D0138358

Limnoecology

Limnoecology: The Ecology of Lakes and Streams

Winfried Lampert
Director Max Planck Institute for Limnology
Plön, Germany

Ulrich Sommer
Professor Institute of Marine Research
Kiel, Germany

Translated by JAMES F. HANEY
Professor of Zoology
University of New Hampshire

OXFORD UNIVERSITY PRESS
New York   Oxford
1997

Oxford University Press

Oxford New York
Athens Auckland Bangkok
Bogota Bombay Buenos Aires Calcutta
Cape Town Dar es Salaam Delhi
Florence Hong Kong Istanbul Karachi
Kuala Lumpur Madras Madrid Melbourne
Mexico City Nairobi Paris   Singapore
Taipei Tokyo   Toronto

and associate companies in
Berlin Ibadan

Library of Congress Cataloging-in-Publication Data
Lampert, Winfried. [Limnoökologie. English]
Limnoecology : the ecology of lakes and streams /
by Winfried Lampert and Ulrich Sommer ;
translated by James F. Haney.
p.   cm.   Includes bibliographical references (p.   ) and index.
ISBN 0-19-509592-8
1. Limnology.   2. Lake ecology.   3. Stream ecology.
I. Sommer, Ulrich.   II. Title.
QH96.57.A1L3513   1996
574.5'2632—dc20   96-33934

9 8 7 6 5 4 3

Printed in the United States of America
on acid-free paper

Contents

Preface to the English Edition

Fresh waters form only a tiny part of the earth's surface. Nevertheless, their importance for drinking water, irrigation, fisheries, aquaculture and tourism are beyond dispute. Limnology, the science of inland waters, provides a necessary scientific basis for the management of lakes and rivers. There is, however, a bigger and more fundamental role for limnology. Even the earliest limnologists believed that they had an important message for ecology. This continuing belief is highlighted by article titles like "the lake as a microcosm" (Forbes 1877) and "copepodology for the ornithologist" (Hutchinson 1951). Lakes, especially, have been considered as little theaters where the great play of the ecological interactions, interactions among organisms and between organisms and their environment, could be studied more easily than anywhere else. Lakes can serve as those little theaters ("microcosms" sensu Forbes) because they are relatively easy to sample, because they have clear-cut boundaries (compared to terrestrial ecosystems) and because field experiments are relatively easy to perform. There has been a continuous flow of information from limnology into other fields of ecology, at all levels of science, starting from methodological problems of quantitative sampling to the highly abstract concepts such as the food chain and the trophic cascade. One of the most prominent limnologists, G.E. Hutchinson, has developed most of the foundations of modern population and community ecology. Conversely, limnology has received inputs mainly from theoretical ecology and less so from experimentalists and field ecologists outside the field of limnology. The exception is marine ecology, which has provided limnology with some important methodological advances. More than any other subdiscipline of ecology, limnology has helped bridge the gap between models of theoretical ecology and experimental research. The idea of "the lake as a micro-cosm" implies, of course, a concept of "unity of ecology". It can be assumed that the same principles and laws govern limnetic, marine and terrestrial communities.

Some time ago we felt our German students needed a text book that demonstrated the close connection between limnology and ecology. In order to clearly indentify the concept of this book we invented the title "Limnoökologie" (limnoecology). The "unity of ecology" structured our book. It is modeled after ecology textbooks with a sequence from more elemental to more complex units: individuals-populations-coupled populations-communities-ecosystems. The physical, chemical, and geological background is rather reduced, compared to other limnology textbooks. We were surprised to see that the demand for such a book was not restricted to Germany. Limnoökologie was first translated into Polish. We received continuing inquiries for an English translation, and eventually Oxford University Press made it possible to produce this volume. We are extremely grateful to James F. Haney, who took the challenge to translate a not very easy text. We thank the numerous colleagues who have given us advice for improvment of the German edition. We have incorporated their suggestions and updated the literature in the English edition. The publishers were very generous to allow us to expand the book where it was necessary to clarify ideas. Finally, we would like to thank John J. Gilbert and Nelson G. Hairston Jr. for their constructuve reviews of the first English draft.

Ploen and Kiel   W.L. and U.S.
July 1996

Preface to the German Edition

Limnology is the science that investigates the structure and function of inland waters. It is a subdiscipline of ecology. We decided to use the newly created word *limnoecology* as a title in order to emphasize this relationship.

In the July, 1990, issue of the journal *Limnology and Oceanography*, Nelson Hairston, Jr., published a commentary in which he complained that the numerous textbooks in ecology and evolutionary biology use examples almost exclusively from terrestrial systems. This is astonishing, considering many of the fundamental concepts of modern ecology can be traced back to the late G. E. Hutchinson, one of the most well-known limnologists. Furthermore, many important concepts have been developed in aquatic ecology in the past 20 years that are important for all of ecology. In our opinion it is an appropriate time to illustrate that many ecological concepts can be better explained with examples from the limnological field than from the terrestrial.

While working on the manuscript, we confirmed from discussions with colleagues and students that there was a considerable need for a textbook on the ecology of lakes and streams and that the concept of our book appeared to fill a gap. Organisms, with their evolutionary adaptations and their physiological and genetic boundaries, are the centerpiece of our approach. We have dealt with the areas of the "classical" limnology, such as physical conditions and water chemistry, only as detailed as absolutely necessary to understand the environmental condition of the organisms. We place special importance on the interactions between organisms from which higher units emerge, such as communities, even though maximization of fitness basically serves only the individual. In the beginning of this book we introduce theoretical concepts and then attempt to explain them with examples from fresh water. Where possible we use mechanistic models that deal with the characteristics of the organisms and allow for *a priori* predictions.

Using this concept we hope to satisfy the need for a textbook on limnoecology as well as present some new developments in ecology, emphasizing evolutionary theory. The latter is important, since "evolutionary ecology" has now become an established concept.

We have tried out parts of this book in our lectures at the Universities at Kiel and Oldenburg. Critical questions from students have helped with the writing of the text. Several of our doctoral students have read the manuscript and provided valuable suggestions. We thank them for this. We give special thanks to Dr. K. O. Rothhaupt, Plön, and Dr. F. Schanz, University of Zürich, who took enormous effort to provide us with constructive criticism on the first draft of the manuscript. Finally, we thank the Publisher for the excellent production of the book and their endless patience waiting for the delivery of the manuscript.

Plön and Oldenburg   W. L. and U.S.
1992

REFERENCE

Hairston, N. G., Jr., Problems with perception of zooplankton research by colleagues outside aquatic sciences. *Limnol. Oceanogr.* **35** (1990) 1214.

Translator's Preface

The science of limnology has become increasingly complex and diverse, driven in part by the rapid development of new tools for investigation. As the directions of limnology are simultaneously pushed toward global and molecular levels, it becomes increasingly difficult to identify the underlying hypotheses that make this a unified science.

My first reading of the German textbook entitled "Limnoökologie" provided surprises. Although I was familiar with many of the examples of aquatic research used in this text, I had never viewed this research from such a coherent ecological perspective. Perhaps as many limnologists, I have taught limnology and ecology as related, but separate sciences. I was also surprised by the authors' presentation of ecological theory through detailed research, rather than by simplified generalizations. The approach is informative and direct. Is there a better way to learn than from specific examples?

Acceptance of the authors' request for me to translate their textbook was made easy by knowing that this book was unique and could provide a useful vehicle for furthering the integration of limnology and ecology. It has not been an easy task, for many of the chapters contain concepts that are complex and difficult to express. It is my hope that the English edition of this book is both readable and challenging. I also hope it will provide students with the education that translating and revising it have afforded me.

Durham, New Hampshire   J. F. H.
August 1996

Limnoecology

CHAPTER 1

Ecology and Evolution

1.1 NATURAL SELECTION

The approach of an ecology textbook can often be recognized by the answer to the question "What is ecology?". We will use Krebs's (1985) definition, "Ecology is the scientific study of the interactions that determine the distribution and abundance of organisms" to stress the point that evolutionary concepts are central to modern ecology.

All organisms interact with their living and non-living environment. They are affected by *abiotic* (physical and chemical) and *biotic* (activities of other organisms) factors. We speak of organisms as existing only under conditions to which they are *adapted*. These adaptations are the result of evolution, the driving force of which is natural selection (Darwin 1859).

Prerequisites for evolution to occur include:

1. Individuals of a population are not entirely alike; they differ, if only slightly, in their appearance, size, ability to feed, or their reproductive capacity.

2. At least a portion of this variability is inheritable. Individuals share part of their genome with their parents and contribute their genes to the next generation.

3. Unrestricted, each population has the potential to colonize the entire earth. This leads to resource limitation and competition. In reality, populations do not increase without limitations. Many of the offspring of each generation die before they reach reproductive age. In a condition of equilibrium a single offspring replaces each parent.

4. Not every parent is replaced by its own offspring. The probability that the offspring of a particular parent will survive depends on how well these offspring cope with the environmental conditions in comparison to the young of other parents. When

the traits responsible for the offspring's success are passed on to the next generation, the populations that carry such traits will increase.

5. The likelihood that the offspring of a particular parent make it to the next generation therefore depends heavily on how well the traits of the offspring fit with the specific environmental conditions. Individuals that are successful in a given environment are generally less successful in other environments.

It is a prerequisite of evolution that the results of natural selection are inherited. Natural selection works through phenotypes and does not alter the genome of the individual. It changes the *gene pool* composition of a population causing the populations to change their characteristics from generation to generation. Population genetics investigates such changes at the level of genes and single characteristics of organisms. It is difficult to observe these changes with present populations because our period of observation is relatively brief and the changes occur slowly. The distribution and occurrence of populations that we see today are the product of previous natural selection.

Evolution and ecology are closely related, as indicated above. Natural selection is the product of a population's interactions with the biotic and abiotic environment. These interactions are the fundamental parameters studied by ecologists. *Ecology investigates the forces driving natural selection*. It is no wonder that modern ecology and evolutionary biology are closely coupled, for many new concepts in evolutionary biology have stimulated recent developments in ecology.

1.2 FITNESS

The essence of natural selection is that individuals produce differing numbers of reproductive offspring. At the molecular level this means that certain genes in the population are copied at different rates. A measure of a genotype's long-term success is *fitness*. Individuals that produce the largest number of offspring over a long period of time have the greatest fitness. Fitness is always a relative measure, since the environmental conditions often do not allow for maximum production. One should consider the fitness of an individual relative to other individuals living in the same environment. The individual that has the most surviving offspring contributes the most to the gene pool, the sum of all genes in a population. Its phenotypic features will increase in the population, indicating it has the highest *relative fitness*.

Two components of fitness are reproductive potential and survivorship. It does not necessarily follow that those individuals with the greatest number of offspring automatically have the highest fitness. It is critical to know how many of the offspring survive. A species can achieve a high degree of fitness by either producing a large number of offspring or by providing offspring with special care. Carp, for example, lay millions of tiny eggs that they cannot later care for. Sticklebacks, however, have only a few large eggs that they lay in a nest guarded by the male. During the critical phase of development mouth-breeding cichlids provide cover and protection for their young in the mouth of the parent, but only for a few of the juvenile fish. All three fish species achieve fitness in different ways.

Availability of a particular resource such as food or a specific chemical element often limits the reproductive capacity. The physiological mechanism to achieve greater fitness for an individual consists of either getting more food or utilizing it better than other individuals. Greater survivorship often requires specialized morphological structures and behavior. For example, storage of reserves inside or outside the body as well as periods of diapause can reduce mortality. Methods of avoiding being eaten include protective armor, other defense mechanisms (e.g., bad taste), and fleeing.

There are certainly many ways to increase fitness. The phenotypic characteristics within a population that will be the most successful are those that contribute to maximizing fitness. Since organisms never achieve the maximum possible fitness in nature, it is better to speak of *optimization*. Organisms become adapted to environmental conditions as a result of optimization of fitness through natural selection. A species will never achieve a perfect adaptation because:

1. *Selection works on the phenotype.* There is never a perfect correlation between genotype and phenotype due to mutations and interpopulation gene flow. The immigration of individuals from other populations that are adapted to somewhat different conditions provides a constant flow of genes into the gene pool. Smaller populations are most likely to be affected by random factors resulting in *genetic drift*. Selection affects only the available phenotypes, since no single population has all of the possible genetic combinations that influence fitness. Selection promotes the best adapted genotypes from those that are available rather than the absolute best genotypes.

2. *Physical, chemical, and biological factors are not constant in the environment.* Long-term and short-term climatic changes constantly require new adaptations. Geological events create new habitats, and the organisms themselves also change the environmental conditions. The accidental immigration of organisms into habitats to which they have historically not had access changes the biological environment for the organisms that are already present.

3. *Organisms must constantly make compromises.* One cannot conclude from an organism's presence that the conditions are optimal for it. For example, animals that live in the deep, cold water in lakes do not necessarily require cold water. Some of them can produce significantly more offspring at higher temperatures, but cannot live in the warm surface waters because of predators, against which they have no defense.

4. *Adaptations often have "costs."* The energy that an organism invests for a particular morphological adaptation such as for defense may be deprived elsewhere in the organism. Adaptation to one selective pressure may interfere with adaptation to another selective factor. For example, when algae develop large colonies, they are protected from grazing by zooplankton, but the innermost cells of the colony may not receive adequate CO_2 and nutrients.

5. *There are structural and physiological limits to an adaptation.* Organisms cannot become, for example, infinitely large or small. Likewise, uptake rates for nutrients and feeding rates cannot increase without limits.

6. *The biological relationships between organisms resemble an "arms race."* A prey organism that protects itself by means of armor produces a selective pressure that

promotes those predators that can best break through the armor, which may, for example, result in the evolution of a stronger jaw. This, in turn, creates a larger selective pressure for stronger armor in the prey. As long as the genetic potential for further change is present, the predator and prey will continually change, but never reach an optimal adaptation.

This last point suggests there are differences in the adaptations to abiotic and biotic factors. Although organisms modify the abiotic environment, the changes in the environment do not represent an adaptive response in return. Adaptation to an abiotic factor is a one- sided process. Adaptation to biotic factors is a two-sided process. Here, species interact and become partners in the evolutionary process in which they co-evolve. In a particular habitat all organisms are exposed to the same abiotic factors; for example, the water in an arctic lake is cold for all the organisms that live there. Since there are a limited number of possible physiological reactions that would result in increased fitness in response to cold temperatures, evolving organisms tend to become more similar. Biological factors, on the other hand, are not the same for all organisms. There are close interactions between some individuals or groups of individuals, whereas others may have only slight or no interactions. Various organisms are therefore exposed to different biological selective factors that have the effect of maintaining differences. Only those organisms that have a sufficiently high fitness can establish themselves as part of a community. The structure of this community, such as which species occur and how they interact, depends on how well these organisms are adapted to the conditions. In this sense, abiotic factors create the "frame," but biotic factors produce the "picture in the frame" (see Section 8.7.3).

1.3 PROXIMATE AND ULTIMATE FACTORS

Ecology is changing from a descriptive science into an explanatory science. There has been a shift to view most aspects of ecology from the standpoint of fitness. The new focus of ecologists is to investigate not only *how* organisms are adapted to their environment, but *why* they have a particular adaptation, which is to ask what is their "adaptive value."

An example may help clarify the distinction. Some planktonic copepods (cyclopoid copepods) have a diapause or resting stage in their life cycle during which they do not undergo further development. This diapause often occurs in the summer. Adult copepods reproduce in the spring. Their larvae (nauplii) and the first juvenile stages (copepodites) develop, but the fourth copepodite stage ceases development and buries itself in the sediment where they pass the summer. They reappear in late fall and complete their development. Spindler (1971) demonstrated experimentally that day length determines the onset of diapause. This explains what controls the behavior, but not why the behavior has developed. One would not expect that longer days represent a disadvantage for fitness. On the contrary, a copepod that does not go into diapause could produce several generations of offspring during the summer, thereby increasing fitness.

Day length is a *proximate factor*, an environmental factor that determines the immediate response of an organism. In addition, there must be one or more inconspicuous
factors that cause the diapause behavior to represent a fitness advantage. The evolutionary response is controlled by *ultimate factors*. It is not always easy to recognize an
ultimate factor, since natural selection acts to eliminate ultimate factors. It is rare that
there is historic evidence that indicates an ultimate factor is still operating in a population. As a rule we only observe a stage that approaches the final result of the adaptation.

The search for ultimate factors is nonetheless one of the most important tasks of the
ecologist. One must develop hypotheses to test which ultimate factors are important.
In our example such hypotheses might appear as:

1. The copepods cannot tolerate the high summer temperatures because of physiological limitations.
2. The larval stages that depend on algae for food are poor competitors and cannot
   succeed in the summer in the presence of other zooplankton that graze on algae.
3. The copepods are susceptible to fish predation and avoid those periods when fish
   are abundant and most active in the lake.
4. The behavior is a historic relic from a time when the copepods lived in small water bodies that dried out in the summer.

Some of these hypotheses can be tested experimentally and others by comparisons
of populations in various lakes. Normally one can only exclude a proposed ultimate
factor rather than prove it is the responsible one (cf. Section 2.1). Hypothesis (1) could
be easily discarded if the copepods could be cultured at short day lengths and at high
water temperatures (see Section 6.8.2). By excluding more and more possible explanations, we reduce the number of possible alternative explanations until, ideally, only
one ultimate factor remains. Even this is not a final proof, for an infinite number of
new explanations could be created.

An understanding of ultimate factors is important, however, for it enables us to make
predictions of how organisms will respond to environmental changes. It may also help
reduce the number of factors that must be investigated and lead to more general principles that apply to all organisms. Ultimate factors may affect more than a single group
of organisms; their effects are often cascading. Returning to the example of diapause,
adult cyclopoid copepods are predatory, feeding on other small zooplankton. When
these copepods enter diapause in the summer, they no longer threaten the small zooplankton, which then can develop populations that become large enough to suppress
certain species of algae that they feed upon. Natural selection directly changes the phenotypes of a single population (in our case, the copepods), but indirectly alters other
populations and communities. Changes in the characteristics of one population create
changes in the environment of those organisms with which they interact.

Ecology integrates at levels ranging from the individual, population, community,
and ecosystem to the biosphere. As complexity increases at each higher level, our ability to find adequate explanations lessens (Krebs 1985). Quite likely we will never be

able to predict the effects of global climate change in terms of theories of evolutionary biology. This, however, may not be necessary, since such questions deal with large groups such as the green plants rather than species or individuals and require special methods to measure the responses and changes. Nevertheless, evolutionary theory is involved at the highest levels of integration, for even the atmosphere is a result of the interactions of organisms and their environment, and it has only developed its present condition after the evolution of oxygen-producing plants.

Current advances in ecology suggest that we will develop general principles of evolutionary adaptations at the level of communities. Communities have unique features that result from their complexity, such as diversity, stability, species structure, hierarchical divisions, and regenerative ability. In contrast to the DNA of individuals, there is no molecular basis for the inheritance of the integrated features of a community. Ultimately, the features of communities must therefore stem from the interactions of individuals. One of the most challenging tasks of ecology is to derive from the characteristics of individuals the principles that regulate higher-level characteristics such as diversity.

CHAPTER 2

Methods of Ecological Research

2.1 HYPOTHESIS TESTING

Ecology, like all other sciences, operates with a repertoire of methodological rules. Methods and approaches have changed as ecology has moved from a descriptive, natural history study to an experimental science.

The early "developmental" stage of a science usually begins with a description of individual observations. In the next phase of "classification," the objective is to organize the multitude of observations. Lake typology (cf. Section 8.5.4) represents the classification stage in limnology that was developed in the first decade of the twentieth century. The classification stage frequently spawns discussions concerning the definition of concepts rather than their cause–effect relationships. Ecology has now reached the stage of "scientific law" that focuses on predictive statements concerning causal relationships.

Predictive laws demand tests of veracity, whereas classification schemes are either practical or impractical rather than true or false. Testability of validity is the tool of scientific theory, a subfield of philosophy. As in almost all areas of philosophy, there is no consensus of opinion amongst the scholars concerning the definition of science and the scientific method. Karl Popper, who has achieved wide acceptance within the natural sciences, suggests there are two fundamental ways to make predictions:

1. *Deduction,* the derivation of a specific prediction from general statements or laws.
2. *Induction,* the inference from many individual observations to a general law that should be valid for similar cases that have not yet been observed.

It is not possible conclusively to prove a statement derived through inductive reasoning, since one cannot test all possible individual cases. However, a single case refuting the observation is sufficient to *falsify* or disprove it. An essential feature of a scientific statement is its falsifiability. A hypothesis must be formulated so that it can be falsified. Nonfalsifiable statements do not contribute to the development of knowledge.

Popper's example with swans provides a graphic illustration of the falsification principle. Before black swans had been discovered in Australia, the statement, "all swans are white," was a legitimate hypothesis, since it could be falsified by a contradictory observation. The statement that "all swans have some color," would be irrefutable, but it would also be nonsense, for it contributes no information.

Chance occurrences are so important in ecology that it is not useful to apply too rigid a rule of falsification. Falsifiable hypotheses can be formulated to include the probability of the occurrence of an event or the confidence limits of an expected value. Such *probabilistic hypotheses* can be falsified only with a sufficient number of observations, rather than by a single observation.

Scientific theories are logically bound complexes that consist of hypotheses and axioms. Axioms are definitive statements such as the "survival of the fittest" principle of Darwin's theory of evolution. Axioms are not falsifiable through direct observations or experimentation. They are subject to the historical testing process through which theories based upon the axioms are tested.

The empirical testing of theories follows the *hypothetical–deductive* method. In this method falsifiable hypotheses are derived from a theory and tested by observation or experimentation. Two theories are often in competition to explain a fact. One must examine the theories, derive the contradictory hypotheses, and decide which is correct. An example taken from limnology is in Box 2.1. This procedure is becoming established in ecology slowly, in part because of tradition and in part because the available methods and theoretical framework are still inadequate to deal with such complex systems of interactions.

2.2 OBSERVATIONS AND RECORDING

Observations of nature, the collection and recording of field data, and their systematic analysis dominated the earlier phase of ecology. The concept of the "balance of nature" was particularly important in the development of the early theories, as ecologists applied inductive reasoning to their interpretation of field data. Even though ecology today is moving toward more experimental approaches, the collection of field data is still important, for they provide tests of hypotheses as well as the monitoring of the environment. The once widely held concept that ecologists should collect as many data as possible in hopes of gaining insights through inductive reasoning, however, is no longer contemporary.

Field data commonly collected by freshwater ecologists include physical parameters, such as temperature, density of water, current velocity, light intensity, concentrations of chemical substances, and the abundance and composition of species. Ecolo-

BOX 2.1	EXAMPLE FOR THE TESTING OF HYPOTHESES

Observation: In many lakes the spring population maximum of phytoplankton is followed by a maximum density of herbivorous zooplankton. At the time of the zooplankton maximum there is a spectacular collapse of the phytoplankton ("clear water stage," see Section 6.4.1).

Correlation: The temporal changes in phytoplankton biomass (the derivative of the phytoplankton curve; dB/dt) is negatively correlated with the density of zooplankton.

Hypothesis: The grazing by zooplankton leads to the clear water stage.

Alternative hypothesis: A deficiency of nutrients causes the collapse of the phytoplankton.

Testing by examination of field data:

1. The production rates of the phytoplankton per unit biomass are so high during its collapse that nutrient limitation can be excluded.
2. The grazing rates of the zooplankton are higher than the growth rates of the phytoplankton.

Conclusion: The alternative hypothesis is falsified, whereas the initial hypothesis has not been disproved.

Testing by field experiments:

1. The unrestricted development of zooplankton in fertilized mesocosms (see Section 2.4) also leads to a clear water stage.
2. When the zooplankton are suppressed through fish predation or removal by a plankton net there is no clear water stage, with or without addition of nutrients.

Conclusion: The alternative hypothesis is falsified, and the initial hypothesis has not been disproved.

---

gists often measure the mass of organisms, *biomass,* to avoid the problem of comparing organisms of vastly different size, where the *number* of individuals has little meaning (one large fish has more effect on its environment than thousands of microscopic algae). Ecologists refer to *concentrations or densities*, rather than absolute numbers or biomass. For example, the number of water fleas per liter or per meter squared is usually more useful than the total number in a lake, to describe general processes rather than events in a particular lake.

The distribution of chemicals and organisms in their environment is not uniform. Organisms are spatially variable or *patchy*, even within a habitat that appears physically homogeneous. Rarely is a total census of a habitat possible. Thus quantitative ecology was confronted quite early with the problem of obtaining a representative sample and with related statistical questions. Even with ideal sampling designs there are serious limitations because of variation due to *sampling errors*. Sometimes ecologists conclude that it is sufficient and more efficient simply to categorize organisms into estimates of frequency groupings such as single, rare, abundant, very abundant, and massive. If possible, such simplified systems should be avoided, for they may have *subjective errors*, resulting from overestimation of conspicuous and anticipated organisms.

Concentration, density, and biomass change temporally. It is important that the frequency of sampling and the length of the period of study are appropriate for the *time scale* of these changes. In lakes, the species composition of the phytoplankton can completely change within a few weeks; therefore, it is not reasonable to describe their changes on the basis of a single sample per year. Samples taken once weekly or once every 2 weeks are usually necessary to describe the seasonal changes in species composition of the phytoplankton. Observations must be made over many years to determine whether the observed changes represent a consistent pattern that is repeatable.

Since finances and personnel are usually limited, ecologists must always make compromises concerning frequency of sampling as well as the degree of resolution from the samples. For example, one must often decide whether to determine the frequency of each species or whether it is permissible to lump the species into taxonomic or functional *group categories*. Sometimes the grouping of species not only results in less work, but also aids in the identification of patterns that would not be so apparent with a more detailed analysis. In certain lakes one can predict, for example, that diatoms will dominate the phytoplankton early in the season, but it is usually difficult to say which diatom species will be dominant.

One of the serious limitations of purely descriptive research is that relatively few statistical tests are appropriate. The seasonal changes in concentration and biomass result from growth and losses. Similarly, the rate of change in the density of a population is the net effect of birth, immigration, emigration, and death rates. For many questions, processes are more important than the empirical measurements such as concentration of chemicals and density or biomass of organisms. Growth and losses occur simultaneously. Thus rates cannot be derived from the static measurements made at different times of the year. The investigation of growth and loss rates requires the isolation of individual processes. Although in situ measurements require manipulation, they are usually considered more descriptive than experimental.

2.3 CORRELATION OF FIELD DATA

The transition from observation and recording to the formulation of laws governing cause and effect often begins with an attempt to find correlations between suspected independent and dependent variables. Changes in biological variables are generally viewed as a function of the abiotic environment or other biological variables. An example for the first case is the attempt to explain the distribution of a species (a biological variable) on the basis of temperature (abiotic variable). In the case of biological variables, the distribution of a predatory species might be related to the distribution of its prey.

Mathematical methods and the widespread use of computers and statistical packages are replacing data analysis "by eye." Many basic problems, however, remain unchanged, despite the improved objectivity. *Correlation analysis* can be used to draw numerical, but not functional, relationships. The *cause–effect relationship* must have been already formulated as a hypothesis. If, for example, the pH of the water is positively correlated with the rate of photosynthesis of the phytoplankton, one cannot conclude whether a higher pH stimulates photosynthesis or whether photosynthesis causes the pH to rise. If we already have formulated the hypothesis that photosynthesis drives the pH up as a result of the release of OH^- ions, a correlation analysis can be used as a test.

A causal relationship can be misleading if both variables are actually dependent on a third variable that has not been considered, rather than on one another. A common mistake is to conclude there is a causal relationship based on trends in a time series. An example of such an *apparent correlation* is the simultaneous decrease in the human birth rate and in the number of storks in the wealthy countries of Europe.

Lags in the reaction of the interdependent variables present another problem in the use of correlation analysis for field data. Increases and decreases in populations often occur after the causes for these changes. Thus the changes in numbers of organisms are actually more closely related to past rather than present conditions. Statistical methods such as cross-correlation analysis take into consideration this phenomenon and can be used to determine *time lags*, but such tests require measurements that are taken frequently and over a long period of time.

Routine statistical tests such as *multiple correlation* can be used to find correlations between many variables, but there are difficulties with the assumptions required for these tests. Thus mathematical analysis cannot substitute for the use of experimental manipulations as a means of revealing causal relationships.

2.4 LABORATORY AND FIELD EXPERIMENTS

Experimentation has been the classic method of the natural sciences since the time of Galileo. Experiments are better suited than pure descriptive science for the falsification of hypotheses and deciding between alternate hypotheses. Critical experiments that can resolve between competing hypothesis contribute greatly to a science and carry much prestige. Unfortunately, they are relatively rare. Not all experiments are conducted with the intention of testing hypotheses. Many experiments simply provide new measurements that can be used in calculations, such as the maximum growth rate of a species. They also have the implicit potential to falsify the basic formulas for which they are being tested, or the experiments may be a part of broader program that leads to hypothesis testing. Unfortunately, many ecological experiments are carried out without a theoretical framework and only for the "fun" of seeing the effects of a manipulation.

The classic construction of an experiment is to hold all factors constant except for one so that the effect of the varied factor can be investigated. Such experiments are most easily carried out in the laboratory rather than in nature. Laboratory experiments with individuals or *clones*, a number of genetically identical organisms, is a recognized method in physiology. They are also very important to ecology since the fitness of an organism or its role in biochemical processes is dependent on its physiological condition and abilities. Most of the results referred to in Chapter 4 are derived from this type of experiment.

Some researchers use this method to investigate processes at the population or community level. Experiments that use artificially composed communities are called *microcosms* (Fig. 2.1a). Two or more populations are placed in containers in which experimental manipulations are made. Microcosms are either open or closed systems regarding the addition of substances, the removal of waste products, and the sampling of organisms. Immigration of organisms into the containers is usually excluded. The early competition experiments of Gause are prototypes of these experiments (see Section 6.1.1).

c

Figure 2.1 Ecological experiments with different size scales: (a) Different sized microcosms in the laboratory. (b) Mesocosm experiments in Schöhsee (Holstein, Germany). These plastic bags are 1 m in diameter and 3 m deep. They contain approx. 2 m^3 of water (Max-Planck Society press photograph). (c) Experiment in which a lake is divided conducted at Neuglobsow (Brandenburg, Germany). The lake was divided into four sections with plastic curtains anchored in the sediment. Each section received a different treatment.

The main criticism of microcosm experiments is that organisms must be contained in an unnaturally small space. It is assumed that biotic interactions such as competition and predation are exaggerated over those in nature, due to the lack of immigration, emigration, and avoidance mechanisms. This criticism is a misunderstanding, for the purpose of a microcosm experiment is to study the rules regulating ecological processes such as competition. It is not the immediate goal to determine how important the process is in nature. Microcosm experiments can reveal, for example, which patterns of seasonal changes in the density of organisms are caused by competition or predation. If similar patterns are seen in field observations, it is the first indication that such processes may be important in nature. It is, of course, a reliable indicator, only when alternative mechanisms that could lead to the same patterns can be excluded.

The size limitations of microcosms make them ideal for small plankton, sessile microorganisms, and small insects, but they are of little value for the study of processes that involve larger organisms such as fish. The advantage of microcosms is that several parallel systems can be used to allow for statistical comparisons. Processes that "blow up" in small microcosms can sometimes be examined in *mesocosms* (Fig. 2.1b). These are sections of the natural environment that are isolated by means of artificial barriers such as transparent plastic sheets. These mesocosms (sometimes called limnocorrals in lake studies) are used to contain (*enclosures*) or exclude (*exclosures*) certain communities or populations. Experimental ponds can be used for similar experiments. Mesocosms and ponds are exposed to natural variation in temperature, light, and wind conditions and therefore do not have the constancy of variables that are necessary for an ideal controlled experiment. One assumes that despite these variations in conditions the effect of the manipulation can be recognized by comparing the unmanipulated (*control*) and manipulated mesocosms. Parallel experiments are needed to exclude the random effects. Experiments of this type typically involve fertilization or the addition or removal of organisms.

In limnology, a more controlled type of mesocosm such as the plankton towers (Lampert and Loose 1992) are rare because of their expense and complexity. These 10-m-high metal cylinders have heating and cooling elements that control the thermal gradients. By controlling the light and wind effects as well as the distribution of chemical factors, these towers have about the same degree of experimental control as a microcosm.

The **manipulation of ecosystems** is the largest, but the least controlled, form of ecological experiment. Small lakes and ponds are best suited because they are isolated systems. Processes requiring large space such as fish population dynamics can be investigated in this type of experiment, but not in micro- and mesocosms. On the other hand, it is not possible to have direct control of the populations of plankton and microorganisms. Ecosystem manipulations characteristically involve addition of chemicals such as nutrients and acid or the stocking of a fish species. In contrast to mesocosms, there is no true control to compare to the effect of the manipulation. Neither the same system before the manipulation nor similar, unmanipulated systems can be viewed as controls, in the strict sense of the word. It is better here to compare many years before and after manipulation or many manipulated lakes with many unmanipulated lakes so that random variations in the unmanipulated systems can be identified. In an attempt to improve the controlled aspects of ecosystem manipulations, lakes are sometimes divided into two or more subunits with a plastic curtain (Fig. 2.1c). Each lake subunit can be manipulated or kept as a control. In a well-known example of this approach, Schindler (1980) demonstrated that phosphorus was the limiting nutrient in Lake 227 in the Experimental Lakes Area (ELA) of Ontario.

A fundamental problem in limnological experiments is that organisms and physical processes cannot be miniaturized. The larger an experiment becomes, the more realistic and "nearly natural" it becomes and the more types of organisms that can be included. At the same time larger size means sacrifices in the ideal experimental conditions. Complete control of experimental conditions is relinquished as one moves from micro- to mesocosms. In moving from mesocosms to ecosystem manipulation one cannot assume that the experimental conditions are identical in the manipulated and nonmanipulated systems. Gain in reality is made at the cost of replication and accuracy.

2.5 MATHEMATICAL MODELS

Relationships expressed in mathematical terms can be called mathematical models. *Models* may be a simple mathematical formula or a system of connected equations that describe the interactions of various components of a system. With an increasing number of components and complexity mathematical models become more difficult to solve analytically. At this point, *simulations* may be used to solve for specific conditions by providing for each parameter an assumed numerical value. After a sufficient number of simulations have been run, one can determine the range of conditions within which a particular result is possible, as, for example, the conditions allowing for the coexistence of populations. *Deterministic* simulation models use a specific value for each parameter, whereas *stochastic* models use the variability around a mean value that is provided.

At first glance, it appears that a simulation model is very similar to an experiment since in each case one manipulates a particular input parameter and then waits for the result. Simulations have advantages in not having the temporal, spatial, and technical limitations of an actual experiment. A fundamental difference between simulations and experiments is that simulations are strongly deductive, since the results are implicit in the formulas and parameter assumptions. Thus simulations cannot test predictions of empirical facts, but they can show the consequences of their own assumptions that are not always intuitive. An impressive example of this was the discovery of deterministic chaos (May 1975) that proved that unpredictable time courses could be created from perfectly deterministic formulas.

Comparisons of simulations and experiments are valuable in testing hypotheses. There may be several reasons to account for discrepancies between the experiment and its simulation:

1. The processes that were simulated simply do not exist or were perhaps masked by processes that were not considered;

2. One or more formulas used in the simulation are unrealistic; or

3. The assumptions concerning the input parameters are incorrect.

One must be careful, especially considering the last point, for especially with complex models it is often possible to achieve the "desired" result with enough "fine tuning" of the parameter values. In such cases the agreement or disagreement of the model with reality has little significance. Models that show major shifts in response to minor alterations of the inputs are suspect, since most ecological measurements have broad error margins.

CHAPTER 3

Special Features of Aquatic Habitats

3.1 EFFECTS OF THE MOLECULAR STRUCTURE OF WATER

3.1.1 Association of Molecules

The unique molecular structure of water underlies many of the features of aquatic habitats. The hydrogen atoms form bonds at an angle of approximately 105°, making the water molecule strongly bipolar. This bipolarity is ultimately responsible for the tendency of water molecules to associate with one another, to dissolve other substances and for biochemical processes.

Water molecules form relatively weak hydrogen bonds with one another, creating associations or "swarms" called *clusters*. An important feature of clusters is that they are dynamic structures, constantly forming and breaking bonds with other water molecules. The number of molecules in a cluster can therefore only be described statistically. The mean number of water molecules in each cluster decreases with increasing temperature. For example, at about 0°C there are on average 65 molecules per cluster, and at 100°C there are only 12. The structure of water changes dramatically as it freezes. The molecules line up in a lattice in which four H atoms surround each oxygen atom in a tetrahedron.

Water has unique characteristics resulting from the arrangement and bonding of water molecules. It is most dense at 4°C. The surface tension of pure water is higher than any other fluid except mercury, allowing for a community including algae and insects to live upon, within, or under the water surface. The high viscosity of water is important for the locomotion of organisms. Also uniquely high are the specific heat, the latent heat of fusion, the heat of vaporization, and the boiling point of water. Water lacking this association of water molecules would have a boiling point at about −80°C; that is, there would be no liquid water on earth, just vapor.

3.1.2 Density Anomaly

Water molecules in ice are widely spaced as a crystalline matrix; thus ice has a relatively low density. Upon melting, the molecules move together more closely and the water has a higher density. The density difference between ice and liquid water at 0°C is 8.5%, allowing ice to float on the water surface. When the temperature increases above 0°C, the cluster size decreases, thereby increasing the density, since smaller clusters can be more densely packed. Simultaneously, there is an increase in thermal expansion that leads to a reduction in the density. At 4°C, thermal expansion fully compensates for the increasing density, resulting in a *maximum density* at this temperature. Above 4°C, the effects of thermal expansion dominate and the density continually decreases as the temperature rises. This peculiarity of having a maximum density at 4°C, rather than at the lowest temperature, is referred to as the *density anomaly* of water. Without the anomaly, the coldest water would sink to the bottom in the winter, eventually freezing the lake solid. Because of the temperature anomaly, water that is 4°C collects near the lake bottom, so that the lake only freezes from above. The surface ice cover insulates the lake against further cooling. Even in the most severe winter there is always water available to aquatic organisms in a deep lake. The temperature of maximum water density is lowered by salt, generally negligible in inland lakes, and by hydrostatic pressure. For every 10 bars of pressure the temperature at which water is most dense drops by about 0.1°C. In a deep lake such as Lake Constance, Germany, with a maximum depth of 250 m, the bottom water temperature is only 3.8°C.

3.1.3 Thermal Features

Water has a high specific heat. At 15°C, 4.8186 kJ are required to heat 1 kg of water one degree Celsius. This is a very high value, only exceeded by ammonia gas (5.15) and liquid hydrogen (14.23). Likewise, water loses heat very slowly, enabling it to store much heat. As a result, large lakes act as climate buffers, and lake water temperatures change slowly over the seasons. Both of these effects have important consequences for the lives of aquatic organisms.

Water transmits heat very poorly. With a temperature difference of 1°C, only 0.00569 J cm^{-1} grad^{-1} s^{-1} flows through a cube. Thus heat transport via molecular diffusion in lakes is negligible. Consequently, heat should actually stay at the location where it was absorbed, such as at the highly lighted surface waters. This is not, however, what happens (Fig. 3.5). Wind and water currents transport the heat by means of turbulent eddy diffusion, but only as deep as its force permits. These physical processes result in characteristic heat gradients of lakes (see Section 3.2.2) that are in turn responsible for chemical gradients and the distribution of organisms (see Section 3.3).

3.1.4 Surface Effects

The association of water molecules is especially noticeable at interface surfaces. The air–water interface is a habitat for organisms as well as a barrier to their distribution. This is possible because of surface tension, which is higher for pure water than for any other liquid except for mercury. Temperature and pollutants strongly affect the surface tension of water. Many organisms utilize the surface "film" by attaching to it or even walking on the surface.

Water molecules not only attract one another (cohesion), but also have attraction for submerged surfaces (adhesion). The degree of adhesion depends on the chemical composition of the surface. A surface is called *hydrophilic* if the cohesive forces are less than the adhesive forces. Taken out of the water a hydrophilic surface would appear wet. When the cohesion between water molecules is greater than the adhesion to the surface, the object is *hydrophobic*. Such surfaces cannot be easily wetted. Water adhered to a hydrophobic surface will form discrete droplets. Many organisms (e.g., water beetles) have hydrophobic surfaces that serve as a physical gill (cf. Section 4.2.2).

3.1.5 Viscosity

The mutual attraction of water molecules results in a resistance in flowing water. This *dynamic viscosity* or, shortened, viscosity, is due to the "internal friction" of water. Flowing water or a body that moves in water must overcome this "syrupiness." This is clearly demonstrated when one attempts to pour a liquid with a high viscosity such as honey. Water has a relatively low viscosity, compared with other fluids. Viscosity is dependent on temperature, as well as on dissolved substances, although the latter have a negligible effect on viscosity in fresh water.

The unit of measure of dynamic viscosity is the pascal second (1 Pa s = 1 kg m^{-1} s^{-1}), normally indicated with the Greek letter μ. This is the force necessary to move a mass of 1 kg, 1 m in 1 s. The viscosity of water decreases with increasing temperature. Water at 20°C has a dynamic viscosity of about 1×10^{-3} Pa s, and at 0°C, 1.8×10^{-3}. Thus warm water is "more fluid" than cold water. *Kinematic viscosity* or v is used for various calculations. It is derived as the dynamic viscosity divided by the density of the fluid (δ): $v = \mu/\delta$ and has as units kg m^{-1}/kg m^{-3} = m^{-2} s^{-1}. The relationship between dynamic viscosity and kinematic viscosity is not entirely obvious. Dynamic viscosity expresses how strongly a "parcel" of fluid will be pulled along in synchrony with its neighbors. Density is an expression of the mass of a "parcel" of fluid and thus its tendency to continue in its line of movement (inertia). The relationship of the two is a measure of the ability to counteract the irregularities in the speed of flow of a fluid.

Viscosity and other related phenomena such as water flow are of fundamental importance in regulating aquatic life, the distribution of heat and matter, passive sinking and active swimming, collection of food by filtration and residing in flowing water. Water can move in *turbulent* or *laminar* flow (Fig. 3.1). In laminar flow tiny particles in water move in parallel tracks that can be visualized by parallel *streamlines*. Turbu-

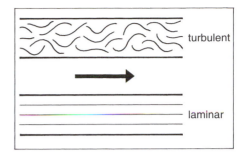

Figure 3.1 Turbulent and laminar flows of water in a pipe. The streamlines describe the paths of the water movement.

lent flow causes the streamlines to be distorted. The entire water mass may move in one direction, but the individual water particles have irregular paths. Flowing water can change abruptly from laminar to turbulent conditions if, for example, the flow velocity increases. This can be demonstrated when a water tap is opened. If the water stream leaving the tap is thin (slow flow), it flows smoothly (laminar flow). When the tap flow is increased, the stream begins to splash (turbulent flow).

Laminar flow is a consequence of the viscosity; closely attracted molecules move synchronously. In turbulent flow the force of inertia is of greater importance than molecular attractions. If you move a rudder in the water, you can see how a parcel of water that is hit continues to move as a result of its inertia. It quickly slows down due to internal friction (viscosity), and after a short distance there is no trace of the turbulence that was created. The type of flow depends strongly on the degree of viscosity and drag forces. The relationship between inertial and viscous forces is characterized by the dimensionless *Reynolds number* (*Re*):

$$\frac{\text{inertial forces}}{\text{viscous forces}} = Re = \frac{\delta U l}{\mu}$$

where U is the speed (m s^{-1}) with which water and an object move relative to each other. It is unimportant whether the water is passing a stationary object or whether the body is moving in stagnant water. l is a characteristic length (m), such as the diameter of a pipe through which the water is flowing or the length of a body in the direction of flow. The density (δ) is removed if the viscosity (μ) is replaced with the kinematic viscosity (ν):

$$Re = \frac{U l}{\nu}$$

When Reynolds numbers are small, the forces of viscosity prevail, whereas large Reynolds numbers indicate inertial forces are more important. Thus one predicts laminar flow with small Reynolds numbers and turbulence at high numbers. The Reynolds number is determined primarily by the speed of movement (U) and the characteristic length (l), since the kinematic viscosity of water varies within a very small range (about

1×10^{-6} m^2 s^{-1}). Higher velocity and larger objects produce larger Reynolds numbers. The characteristic length is difficult to determine as it depends on the length as well as the shape of the object. This inaccuracy is of little consequence in ecology, for estimates of Reynolds numbers are usually given simply as orders of magnitude because they range so widely in nature. Some examples are:

	Re
Large swimming whale	10^9
Trout in a stream	10^5
Escaping zooplankter	10^2
Ciliate swimming	10^{-1}
Filtering setules on a zooplankter	10^{-3}

Very small organisms live in a viscous environment, and large organisms live in a turbulent environment. One can ignore turbulent flow when Reynolds numbers are much smaller than 1. Very small organisms such as algae and bacteria are always surrounded by laminar flow (Fig. 3.2A). An algal cell may be transported in a turbulent water parcel, but there is laminar flow in its immediate surroundings. This has important consequences. In laminar flow an object is encompassed with a layer of water with a very low flow velocity. This *boundary layer* contains water molecules that always accompany the object. The width of the boundary layer increases with increasing wa-

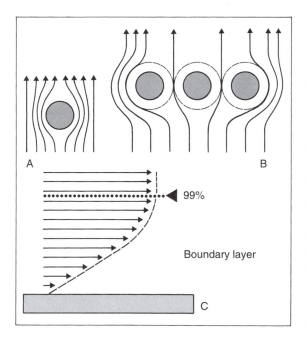

Figure 3.2 Flow at very low Reynolds numbers. (A) Laminar flow around a small particle (e.g., a sinking alga). (B) Cross section of flow across a filtering structure. The boundary layers around each cylinder are symbolized by dashed lines. Pressure is required to force the water through the pores of the filter. If there is insufficient pressure the water will flow around the filter. (C) Formation of a Prandtl boundary layer when water flows across a fixed surface (e.g., a rock). The length of the arrow symbolizes the flow velocity. It is zero exactly at the stone's surface.

ter velocity and with decreasing size of the object. If an organism removes a molecule from the boundary it can be replaced from the surrounding water by diffusion. This is a very slow process. The concentration gradient in the boundary layer will determine how rapidly the nutrient deficiencies are replaced by diffusion. A cell does not benefit by swimming about in a medium with nutrients equally distributed throughout, for it simply takes with it the nutrient-poor boundary layer. Diffusion remains the only way to supply more nutrients to the layer and is thus the factor that ultimately limits the rate of uptake by the cell. The advantage of changing location for the cell is that it may end up in a region of higher nutrient concentration where there is a steeper nutrient gradient between the boundary layer and the environment.

The common argument that an algal cell sinking slowly through the mixed surface layer of a lake will take up nutrients more easily than a cell that does not sink is correct only if it is a very large algal cell. Even then diffusion is the dominant process, since laminar flow predominates around a sinking algal cell. A diatom with a radius of 10 μm achieves a sinking speed of about 10 μm/s (85 cm/day) (see Section 4.2.6 for calculation of the sinking speed), resulting in an increase in nutrient diffusion through the cell surface of about 10%. A green alga sinking more slowly or a diatom with only a 5 μm radius will have almost no significant improvement in nutrient uptake ($<$ 1%). For very small cells active locomotion is more important. For example, swimming at rates up to 100 μm/s increases the nutrient diffusion for a 1 μm flagellate about 5%, for a 5 μm flagellate 30%, and for a flagellate with a 10 μm diameter about 50%.

The boundary layers of closely positioned small bodies can affect one another. For example, the boundary layers around the very fine filtering structures of filtering organisms, viewed as a system of cylinders, can overlap (Fig. 3.2B). Although such a filter has many holes, the water cannot flow through them unless the necessary pressure is applied.

Even when water flows over a flat surface a boundary layer develops. Directly on the surface the flow velocity is zero, and it increases with distance from the surface to a maximal speed. This layer around fixed surfaces in which the flow velocity is reduced is called the "Grenzschicht" or *Prandtl boundary layer*. Its specific thickness cannot be determined precisely, since the flow velocity increases continuously with distance from the surface. Its thickness is operationally defined as the distance from an attached surface at which the flow velocity reaches 99% of its speed away from all objects. The further the surface faces away from the flow, the thicker the boundary layer. At the same time the velocity gradient at right angles to the surface becomes lower (Fig. 3.2C). The larger the distance from the edge facing the current, the thicker the boundary is. The boundary layer thickness (σ) can be calculated more precisely as:

$$\sigma = 5\sqrt{\frac{x\mu}{\delta U}}$$

The thickness depends primarily on the distance from the surface onto which the water is flowing (x) and the flow velocity (U), since the kinematic viscosity (μ/δ) does not change much. The boundary layer around rocks in a stream can extend several millimeters. By treating x as though it were the characteristic length in the Reynolds number at location x (Re_x), we get:

$$Re_x = \frac{Ux}{\nu}, \qquad \frac{\delta}{x} = 5\sqrt{\frac{1}{Re_x}}$$

Thus the relative thickness of the boundary layer (σ/x) depends only on the Reynolds number.

The Prandtl boundary layer is especially important for stream organisms since it allows them to avoid contact with the full force of the flowing water. Algal mats growing on the surface of rocks in flowing water can remain in the boundary layer as long as they do not become too thick. Many stream animals have morphological features and behaviors that are adapted to keeping them within the boundary layer. For example, many insect larvae that graze on algae on the surface of rocks are flattened. At high stream velocities they press themselves tightly against the rock to take advantage of the thin layer of quiet water. Flattened insects are actually not forced against the rock surface, since water flowing over the back of the animal pulls the animal outward, like the airfoil of an airplane wing. Therefore, these animals need devices such as claws and suction cups to hold on to the rocks.

Stream animals must often make compromises. For example, stonefly larvae lacking elaborate external gills would benefit by being further from the rock in the faster-flowing water where the oxygen exchange is better, but this would expose the insect to greater water turbulence. Filter-feeding black fly larvae must keep their filtering appendages in the flow to supply them with food particles drifting in the water, while securing their position with specialized hooks and adhesive silk. A flattened form does not always indicate an adaptation to the boundary layer. Turbularians, for example, are quite flat yet live under rocks and in cracks. Other animals may live in fast-flowing water by reducing their resistance by streamlined body forms (cf. Section 4.2.5).

3.1.6 Water as a Solvent

Water has a high dielectric constant ($\epsilon = 80$ at 20°C) because of the asymmetrical structure of the water molecule. Water thus breaks up or dissociates heteropolar bonds, making water an excellent solvent. Solution and transport of gases and ions are very important environmental factors for aquatic organisms.

Dissolved gases
Gases in water come from the air or from the metabolic activities of organisms. The most important gases dissolved in water and their origin are:

Oxygen	O_2	Atmosphere, photosynthesis
Nitrogen	N_2	Atmosphere, bacterial activity
Carbon dioxide	CO_2	Atmosphere, respiration

| Hydrogen sulfide | H_2S | Bacterial activity |
| Methane | CH_4 | Bacterial activity |

A diffusion equilibrium develops between a specific gas and water. The amount of gas that dissolves depends on the solubility constant for the specific gas and the pressure according to Henry's law:

$$C_s = K_s P_t$$

where C_s is the amount of gas that dissolves under the given conditions, K_s is the solubility coefficient for the given temperature, and P_t is the partial pressure (e.g., for oxygen for standard conditions in the atmosphere it is 0.21).

Water that contains the amount of a gas it can hold through the solubility equilibrium is said to be saturated. Equilibrium conditions are rare, however, in nature. The rates of consumption and production of gases in lakes exceed the rate of gas exchange with the atmosphere, leading to conditions of supersaturation or deficit. The concentration of a gas present expressed as a percentage of the concentration resulting from equilibrium conditions is called the *relative saturation*. The relative saturation of a gas can change as a result of changing conditions, even though the concentration has not changed. The saturation concentration is determined by pressure, temperature, and salt concentration, although the latter is negligible in freshwater. For example, oxygen saturation in water at standard pressure and 20°C is 9.09 mg/l (100%). If the sun were to warm this water to 22°C, the relative saturation would become 104%, since the saturation value at 22°C is only 8.74 mg/l. Air pressure also affects the saturation concentration. A gas concentration equivalent to 100% saturation at sea level would have a saturation of 135% in a lake at an altitude of 2500 m.

Gases are very soluble deep in a lake due to the great hydrostatic pressure. Since diffusion is very slow, most atmospheric gases exchange at the surface and then are transported elsewhere. During fall circulation, for example, water masses move to the surface, exchange with the atmosphere, and are then moved by wind action to the bottom of the lake. Thus the effects of hydrostatic pressure are usually ignored, and the relative saturation is calculated relative only to the water surface. If gases are produced at depth, such as through bacterial activity, considerable amounts can be dissolved because of the great pressure. An impressive illustration of this effect was seen in the Black Forest lake, Schluchsee (Germany), which was lowered 30 m for repairs to its dam. Suddenly, large quantities of gas bubbled through the lake, the result of reducing the hydrostatic pressure to about one-half, thereby causing the sediments to become highly supersaturated with gas (N_2, CH_4, and CO_2). Fish populations concentrated in the remaining lake were destroyed as the sediments riled up and the noxious gas fizzed out of the lake like champagne bubbles, scavenging the dissolved oxygen.

Carbonate–bicarbonate–carbon dioxide equilibrium
Carbon dioxide has a special ranking among the dissolved gases, for it does not follow Henry's law; there is normally much more CO_2 in water than expected. This is because CO_2 occurs in chemical forms in addition to a free gas. When CO_2 dissolves

in water, a small proportion (less than 1%) hydrates to carbonic acid.

$$H_2O + CO_2 \rightleftharpoons H_2CO_3$$

Some of the carbonic acid dissociates to bicarbonate and H^+ ions.

$$H_2CO_3 \rightleftharpoons HCO_3^- + H^+$$

This leads to a decrease in pH. Now a second dissociation step results in the liberation of another proton.

$$HCO_3^- \rightleftharpoons CO_3^{2-} + H^+$$

The degree of dissociation of carbonic acid depends on the pH (Fig. 3.3). At a pH of 8 there are almost exclusively bicarbonate ions present. If the pH shifts to the alkaline side, the equilibrium moves more and more toward the carbonates. When the pH is very low, free CO_2 and carbonic acid predominate. This pH relationship is very important since most aquatic plants can only utilize CO_2 and bicarbonate for photosynthesis (cf. Section 4.3.6).

In most natural lakes the carbonic acid can bind with alkaline earth metals and alkali metals to form insoluble salts. This disturbs the equilibrium, allowing new CO_2 to diffuse into the water, resulting in a greater than expected amount of dissolved CO_2. The calcium ion and the calcium–carbonic acid equilibrium are especially important in fresh water:

$Ca(HCO_3)_2$	$\rightleftharpoons$	$CaCO_3$	+	H_2CO_3
calcium bicarbonate		calcium carbonate		carbonic acid
(highly soluble)		(low solubility)		

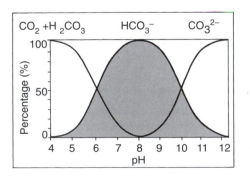

Figure 3.3 Relative proportions of the various forms of CO_2 at different pH values.

[24]

This equilibrium requires that a small quantity of carbonic acid always be present. If it is removed as, for example, by a biological activity, then the bicarbonate changes to carbonate, which precipitates because of its low solubility. In calcium-rich lakes aquatic plants can become encrusted with a layer of insoluble $CaCO_3$ that precipitates when CO_2 is removed during periods of high photosynthetic activity. One can see in many lakes a cloudiness of the water caused by the calcite crystals. A spectacular example of such "biogenic decalcification" are the Plitwicz lakes in Croatia. Over the years, the aquatic moss deposited calcium barriers 3–50 m high, damming 16 lakes, and causing them to cascade into one another.

If CO_2 is added to the system, as through respiration of aquatic organisms, then the carbonate changes to bicarbonate until the excess CO_2 has been used. If , for example, biological activity produces CO_2 in excess of the equilibrium, causing an increase in acidity, we speak of "aggressive carbonic acid," since this can cause corrosive problems in water lines.

The carbonate–bicarbonate–carbon dioxide equilibrium is primarily responsible for the buffering capacity of a lake, that is, its ability to receive H^+ or OH^- ions without changing its pH. *Alkalinity* is a measure of the buffering capacity relative to acids. Although other ions can contribute to buffering capacity, the carbonate–bicarbonate–carbon dioxide equilibrium plays a major role in most waters. If protons (H^+) are added to the water, they react with carbonate (CO_3^-) to form bicarbonate (HCO_3^-); thus they disappear from the system and the pH remains stable. The opposite will happen when OH^- ions are added. The more calcium in the water, the more bound carbonic acid will be present and the more H^+ or OH^- that can be added without changing the pH. Calcium-poor lakes are weakly buffered and are usually slightly acidic. High rates of photosynthesis in such lakes can raise the pH to about 9 by removing CO_2 and even to a pH of 11 if bicarbonate is used by the plants. This can lead to dramatic daily cycles of pH change in poorly buffered lakes. Calcium-rich lakes usually have a pH of 7–8. When CO_2 is removed from this water, the equilibrium shifts to insoluble $CaCO_3$, releasing CO_2 from the bicarbonate into the water (cf. Section 4.3.6). The pH remains constant until all the CO_2 has been used up. High photosynthetic activity can nevertheless lead to pH fluctuations here, since there is a time lag in the recovery of the equilibrium.

The buffering capacity of a lake depends on the geology of the lake basin and watershed. Lakes in regions with siliceous bedrock (e.g., Scandinavia, Northeastern United States, and Eastern Canada) receive little calcium and are therefore poorly buffered. An addition of acids, such as sulfuric and nitric acids from acid rain, depresses the pH rapidly in these naturally acidic lakes. Calcium-poor bogs have pH values as low as 4.5, an acidity about 1000 times that of neutral water, due to their humic acids. Volcanic activity can result in extremely low pH values. For example, a crater lake in El Salvador had a pH of 2 due to sulfuric acid produced by fumaroles. There are also examples of naturally alkaline lakes that may have a pH over 9, such as in the soda lakes in which sodium carbonate (Na_2CO_3) replaces calcium carbonate. Most of these lakes are closed basins lacking an outlet located in arid regions. Well-known examples are Lake Nakuru (Vareschi and Jacobs 1985) in Kenya and Lake Lenore and Soap Lake in Washington, studied extensively by W. T. Edmondson (Edmondson 1991).

Ions and polar molecules

The most important anions in water, according to their abundance, are CO_3^{2-} and HCO_3^-. Sulfate (SO_4^{2-}), chloride (Cl^-), and nitrate (NO_3^-) are less abundant. Of the cations, calcium (Ca^{2+}) is the leader, followed by magnesium (Mg^{2+}), sodium (Na^+), and potassium (K^+). These ions come from the weathering of rock, except for a small quantity that enters lakes as precipitation. The most soluble salts such as NaCl and Na_2CO_3 were the first to dissolve during geological times and thus are now primarily found in the oceans. Less soluble salts including $CaCO_3$ are still being eroded. Thus the salt composition of lakes and oceans are fundamentally different.

The ions that are most abundant are not necessarily biologically most important. Many ions such as essential nutrients are biologically important because they are so rare, and they therefore can limit biological production. Phosphate, silica, nitrate, ammonia, and iron are examples of ions that limit the productivity of lakes.

In addition to inorganic ions, polar organic molecules that originate from biological activities also dissolve in water. Substances that make up the dissolved organic matter (DOM) in lakes are products of metabolism and decomposition of dead organic material. Lakes contain, as a rule, a wide array of dissolved organic substances. They are difficult to identify and are usually simply described as total dissolved organic carbon (DOC). DOC varies greatly with time of year and from lake to lake. Dissolved organic substances can provide an important energy source for microorganisms, but the majority of the DOC decomposes very slowly. This *refractory* (resistant) material represents the "leftovers," after the *labile* or rapidly transformed portions of the DOC have been used up. The dissolved organic carbon is a dynamic pool, with a constant inflow and outflow of DOC. The slow decomposition of dissolved carbon gives this pool a stabilizing effect on the carbon cycle of a lake. Small amounts of dissolved organic matter such as exoenzymes and signal chemicals can have important regulatory effects in natural waters.

3.2 VERTICAL GRADIENTS

3.2.1 Light

Almost all available energy on earth comes from the sun's radiation. The total radiation that reaches the earth's surface consists of direct sunlight and diffuse skylight. This *global radiation* includes wavelengths from 300 to 3000 nm, since the atmosphere absorbs some of the shortest wavelengths. It can be divided into three categories, each with differing effects:

300–380 nm	Ultraviolet; damaging effects on organisms
380–750 nm	Visible radiation; includes the photosynthetically active radiation (PAR; 400–700 nm)
750–3000 nm	Infrared radiation; produces heat

[26]

When radiation reaches a lake, a small amount is reflected, and the remainder penetrates the water, where it is absorbed. The portion that is reflected depends on the angle of declination of the sun, the wavelength, and the surface wave conditions of the water. In Middle Europe, for example, about 3% of the direct sunlight is reflected in the summer and 14% in the winter, and about 6% of the diffuse skylight is reflected. Reflection can increase to 30–40% when there is strong wave action.

Since light always comes from above in a lake and is always absorbed as it passes through the water, there is a vertical light gradient in every lake that has profound effects on the production and life in the lake. Light that penetrates into the lake is either scattered, absorbed as heat, or transformed into other energy sources. In photosynthesis, for example, it is stored as reduced carbon. The degree of light retention by a layer of water is referred to as the *light extinction*, and the amount of light passing through as the *light transmission*.

The concept of light extinction, strictly speaking, only applies to the decrease of monochromatic light with parallel light beams in pure solutions (Lambert–Beer law). Obviously, these conditions are not met in lake water, where light is polychromatic and suspended particles absorb and scatter light. The reduction of light intensity due to absorption and refraction of light is called the *vertical light attenuation*. It can be mathematically approximated by the Lambert–Beer law.

Light intensity does not decrease linearly as it passes through lake water, but rather as a fixed proportion of the light still remaining at each depth. This leads to an exponential decrease in light with depth:

$$E_d(z) = E_d(0)e^{-k_d \cdot z}$$

$$k_d = \frac{\ln E_d(0) - \ln E_d(z)}{z}$$

Where $E_d(0)$ and $E_d(z)$ represent the light intensities at the surface and at depth z, and k_d is the *vertical attenuation coefficient*. A higher k_d indicates light is absorbed more rapidly and the vertical light gradient in the water is steeper (Fig. 3.4). k_d can be calculated from light intensities measured at two depths with an underwater photocell (cf. Box 3.1).

Each wavelength that penetrates through the water has a different attenuation coefficient. The different absorption of light at different depths, depending on the wavelength, results in changing color with depth. Red light is absorbed most rapidly in pure water, and blue light has the greatest transmission. In pure water, about 65% of the red light (720 nm) is absorbed in the first meter ($k_d = 1.05$). In contrast, only 0.5% of blue light (475 nm) is absorbed in the same distance ($k_d = 0.005$). A diver, for example, sees no red hues below only a few meters of water. Dissolved substances and particles are also very important in determining light transmission in lakes. Algae shift the absorption maximum to the longer green wavelengths, and dissolved humic substances shift to the shorter yellow wavelengths. Thus both the light intensity and the spectral composition of the light changes with depth. This is of great importance for photosynthesis, which functions only within the range of 400–700 nm.

[27]

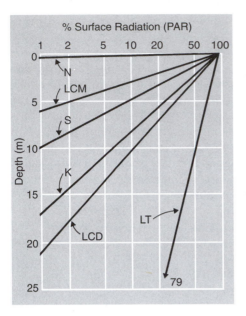

Figure 3.4 Decrease in the photosynthetically active radiation (PAR, 400–700 nm) with depth in different lakes. N, Lake Nakuru in Kenya; mean value for 1972 (k_d = 17.1); from Vareschi (1982). LCM and LCD, Lake Constance in May and December (k_d = 0.768 and 0.219); from Tilzer et al. (1982). S, Schöhsee (Holstein) in June (k_d = 0.461). K, Königsee, mean value for 1979 (k_d = 0.271); from Siebeck (1982). LT, Lake Tahoe (California) before eutrophication in 1970 (k_d = 0.058); from Tilzer et al. (1975).

BOX 3.1	LIGHT: UNITS OF MEASURE

Light can be expressed as energy or as mol quanta. These units cannot be directly converted, for the energy content of a light quantum is different for different wavelengths. The following units are generally used:
Radiation energy:

$$1 \text{ J} = 1 \text{ W s} = 0.2388 \text{ cal}$$

Energy flow density:

$$1 \text{ W m}^{-2} = 1 \text{ J s}^{-1} \text{ m}^{-2}$$

Photon flux density:

$$\text{mol m}^{-2} \text{ s}^{-1} \text{ or } \mu\text{E m}^{-2} \text{ s}^{-1} \text{ (E for Einstein)}$$

An approximation for PAR (400–750 nm):

$$1 \ \mu\text{E m}^{-2} \text{ s}^{-1} = 0.2 \text{ to } 0.25 \text{ W m}^{-2}$$

Occasionally the older term *illumination intensity* is still used with the unit of lux (lx), although it is usually not a useful measure and should be avoided in limnological research:

1 W m^{-2} = approx. 95 lx at 10 degrees sun elevation, approx. 120 lx at 50 degrees sun elevation and about 140 lx with heavy cloud cover.

Figure 3.4 compares the vertical transmission of photosynthetically active radiation (PAR) in lakes with contrasting concentrations of algae and turbidity. The light curves are straight when plotted on a semilogarithmic graph, since the light is absorbed exponentially with depth (cf. Fig. 3.5). One extreme curve is for Lake Nakuru in Kenya, which has high turbidity due to enormous densities of *Spirulina*, a cyanobacterium, and which supports large populations of flamingos. The other extreme is for the exceptionally clear Lake Tahoe in California. One can read from the graph the depth at which 1% of the light is still available. This depth is often used as a rough estimate of the lower boundary of the euphotic zone, the region in which there is a positive energy balance due to photosynthesis (cf. Section 4.3.5). This depth can change dramatically, depending on the time of year, as illustrated with the curves for Lake Constance in May and December. As an alternate to actually measuring the light intensity at different depths, one can estimate the transparency of lake water by determining the visibility of a standard white disk (Secchi disk) lowered into the lake (cf. Box 3.2).

One can easily see that there is a wide variety of colors of lakes. Such *water color* depends on the spectral range of the light reflected from the water surface (e.g., the color of the sky) and the color of the light that leaves the water after having penetrated it. Some of the light is scattered by water molecules and particles so that it eventually leaves the water. Light energy is lost indirectly as a result of scattering, which increases the distance traveled by light, thereby increasing the amount of light absorbed. Scattering is also selective, depending on wavelength; short-wavelength light is scattered more than long wavelengths. Pure water appears blue, since blue light is scattered most strongly and also has the greatest transmission. Chlorophyll content and humic substances provide green and brown water colors. Turbidity does not change the color, but it does reduce the color intensity.

Ice cover can strongly change the optical characteristics of a lake. Clear ice has the optical characteristics of distilled water and allows many algae to grow attached to the underside of the ice. Air bubbles, and most important, snow on top of the ice, strongly affect the light transmission. A 20-cm-thick dry snow cover can absorb and reflect 99% of the incident light, thereby seriously limiting the photosynthesis under the ice.

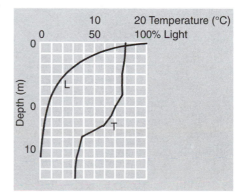

Figure 3.5 Depth profiles of light (L) and temperature (T) in a lake (Schöhsee, Holstein, Germany, June 7, 1983). There is a continuous decrease in light with depth from the surface; the temperature is uniform in the upper water layer due to turbulence (eddy diffusion).

BOX 3.2	A SIMPLE DETERMINATION OF TRANSPARENCY

The *Secchi disk* is a simple method for determining the *Secchi disk transparency* or depth of the maximum visibility in the water. The Secchi disk transparency is the water depth at which a white disk, usually 20 cm in diameter, disappears from the view of the observer at the water surface. It is a measure of half the distance that the light travels to the disk and back. The Secchi disk is a relatively reliable and often used estimate for the rapid measurement of the light conditions in a lake. It is remarkably useful despite the many factors that influence it such as the contrast between the disk and the environment, the light conditions, the visual acuity of the observer, and the diameter and reflectivity of the disk. Parallel measurements of different observers are surprisingly similar. Some typical visibility depths for different lakes are:

Extremely clear Crater Lake (U.S.)	40 m
Lake Constance (Germany) at various seasons	1.5–12 m
Schöhsee (Germany) in summer	5 m
Nutrient-rich northern German lakes in summer	<1 m
Carp ponds	20 cm
Lake Nakuru (Kenya)	5–10 cm

3.2.2 Temperature

Most of the radiation entering a lake, especially the long wavelengths, is absorbed near the surface and transformed into heat. Since molecular diffusion can be ignored (cf. Section 3.1.3), one should assume that the heat will remain where it was absorbed. We would therefore predict that the temperature will have an exponential decrease with depth just as with light. This, however, is not what happens (Fig. 3.5). The reasons for this are the density anomaly of water and the effect of the wind.

Since water is most dense at 4°C, it can become lighter either by cooling or by warming. Less dense water is buoyant compared to water that is more dense. From 4 to 0 water has a density difference of about 0.13 and 1.77 g/l between 4 and 20°C. The density difference gets larger with increasing temperature, and between 24 and 25°C it is 30 times as great as between 4 and 5°C. Most of the exchange of heat between lakes and the environment takes place through the water surface. The lake is warmed by solar radiation and cooled through radiational cooling (e.g., at night) and evaporation. As cooling surface water becomes more dense it sinks until it encounters water of even greater density. At temperate latitudes there are usually periods in the spring and fall during which the surface water is cooled to 4°C. Therefore, the deep water temperatures in lakes of sufficient depth is always near 4°C. One exception is high-altitude lakes that are extremely cold. In tropical lakes the surface water never cools to 4°C, so the deep water temperatures are higher, and in polar lakes that are always frozen the temperatures can be even colder.

The wind produces turbulence and currents at the water surface that mix the shallow water. When the warmest water is "floating" near the surface, the wind cannot easily mix it with the underlying cold water. This produces a resistance to the wind effect that is proportional to the density difference between the upper and underlying water. A well-mixed layer is formed in the summer that extends to a depth at which the force of the wind is equal to the positive buoyancy of the warmer water.

There is an increase in the stability (the resistance to mixing of two water masses) from the deep, cold water to the shallow, warm water, since the density changes per degree centigrade are greater at high temperatures than at low temperatures. Wind can only attack the surface, and its power dissipates rapidly with depth allowing a relatively sharp boundary between the mixed surface waters and the colder deep water. This gives rise to the typical temperature profile of a *stratified* lake with two separate water masses, the warm *epilimnion* and the cold *hypolimnion* (Fig. 3.6).

A region of greatest change in temperature, known as the *metalimnion,* separates these two layers. The metalimnion is not easily defined. If one follows the original definition of Birge (1897), the upper and lower boundaries of the metalimnion are located where the temperature difference is at least 1°C per meter. One can also lay an imaginary plane through the lake at the depth of the greatest relative temperature change (Hutchinson 1957). This plane dividing the lake into two levels is called the *thermocline.* The terms *metalimnion, thermocline,* and *Sprungschicht* (German) are often used as synonyms.

When the temperature of the epilimnion approaches 4°C in the spring, density differences between the water layers have almost dissipated. At this time a strong wind is sufficient to mix the lake to bottom, thereby initiating *spring circulation.* The lake then becomes *homeothermic,* with the same temperature from the surface to the bottom.

Increasing sunlight surface warms the surface water and the lake begins to stratify. At first the stratification is so weak that a brief episode of wind can easily destroy it, but a stable epilimnion develops after a short time. The lake then enters the period of

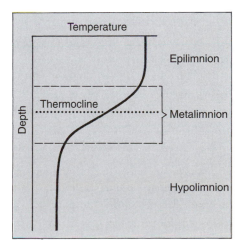

Figure 3.6 A typical summer temperature profile in a lake at temperate latitudes.

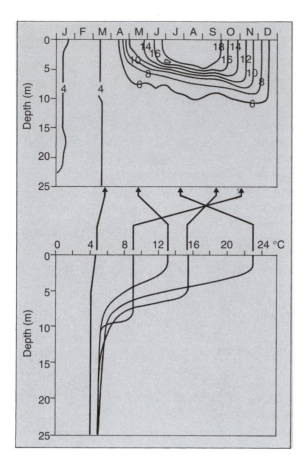

Figure 3.7 Temperature conditions in Plußsee (Holstein, Germany) in 1986. (Top) Presented as isotherms (lines of equal temperature). (Bottom) Temperature profiles from 5 selected dates (data from H. J. Krambeck).

summer stratification. The stratification becomes increasingly stable throughout the summer. The metalimnion and the thermocline deepen throughout the summer and into the fall (Fig. 3.7). Thus the greatest depth of the thermocline does not coincide with the highest water temperatures. For example, in Lake Constance, the thermocline is deepest in October.

By late autumn the epilimnion has cooled down and the wind can once again completely mix the lake. Following *fall circulation* the lake is homeothermic for the second time that year. Lakes that have an ice cover develop a reverse thermal or *winter stratification* with colder, less dense water directly under the ice floating on top of deep, warmer, and more dense water at about 4°C. An epilimnion cannot develop, however, for the water is protected from wind stress by the ice cover.

This change between stratification and circulation that takes place on an annual cycle in temperate latitudes can occur daily in shallow, tropical lakes. Very small temperature differences can sustain stable thermal stratification in warm, tropical lakes, since the relative density differences per degree temperature change is greatest at high temperatures. These lakes stratify during the day as the surface waters are heated, and each night, as the surface waters cool, the wind destratifies them.

Thermal stratification and heat exchange depend on solar radiation and wind. In addition to climate, the size and wind exposure of a lake, and occasionally water inflow, are the major factors that determine the type of circulation. The wind cannot mix a small, protected woodland pond as well as a large lake with a long axis oriented parallel to the wind. Lakes can be categorized according to the degree and frequency of their *circulation patterns* (Box 3.3).

The epilimnion is referred to as the *mixed layer*, implying that it is a thermally homogeneous water mass. Literally, this means that the lower limit of the epilimnion extends to the depth of the lowest temperature measured at the surface during a 24 h cycle, since a water parcel cooled at night at the surface could theoretically sink to that depth. However, investigations with highly sensitive temperature sensors show that the epilimnion is not completely homogeneous, especially on calm days. Even very slight temperature differences between the shoreline region and the open water can result in density currents that produce a *fine structure* in the thermal stratification that can be of great importance for the distribution of aquatic organisms (Fig. 3.8).

BOX 3.3   CIRCULATION PATTERNS IN LAKES

Amictic lakes never mix, since they are permanently frozen. Such lakes are found in arctic and antarctic regions and at very high altitudes.

Meromictic lakes mix only partially; the deep water layers never mix either because of high water density caused by dissolved substances or because the lake is so protected from wind effects.

Holomictic lakes mix completely and are classified according to the frequency of circulation.

Oligomictic lakes do not mix every year. Because such lakes are usually large and have a large heat storage capacity, whether or not they mix completely depends largely on the specific climatic conditions.

Monomictic lakes mix only once each year, either in the summer or in the winter (c.f. Section 3.3.4).

Cold monomictic lakes are found in polar regions. They thaw, but rarely reach temperatures above 4°C, and mix in the summer.

Warm monomictic lakes mix in the winter, since they cool down to about 4°C, but do not freeze over. An example is Lake Constance in S. Germany, which on the average freezes over about once every 33 years because of its large size.

Dimictic lakes mix twice a year (in spring and fall) and are the most common lake type at temperate latitudes.

Polymictic lakes mix frequently and sometimes even daily. These are usually shallow tropical lakes or shallow lakes at temperate latitudes with great wind exposure.

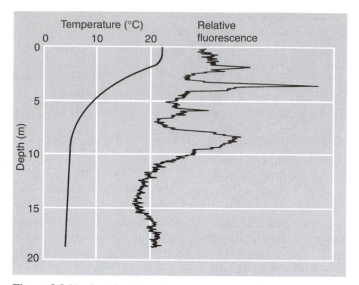

Figure 3.8 Vertical fine layering of organisms indicating that the "mixed zone" is not actually homogeneous. The fluorescence is caused by pigments (e.g., chlorophyll). Plußsee (Holstein), July 9, 1987 (data from A. Baker and C. Watras).

3.3 OTHER VERTICAL GRADIENTS

3.3.1 Oxygen

Most organisms require oxygen for their metabolic activities. An adequate supply of oxygen is never a problem for air-breathing organisms, but in aquatic environments sufficient amounts of oxygen are not always available. The supply of oxygen in water comes from exchange with the atmosphere or from photosynthesis by green plants and cyanobacteria (blue-green algae). Oxygen enters only in the upper waters, since photosynthesis is dependent on light and atmospheric exchange occurs at the water surface. Photosynthesis produces organic matter and releases oxygen (cf. Section 4.3.5), whereas aerobic respiration consumes organic matter and uses oxygen. Oxygen production usually predominates in the light, and oxygen consumption predominates in the dark. Lakes can be divided into two regions based on these processes: the lighted *trophogenic zone*, in which organic matter is synthesized and oxygen produced, and the *tropholytic zone*, where organic matter is decomposed and oxygen is consumed.

The trophogenic zone often, but not always, corresponds to the epilimnion. Much of the organic matter that is produced in the epilimnion eventually sinks to the deep water, but the oxygen produced remains in the epilimnion. Thus there is a sharp division between O_2 production and O_2 consumption. Addition of oxygen from the atmosphere and loss to the atmosphere, when the water is supersaturated, occurs when the surface waters are disturbed by the wind. This oxygenates the entire water mass during circulation periods, whereas during the stagnation periods only the epilimnion can exchange gases with the atmosphere and the hypolimnion has no atmospheric exchange. The oxygen available for decomposition of organic matter that sinks into the deep water must therefore come from the supply obtained during the periods of circulation. The hypolimnion constantly removes this supply of oxygen.

The concentration of oxygen in the deep waters depends on several factors:

1. The circulation pattern determines how often the supply of oxygen is renewed.

2. The size of the oxygen supply can be estimated from the volume of the hypolimnion times the oxygen concentration immediately after circulation. A deep lake has a larger hypolimnion than a shallow lake with the same surface area. The deep lake can therefore store more oxygen and has a trophogenic to tropholytic zone ratio that is more favorable for oxygen balance.

3. The quantity of decomposable matter that sinks into the bottom water is a function of the production in the epilimnion. In lakes with low productivity, most of the organic matter has been decomposed before it reaches the bottom water and therefore does not use up the oxygen there.

4. The rate of decomposition depends on the temperature. Decomposition occurs much more rapidly at 25°C in the hypolimnion of a tropical lake than at 4°C in a lake at a temperate latitude. In polymictic lakes in the tropics loss of oxygen during the daily temperature cycle is enhanced by the lower solubility of oxygen in warm water.

Both *productivity* and *morphometry* are critical determinants of the oxygen balance in a lake. For lakes with the same productivity, those lakes with a large hypolimnion decompose organic matter with little effect on the dissolved oxygen concentration, whereas lakes with little depth may have complete loss of oxygen. Partially decomposed organic matter that settles to the bottom of a lake forms sediments that are rich in organic matter. The water layer in contact with the sediments then also loses oxygen, for the decomposition processes continue within the sediments. Lakes with a large ratio of sediment surface to lake water volume tend to have a great loss of oxygen to the sediments. Loss of oxygen in the hypolimnion begins with the onset of stagnation, but in very productive lakes there can also be considerable loss of O_2 in winter under the ice. The longer the stagnation period, the greater the oxygen loss in the deep water.

The annual cycle of vertical changes in oxygen in a lake are closely tied to the pattern of circulation (Fig. 3.9). Every lake has its own peculiar oxygen profile (Fig. 3.10). Deep *holomictic* lakes with low productivity retain throughout the summer the same oxygen curve that they had immediately after spring circulation. In these lakes the water is saturated with oxygen from the surface to the bottom. The absolute concentration of oxygen may be somewhat less in the surface waters due to the lower saturation limit in the warmer water (cf. Section 3.1.4). Such oxygen profiles are called *orthograde* curves. In productive lakes where the oxygen concentration often decreases to zero in the hypolimnion, the resulting oxygen curve is described as *clinograde*. Clinograde oxygen curves can also develop minima and maxima near the metalimnion. For ex-

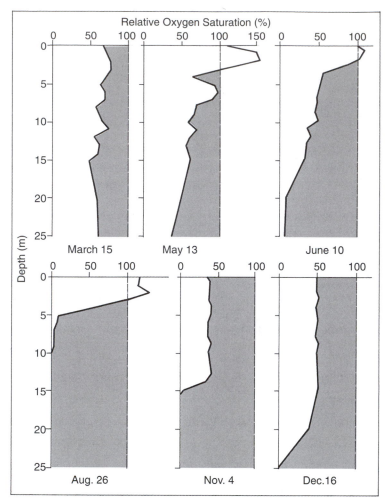

Figure 3.9 Depth profiles of dissolved oxygen (relative saturation) in a eutrophic lake (Plußsee, Holstein, Germany) throughout 1986. The shaded area clearly indicates the region of subsaturation. After ice out the lake becomes mixed homogeneously, but already has an oxygen deficit. The oxygen concentration in the deep water continually decreases during the summer stagnation period, whereas a supersaturation of oxygen develops in the surface waters. In the fall oxygen is once again mixed in from the surface (data from H. J. Krambeck).

ample, high bacterial activity may develop in accumulations of dead organic matter that is retained by the sharp density gradient. This produces a metalimnetic oxygen minimum or *negative heterograde curve*. On calm, sunny days a productive lake may have a supersaturation of oxygen in the epilimnion. Lack of mixing in the lower epilimnion prevents escape of the excess oxygen into the atmosphere, and an oxygen maximum or *positive heterograde curve* develops.

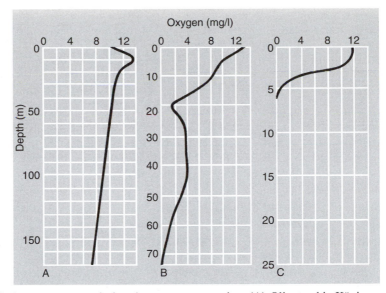

Figure 3.10 Typical oxygen curves during the summer stagnation. (A) Oligotrophic Königsee (July 5, 1980) has a nearly orthograde oxygen curve (from Siebeck 1982). (B) The deep, eutrophic Bieler Lake in Switzerland (Oct. 11, 1976) has a negative heterograde oxygen curve (from Tschumi 1977). (C) The eutrophic, wind-protected Plußsee in Holstein (Sept. 4, 1989) has a strongly clinograde curve (data from H. J. Krambeck)

The vertical distribution of oxygen in a lake is one of the most important abiotic factors for freshwater organisms. Aquatic life in lakes has evolved many adaptations to the wide range of oxygen conditions in a lake. For example, for some species the lack of oxygen in the deep water limits their vertical distribution, whereas for other species that can tolerate periods without oxygen (*anoxibiosis*), these oxygen-devoid regions represent a refugium from predators. Many organisms have life cycles adapted to the predictable changes in oxygen conditions. *Obligatory anaerobic* microorganisms can only live in regions that are free of oxygen. The composition of aquatic communities is highly dependent on the oxygen conditions in a lake.

3.3.2 pH

Vertical differences in biological activities in a lake lead to the development of vertical gradients and temporal changes in pH. Three major processes that affect the pH are photosynthesis, respiration, and nitrogen assimilation. The effects of photosynthesis and respiration on the pH depend largely on the carbonate–bicarbonate–carbon dioxide equilibrium (Section 3.1.6). The simplified formulae for photosynthesis is as follows, depending on the prevailing form of dissolved inorganic carbon (DIC), either CO_2 or bicarbonate:

$$6\ CO_2 + 6\ H_2O \rightleftharpoons C_6H_{12}O_6 + 6O_2$$

$$6\ HCO_3^- + 6H^+ \rightleftharpoons C_6H_{12}O_6 + 6O_2$$

Notice that no protons are used when CO_2 is taken up (assimilated) during photosynthesis, whereas one proton per atom of carbon is used when bicarbonate ions are used for photosynthesis. The reverse is true for respiration. The result is that when the pH is less than 6.3 and only CO_2 is present, respiration and photosynthesis have no effect on the pH. At higher pH values, when other forms of inorganic carbon are available, photosynthesis and respiration alter the uptake and release of protons. The exchange of protons first affects the alkalinity or acid-neutralizing capacity of the water. The effect that proton addition or removal has on the pH depends on the buffering capacity. The vertical gradient of pH that develops in a stratified lake (usually higher in the epilimnion) depends on the initial pH and the buffering capacity.

Nitrogen assimilation by aquatic organisms also can affect the pH of a lake. If ammonium ions (NH^{4+}) are used as a source of nitrogen, an equivalent number of protons must be released to maintain the balance of charges. When nitrate (NO^{3-}) is assimilated, in contrast, an equivalent number of protons must be removed. This generally applies for other ions that are taken up, but these are quantitatively unimportant. Since nitrogen assimilation is not as important as carbon assimilation, its effect on pH is most important at low pH.

3.3.3 Redox Potential

Many chemical and biochemical transformations in water are *redox reactions* or reactions in which there is transfer of electrons. The *electron donor* is referred to as the reducing agent and the *electron acceptor* as the oxidizing agent. In a redox reaction, the electron transfer changes the original reducing agent into an oxidizing agent, and the original oxidizing agent into a reducing agent. The direction and speed of the reaction depends on the amount of energy that is released. The more the equilibrium, $Red \rightleftharpoons Ox + e^-$, moves to the oxidation side, the more electronegative the redox potential becomes.

Photosynthesis and respiration can also be described in terms of redox reactions. In photosynthesis, CO_2 is the original oxidizing agent and H_2O is the original reducing agent. The oxidation level of carbon is reduced during photosynthesis ($+IV$ to 0), resulting in a reduced organic compound and oxygen as the terminal electron acceptor. Other biologically important elements also undergo redox changes (N, S, Fe, but not P).

Oxidation levels of biologically important elements:

C ($+$ IV): CO_2, HCO_3^+, CO_3^{2-}

C (0): C, CH_2O

C ($-IV$): CH_4

N (+V): NO_3^-

N (+III): NO_2^+

N (0): N_2

N (−III): NH_3, NH_4^+, $-NH_2$

S (+VI): SO_4^{2-}

S (−II): H_2S

Fe (+III): Fe^{3+}

Fe (+II): Fe^{2+}

The free electrons in a solution can be captured by an electrode placed in a redox system. A potential develops on the electrode that corresponds to the capability of the solution to oxidize or reduce, which is the redox potential (Eh) of the solution. The Eh is expressed in volts and is measured with a standard hydrogen electrode. Eh is often standardized to a pH of 7 (E_7), since Eh decreases 0.058 V for each increase of one pH unit.

Oxygen is especially important for the redox potential. Oxygen dissolved in water follows the reaction:

$$H_2O \rightleftharpoons 1/2\ O_2 + 2\ H^+ + 2e^-$$

The theoretical redox potential of water saturated with oxygen at pH = 7 and 25°C is 0.8 V. Redox potential is very insensitive to changes in the oxygen concentration per se. For example, a 99% decrease in oxygen concentration decreases the E_7 only about 0.03 V, as long as there are no increases in reducing substances associated with the decrease in oxygen.

The theoretical E_7 value of 0.8 V holds only for ideal conditions in a chemical equilibrium, where all redox reactions are reversible. Such conditions do not occur in natural waters. In an oxygen-rich epilimnion the redox potential is between 0.4 and 0.6 V, since the chemical adjustment to the equilibrium is slow and photosynthesis works against such chemical adjustments. When oxygen is not present, the reducing substances become noticeable. Thus lakes with an anaerobic hypolimnion generally have strong vertical gradients in redox potential. Reducing ferrous ions (Fe^{2+}) and organic matter decrease the redox potential in anaerobic regions. There is usually a rapid change in Eh at the sediment–water boundary, where the values in the sediments may reach −0.2 V and in the overlying water seldom drop below 0 V.

Since the redox potential has a strong influence on the solubility of elements, its seasonal changes in the hypolimnion and in the sediment–water interface are especially important in the nutrient cycling in lakes. Iron, for example, plays a key role in the availability of phosphorus (cf. Section 8.3.5). Lower oxygen concentration and the resulting sinking redox potential lead to a series of reductions in redox pairs, as shown in Table 3.1.

Iron is essentially insoluble as Fe^{3+} above an E_7 of 0.3 V, but as the redox drops below this level the iron changes to soluble Fe^{2+}, which can resolubilize the phos-

TABLE 3.1

Redox Pair	Redox Potential E_7 (V)	Corresponding Oxygen Conc. (mg/l)
$NO_3^- - NO_2^-$	0.45–0.40	4.0
$NO_2^- - NH_3$	0.40–0.35	0.4
$Fe^{3+} - Fe^{2+}$	0.30–0.20	0.1
$SO_4^- - S^{2-}$	0.10–0.06	0.0

phorus that was bound in the sediments. If the redox potential drops further, however, sulfate transforms into sulfide, which then combines with iron to form an insoluble, black iron sulfide that precipitates out of solution. As oxygen concentrations decline and reducing substances build up, phosphorus, the element most frequently limiting algal growth, is first dissolved, then precipitates out.

Reducing substances can be chemically oxidized by oxygen, but in nature it is often accelerated by biological processes such as respiration, nitrification, and bacterial oxidation of sulfur and iron (cf. Section 4.3.8).

3.3.4 Meromixis as a Special Case

Some lakes never mix completely at any time during the year. Such lakes are called *meromictic*. The depth below which no mixing occurs is the *monimolimnion*, and the upper water layer that undergoes the circulation according to the climate of the region is the *mixolimnion*. A strong vertical chemical gradient or *chemocline* separates these two zones.

Meromixis develops when the deep water of a lake becomes so heavy due to dissolved substances that its density cannot be equaled by the cooling of the surface water. An increase in salt content of 10 mg/l produces the same increase in density as a decrease in temperature from 5 to 4°C. In a clear lake with a high salt content in the deep water, the monimolimnion may be warmed with solar radiation well above 4°C without mixing, even when the surface waters cool to 4°C.

Different types of meromixis are distinguished on the basis of the source of salt in the monimolimnion: crenogenic meromixis from deep water springs, ectogenic meromixis from occasional delivery of ocean water in coastal lakes, and biogenic meromixis from minerals released by decomposition in the deep water and released from the sediments. Normally, the accumulation of decomposition products during summer stagnation are not sufficient to prevent circulation in the fall or winter. In very deep lakes with relatively small surface areas, inadequate cooling or wind in the fall or winter may result in the lack of a complete circulation. If this happens repeatedly, dissolved substances may accumulated to large enough concentrations to make the lake permanently meromictic. Almost all extremely deep equatorial lakes, such as Lake Tanganyika and Lake Malawi, have this type of meromixis.

A meromictic lake has the same type of vertical chemical gradient as a eutrophic, holomictic lake with an anaerobic hypolimnion. Even unproductive meromictic lakes tend to have an anaerobic monimolimnion, due to the long-term accumulation of decomposing organic matter. The stability of the chemocline also results in much steeper vertical chemical gradients of oxygen and other related substances than in comparable holomictic lakes.

3.4 RUNNING WATER

3.4.1 Flow

Flow is the primary factor that distinguishes streams from lakes. The constant mixing in streams eliminates all vertical gradients, except for light. The unidirectionality of stream flow is critical. For small organisms it means that, unless they can attach themselves to a secure structure, they will be transported downstream, with no possibility that they may be carried back to their original location simply by chance. Plankton that are so characteristic of lakes cannot exist in streams. "Stream plankton" are found only in large, slow-moving streams, with lakelike embayments and impounded regions where the washout rate is slower than the reproductive rate of the plankton population. These regions continually supply the stream with plankton (Reynolds et al. 1991).

Directed flow also means that resources such as nutrients for algae and food particles for animals (see Section 4.3.1) that are not utilized immediately are lost for an organism that is not moving, since the resources never return. On the other hand, new resources are continually supplied at the same rate from upstream. Thus at any point in the stream there is an equilibrium between incoming and outgoing matter.

Flow is a very powerful selective factor, to which stream organisms must be adapted. They must, for example, be able to withstand the shear forces of the water, to attach and utilize the boundary layers (see Section 3.1.5), and, at the same time, to extract sufficient nutrition from the water current. Fluctuating water flow is another strong selective force in streams. Extreme *high water* (floods) can mechanically disturb the stream bottom and have disastrous effects on the populations there. The structure of the stream bottom is rapidly altered by gravel and stones rolling downstream, thereby destroying the habitats of the organisms.

Stream flow is also responsible for structuring the stream bed and its colonization by organisms. Water currents sort particles according to their size and weight, with more rapid flow transporting the largest particles. Thus, in rapidly flowing streams, the bottom consists primarily of very coarse stones, whereas the bottom in quiet habitats is made up of sand and silt deposits. The effects of velocity on the bottom type can be roughly divided as shown in Table 3.2.

Normally one associates high stream velocities with headwater rivers in the mountains. This is correct, although the lower reaches of large streams also have high stream velocities. Stream velocity depends on the incline and smoothness of the substrate, but also on the cross-sectional area of the stream. Velocity increases with increasing gra-

TABLE 3.2

Velocity (cm/s)	Bottom Characteristics of the Water Body
3–20	silt
20–40	fine sand
40–60	coarse sand–fine gravel
60–120	small, fist-sized stones
120–200	larger stones

dient and size of the stream. In high mountain regions there is a steep incline, but streams tend to be small; in lowlands there is minimal drop, but there is a large quantity (cross-sectional area) of flowing water. Large rivers such as the Rhine River in Germany have a high velocity, even where there is little incline. Although one cannot see the effects of sediment transport since the banks of the Rhine have been built up, if a hydrophone is lowered into the river, one can hear the noise of gravel moving along the bottom.

The hydraulic conditions in a stream are determined by its morphology. Since the morphology changes according to its location and tributaries, it is often helpful to describe a stream in terms of its position or *order* in the hierarchy of its tributaries. A first-order stream is a *headwater* stream with no tributaries. After two headwater streams unite, they become a second-order stream. The juncture of two second order streams produces a third-order stream, etc. (Fig. 3.11). The world's largest river systems cover 11–12 stream orders. Although stream order does not say anything about the length, size, and watershed area of a particular stream, there are consistencies that

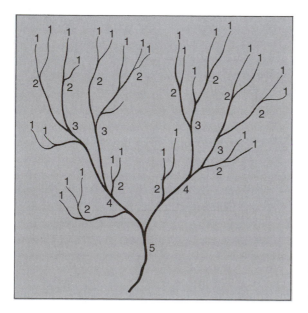

Figure 3.11 The hierarchy of a stream system. The numbers indicate the "order" of the stream.

allow some generalizations. For example, when one moves to a lower order, there are about three to four times more streams and each is on the average less than one-half as long and drains one-fifth of the watershed area (Hynes 1970).

3.4.2 Temperature

Because of the great turbulence and continual mixing, streams do not have a thermal stratification comparable to that found in lakes. Instead, streams typically have longitudinal temperature gradients from origin to end. Springs have very constant temperatures. The temperature at the point of emergence of a spring is determined by the average annual temperature of the drainage area and varies annually only a few tenths of a degree Celsius. In northern Germany this is usually around 8°C. As the water moves downstream its temperature gradually approximates the mean air temperature. This results in gradually increasing temperature downstream in the summer and decreasing temperatures as one moves from the stream source in the winter. Superimposed on these gradual temperature changes, however, are daily variations in temperature, the magnitude of which depends on the size of the stream. A small stream with a small volume of water is rapidly warmed by the sunlight, reaching it highest temperature in the afternoon and then cooling until it reaches it lowest temperature late at night. Small streams can vary 6°C or more in the summer (Hynes 1970), whereas large rivers with a correspondingly large heat capacity have minimal daily temperature fluctuations.

There is a characteristic pattern of temperature changes in the summer related to the fact that streams usually become larger as they progress downstream: with distance downstream, the mean temperature increases, the amplitude of daily changes decreases, and the annual temperature differences increase. This pattern is easily explained by temperature changes from the source to the mouth of the stream, increasing in the summer and decreasing in the winter. Temperature variations in tropical streams are usually much smaller than in streams at temperate latitudes.

Streams that pass through lakes become warmer in the summer as a result of the cooler stream water entering the lake sinking and being replaced in the outlet stream by warm, epilimnetic water.

3.4.3 Oxygen

The oxygen content of unpolluted mountain streams is near 100% relative saturation, due to the constant exchange of gases between the atmosphere and water enhanced by the turbulence. There are characteristic patterns of oxygen changes as one moves downstream. Spring water is often low in oxygen, since it has been underground for a long time. This deficit is quickly overcome by atmospheric oxygen in a rapidly flowing headwater stream. In streams of intermediate stream order the processes of oxygen production and consumption become more noticeable, since there are lags in the exchange of gases with the atmosphere. For example, attached algae and rooted aquatic plants produce oxygen during the day, and oxygen is consumed by decomposition of imported organic matter such as leaf litter. This leads to daily fluctuations in the oxygen concentration with highest oxygen concentrations in the afternoon, due to photosynthesis, and lowest concentrations at night. Such daily changes are primarily under biological control, since physical fluctuations such as temperature, which is cooler at night and warmer in the day, would have the opposite effect on gas concentrations. Physical factors are nonetheless still important in this example, as oxygen is gained from the atmosphere at night and excess supersaturated oxygen is lost to the atmosphere during the day.

The relative importance of atmospheric exchange lessens as one approaches the mouth of a river for the massive volume of water below the surface has less frequent contact with the air. Loading of organic matter, and resulting oxygen depletion, increase with stream order. In slow-flowing, dammed regions, however, planktonic algae may flourish, leading to supersaturation of oxygen. Photosynthesis by phytoplankton is unimportant in the subsurface waters of undammed rivers, for the algae are constantly forced down into the darkened depths where they cannot photosynthesize. Such streams usually develop oxygen deficits. This effect is exaggerated where these rivers collect waste water containing even more organic matter and causing more oxygen consumption. The Rhine River is a good example. In the 1970s, when the Rhine was seriously polluted, the oxygen content in the river was frequently less than 3 mg/l, thereby threatening the fish populations. Following the construction of water treatment plants, loading of decomposable matter was reduced considerably. As a consequence, between 1982 and 1987, the lowest measured oxygen concentration in the Rhine was 5.3 mg/l.

3.5 PREDICTABILITY OF ENVIRONMENTAL CONDITIONS IN WATER

The physical–chemical characteristics of water discussed in the preceding chapter make water bodies unique habitats. *Water buffers* against many of the fluctuations in the environment making living conditions more constant and predictable. For example, on a summer day the ground temperatures in a dry upland habitat may easily differ by 30°C between day and night, but in the epilimnion of a lake rarely exceed 2°C.

Thus, aquatic organisms have evolved many adaptions in response to biological factors rather than physical factors. This makes aquatic systems especially useful for the study of interactions of organisms. The following chapters will focus on these interactions.

The predictability of future conditions is important for an adaptation to evolve. Such predictability is high in aquatic systems, but in streams is limited by occasional flood events. Lakes are especially well suited for the study of predictable patterns and phenomena.

1. They are relatively well-defined systems. The term *ecosystem* is often an arbitrarily divided piece of the landscape, but with lakes, the land–water boundary also rep-

resents for many organisms a functional boundary that separates terrestrial and aquatic habitats. A lake is a relatively isolated system, despite many influences and exchanges of water with the surrounding land.

2. Lakes are not extreme habitats in regard to abiotic factors. There is never a shortage of water, for example, which is often a limiting factor in terrestrial habitats. The temperature is never below 0°C and rarely exceeds 30°C. There are differences in the chemical characteristics of lakes, but these reach extreme levels in only exceptional cases, such as acid-stressed lakes.

3. Abiotic factors are both temporally and spatially more predictable than in terrestrial systems. The high heat capacity of water dampens temperature changes. Maximum water temperatures in lakes show some variation in response to climatic changes, but seasonal changes in lake temperatures change slowly, and day-to-day variations are small. Lakes have predictable vertical gradients of temperature and light that, in turn, regulate the processes of production and decomposition. Anoxic conditions in the hypolimnion are extreme, but they develop slowly and predictably. Lakes generally do not develop an anoxic hypolimnion one year and not in the following year.

4. Lakes have a certain "island character," despite the wide distribution of many aquatic organisms. Density-dependent processes are important in the interactions of organisms, for relatively few species can emigrate when their population gets too dense.

This does not imply that lakes do not change. The appearance of lakes can change from year to year. Massive blooms of a particular species of algae need not occur each year, and the species composition of algae also may change. We can assume that such changes are not the direct result of variations in abiotic factors. In contrast to terrestrial habitats, water maintains a relatively constant "framework" of environmental conditions that is "filled" with biotic interactions. This makes it easier for the ecologist to exclude changes due to random variation, and to find the natural laws responsible for the observed changes.

REVIEW QUESTIONS FOR CHAPTER 3

1. Calculate the Reynolds number for several moving aquatic organisms. Assume water to have a viscosity of $\mu = 1 \times 10^{-3}$ Pa s. (a) A flagellate 5 μm long swimming at 1 m h^{-1}; (b) a diatom of 50 μm size sinking at 1 m d^{-1}; (c) a small fish 10 cm long swimming at 1 m s^{-1}; (d) a large fish 1 m long swimming at 3 m s^{-1}.

2. Plot the thickness of Prandtl's boundary layer as a function of flow velocity for distances (x) of 1, 5 and 10 cm, respectively.

3. Hard waters have higher concentrations of calcium and magnesium ions than soft waters. In which type of lakes do you expect higher concentrations of dissolved, inorganic carbon (DIC $= CO_2 + H_2CO_3 + HCO_3^- + CO_3^{2-}$) and why?

4. It is generally assumed that the compensation depth for phytoplankton photosynthesis equals the depth where ca. 1% of the surface intensity is measured. Calculate the vertical extinction coefficients for lakes with compensation depths of 3, 10, and 30 m.

5. Each of the following large lakes represents one of the three circulation patterns (dimictic, warm monomictic, oligomictic): Lake Vaettern, Sweden, Lake Constance in Germany, Lago Maggiore in Italy. Which lake has which circulation pattern?

6. Which type of lakes and in which season of the year do you expect a hypolimnion devoid of oxygen?

7. Lake Tanganyika in Africa is ca. 1400 m deep. It is oligotrophic, but it has an anaerobic deep water zone. How is that possible?

CHAPTER 4

The Individual in Its Habitat

Ecological requirements of organisms and the adaptations that allow them to meet these requirements are properties of the individual. Likewise, the foremost unit of selection is the individual as carrier of the genome. Physiological ecology investigates how the capabilities of an individual affect its fitness. The individual is not in all cases a well-defined unit, as, for example, colony-forming protists or plants such as reeds that are connected by rhizomes.

A certain minimal size is needed in order to get measurements of physiological activities and characteristics of individuals. For organisms smaller than about 1 mm, such as rotifers, protozoa, microalgae, and bacteria, such measurements are usually made on cultures consisting of experimental populations made up of many individuals. Ideally, clonal cultures are used to minimize variability, but even then genetic identity can only be ensured if clones are used soon after isolation so that mutations are not noticeable. Individual variability, however, cannot be identified from measurements made on experimental populations.

Measurements of the physiological capabilities of microorganisms are often measured in units that are only meaningful when applied to the level of populations. Growth rate of populations is an example, as well as migration movements of small organisms that are usually estimated by analyzing the distribution of a population, rather than observing the movements of each individual. Thus, unavoidably, concepts will be used in this chapter that more appropriately belong in the chapter on populations.

4.1 REQUIREMENTS OF THE INDIVIDUAL

4.1.1 Ranges of Tolerance and Optimality

For some time it has been customary to assign organisms a specific range of tolerance and optimal range for environmental factors such as temperature, pH, stream velocity, and availability of resources. It is often assumed that there is a unimodal curve that describes the response to a gradient of an environmental factor for activities such as metabolic rate, individual growth rate, and population growth rate. The highest point on this curve is then designated as the *optimum* condition for that particular environmental factor. The two points where the lines cross the zero level are the *minimum* and the *maximum*, and the distance between these forms the *tolerance range*. The shape of the curve is called the *reaction norm* (Stearns 1989). Species with a wide range of tolerance are traditionally referred to as *euryotes*, and those with narrow tolerances are *stenotes*. It is important to note that optimal curves do not describe the response of organisms to all environmental factors: a saturation curve applies to the response to resources (cf. Section 4.3), and for toxic substances the optimum is at zero.

Distinctions are usually made between "physiological" and "ecological" optima and tolerance ranges. The former relates to physiological functions that can be determined through experiments with individual organisms or pure cultures. The latter is inferred from distributions of organisms in nature and thus includes the effects of biotic interactions such as competition and predation (cf. Chapter 6). Neither of these approaches is adequate. The limits of tolerance and the optimal range for a species should be determined by examining the relationship of each activity to the actual environmental conditions encountered by the individual.

Extreme conditions can usually be tolerated for short periods. Thus the lethal range for a particular factor is the broadest definition of the tolerance range. The range in which a species has sufficient food to cover its energetic and material costs is narrower, and the range of conditions in which a species can successfully reproduce is even more restricted. All three of these ranges of tolerance, ranges for survivorship, nutrition, and reproduction, are determined by the physiological characteristics of the individual organism. Since they do not involve interactions with other individuals, these are physiological categories of tolerance and optimal ranges.

Through their metabolic activity, organisms alter their own environment and become part of the "environment" of other organisms (cf. Chapter 6). Predators, parasites, or competitors may prevent a species from establishing a stable population in a given habitat, even though the conditions are within the physiological tolerance range for reproduction. It is common that the distribution of a species is considerably more restricted than would be predicted by the physiological tolerance range, even when the population-limiting effects of historical influences such as the distributional history and geographical barriers are excluded. Biological interactions can shift the conditions for the population distribution maximum away from the conditions for the physiological optimum. For example, many aquatic animals live in cold, deep lake water, even though their physiological optimum is near 20°C, a result of the threat by predators in the warmer, shallow regions of the lake.

4.1.2 The Niche

The *ecological niche* is an often-used concept that is related to tolerance range; its meaning has changed considerably since it was introduced by Grinnell (1917). The modern definition can be traced back to Hutchinson (1958), who emphasized that organisms have ranges of tolerance for many environmental factors, rather than for only a single factor. Each environmental factor corresponds to an axis in a hypothetical multidimensional coordinate system. The niche is accordingly an *n*-dimensional *hypervolume* within this coordinate system. Figure 4.1 illustrates two niche dimensions using this approach. Because of the limitations of such a graphic presentation, a maximum of one additional dimension could be included in such a graph. Hutchinson makes the distinction between *fundamental* and *realized* niches, which are analogous to physiological and ecological tolerance ranges. The fundamental niche is thus the hypervolume, within which a species can occur without the effects of biotic interactions. The realized niche is the restricted niche into which the species is driven by competition and predation. The same distinctions apply to the fundamental niche as were made for the physiological tolerance range. This concept will be taken up again in the section dealing with interspecific competition (cf. Section 6.1).

4.1.3 Behavior as an Adaptation to Environmental Variability

The environment of an organism is not constant; it changes, for example, with the time of year. Also, within the life cycle of a species, the environmental pressures and the tolerances of the organism can change. Young stages may have a different niche from

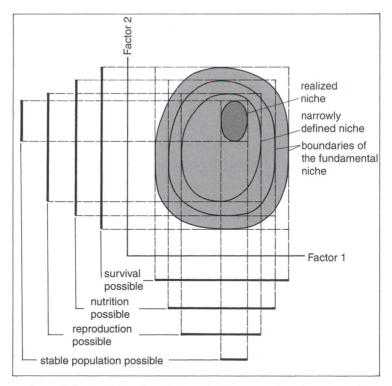

Figure 4.1   A comparison of the Hutchinsonian niche and tolerance regions for two environmental factors.

the adults. Thus it is sometimes necessary that a species divide its life cycle so that the tolerance range of a particular life stage is adapted to its environmental conditions.

The development of a resistant *diapause* or resting stage is the simplest form of adaptation of a life cycle to periodic appearance of detrimental or lethal conditions. Many aquatic plants lose their stems and leaves in the fall and overwinter as a perennial rhizome in the sediments. Insects can overwinter as eggs laid in the fall, but that hatch in the spring. They can also tolerate unfavorable conditions in resting stages as larvae and pupae. Diapause stages are also common within the plankton, as, for example, the zygotes of the green algae, the akinetes of the cyanobacteria (blue-green algae), the statospores of chrysophytes, the ephippia of cladocerans (cf. Fig. 5.13), and the resting eggs of copepods and rotifers.

Various changes in the environment can initiate the formation of resting stages, as, for example, changes in water temperature and day length or reduction in available nutrients. Nutrient decreases often cause the a well-defined pulse of resting stages at the end of the growing period (e.g., the dinoflagellate *Ceratium hirundinella*, Fig. 4.2), commonly referred to as *exogenous controlled encystment*. Golden algae (*Chrysophytes*), in contrast, form statospores that are entirely regulated by the population density, a process called *endogenous controlled encystment*. The golden algae therefore produce the most resting stages at the time of maximum vegetative cell density.

The realized niche is always smaller than the fundamental niche, giving mobile species a certain amount of flexibility when biotic or abiotic conditions around them change. Such species can take advantage of the spatial heterogeneity in environmental conditions by actively avoiding unfavorable conditions and actively searching for beneficial conditions, shifting in this way its realized niche. When, for example, the oxygen concentration decreases in the hypolimnion of a lake, many organisms are placed near the lower limit of their oxygen niche dimension. By moving into the shallow water near the shore they could survive, as long as their tolerance to temperature was not exceeded. It is interesting to consider why such species do not simply stay in the shallow regions after the lake has mixed and reoxygenated the deep water, rather than return to the deep water. It could be that the species returns to the deep water because it has a lower temperature optimum, even though it can tolerate the higher temperature. It could also be that the species moves to the deep water to avoid a higher mortality in the shallow water, such as the presence of predators. The last explanation is most likely. For example, diving ducks that feed on the zebra mussel can dive only a few meters deep (Hamilton et al. 1994), an example of a mortality factor that is limited to the shallow water.

The colonial, flagellated green alga, *Volvox*, has a *vertical migration* in the African reservoir, Cahora Bassa, that represents a *diel* or daily shift in the realized niche. The upper euphotic zone has scarcely any nutrients, but has the light needed for photosynthesis. In the deep water there is an accumulation of phosphorus, essential for cell growth, but, because of the high turbidity, there is inadequate light. To resolve this problem, *Volvox* migrates upward in the morning and downward in the evening. Most of the *Volvox* population stays in the euphotic zone during the day and during the night, when there is no light, moves into the phosphate-rich bottom water of the reservoir (Sommer and Gliwicz 1986).

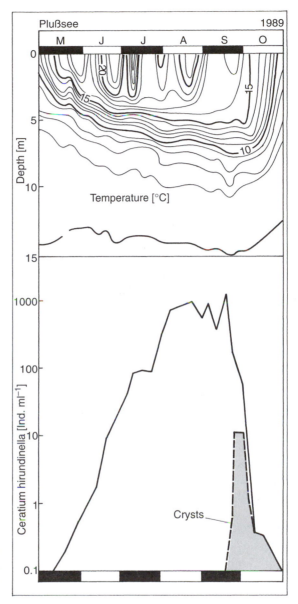

Figure 4.2 The exogenously controlled encystment of the dinoflagellate *Ceratium hirundinella*. The total concentration (vegetative cells and cysts) and density of the cysts (shaded area) in Pluβsee in comparison to thermal stratification.

Often, the ultimate factor responsible for the production of resting stages or changes in behavior can be found in the interactions between organisms, rather than in the abiotic factors (cf. Section 2.3). Many examples of adaptations in life histories and behavior that lead to shifts in the realized niche will be presented in later chapters. These adaptations often enhance the ability of an animal to gather food and avoid predators.

4.2 ABIOTIC FACTORS

4.2.1 Temperature

Aquatic organisms are generally faced with a narrower range of temperature fluctuations than terrestrial organisms. With the exception of hot springs, which may have temperatures near the boiling point, most lakes have temperatures somewhere between 0 and about 35°C, although the range of temperature changes in any single lake is usually smaller. Temperatures also change more slowly in aquatic systems than on the land. For one thing, aquatic organisms do not have to be able to withstand freezing. Thus the lethal limits of temperature are less important in aquatic than in terrestrial systems.

One can assume that heat tolerance is of primary importance to the distribution of organisms in hot springs: thermophilic bacteria can withstand temperatures up to 90°C, thermophilic cyanobacteria up to 75°C, and eucaryotic organisms up to a maximum of 50°C. Other than geothermically heated waters, there are only a few instances where heat resistance is as an important factor. One such case is the diatom *Asterionella formosa,* which does not occur in tropical lakes because it cannot survive temperatures greater than 25°C.

The absence of a species at higher temperatures does not mean that the species is necessarily limited by temperature. It is also possible that other factors that are correlated with temperature falsely suggest temperature dependency. For example, a species may not inhabit warm regions of a lake because the warmer water contains less oxygen. As another example, the scarcity of algae when the epilimnion is cooler might be caused by the deeper mixing depth associated with cooler waters, which in turn gives the algae shorter exposure times to high light for photosynthesis, limiting its growth and population size. Indirect effects of temperature may also lead to shifts in the competitive advantages of a species (Section 6.1).

Species that have a wide range of tolerance and a flattened optimum curve for temperature are called *eurytherms,* whereas species that live only within a narrow set of temperatures and have a steep optimum curve are called *stenotherms.* It was once common to classify organisms as warm- or cold-stenotherms, simply on the basis of their distribution in nature. Physiological tests have demonstrated such designations are not always meaningful.

Some aquatic animals cannot live at higher temperatures and thus retreat into the cool depths of the lake in the summer, as long as there is sufficient oxygen there. For example, the rotifer *Filinia hofmanni* cannot survive at temperatures above 10°C. The freshwater mysid shrimp *Mysis relicta* has a distribution that strongly suggests it is a "relict" from the glacial periods. It is assumed that with the warming climate since the retreat of the last continental glacier 10 to 15 thousand years ago *Mysis* was able to survive in only a few cool mountain lakes, in the hypolimnion of deep lakes and in lakes at northern latitudes. The distribution of *Mysis relicta* suggests that it can exist only at temperatures cooler than 14°C. This contention is supported by laboratory experiments that show that *Mysis* can tolerate higher temperatures, but, at most, only for a few hours. Similarly, there are examples of tropical species that can live and reproduce only above a certain minimal temperature.

Temperature is also important within the lethal limits, since it regulates the speed of the chemical, and ultimately, therefore, the biochemical and physiological processes. *van't Hoff's Law* describes this relationship within the temperature range of biological reactions. For every increase in temperature of 10°C, the rate of the reaction will increase by a factor of 1.5 to 4. This factor is called the Q_{10}.

Activity rates begin to decrease above the *temperature optimum*. Biologically important activities such as feeding may depend on many biochemical reactions. Most enzymes are stable within only a limited temperature range. If each of the individual reactions has a different Q_{10} value, then the physiological subprocesses can come into disequilibrium at temperatures outside the optimum (Hochachka and Somero 1984). Thus the relationship between temperature and biological activities is best described as unimodal curves (i.e., with a single maximum, Fig. 4.3a). A decrease in activity above the maximum is usually more rapid than the increase in the activity rate at suboptimal temperatures. Biological time periods, such as development times of eggs and generation times, become shorter with the speeding up of physiological processes and are minimal at the temperature optimum. Sometimes, however, this optimal temperature is outside the temperature range at which the organisms are actually found in nature, as, for example, the egg development time for some zooplankton species (Fig. 4.3b). On the other hand, some processes, such as photosynthesis, are relatively unaffected by temperature within a broad range. The light-limited rates of photosynthesis are regulated by temperature only below 4°C, whereas light-saturated photosynthetic rates are enzymatically controlled and are more dependent on temperature.

The direct effects of temperature on the ecology of species has often been overemphasized. The zonation of turbellarian flatworms provides a good example of this. There is a characteristic replacement of species along the course of a stream, where *Crenobia alpina* inhabits the upper reaches and species of the genus *Polycelis* live in the middle and lower regions. For a long time this was cited as evidence for the effect of temperature on the distribution of species, for the temperature tends to increase as one moves downstream (cf. Section 3.4.2). Differences in the maximum temperature tolerance were demonstrated for various species in the laboratory. The lethal temperature for *Crenobia alpina* is at 12 and 16°C for *Polycelis felina*, which occurs in lower reaches of streams in Great Britain, and 26°C for other *Polycelis* species. More recent experiments with diets and transplantation of species have demonstrated that biotic factors such as competition and direct interactions (Chapter 6) also influence the distributions of these species. Differences in temperature tolerance explains why *C. alpina* does not occur in the lower reaches, but not why *Polycelis* does not invade the upper regions of the stream (Reynoldson 1983).

4.2.2 Oxygen

Uneven distributions of oxygen conditions are common in aquatic environments (cf. Section 3.3.1). Oxygen deficiencies and even anoxic conditions are found in the deep waters of eutrophic lakes as well as in organically polluted streams. Ground water and spring water are often low in oxygen. Supersaturation of oxygen can be caused by high rates of photosynthesis during the day in nutrient-rich waters, resulting on sunny and

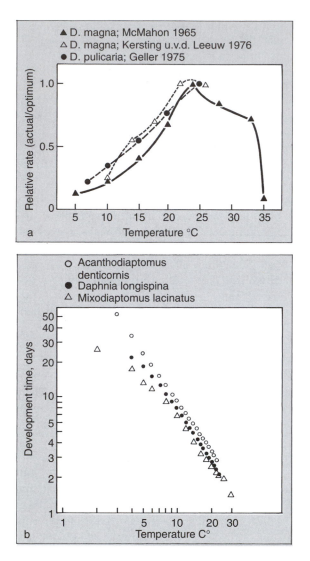

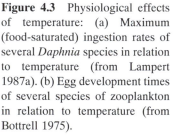

Figure 4.3 Physiological effects of temperature: (a) Maximum (food-saturated) ingestion rates of several *Daphnia* species in relation to temperature (from Lampert 1987a). (b) Egg development times of several species of zooplankton in relation to temperature (from Bottrell 1975).

windless days in oxygen saturation of 200% or more. Supersaturation may create a problem for aquatic animals. For example, gas bubbles may attach to the carapace of zooplankton, causing them to float and attach to the surface film. Occasionally, fish larvae in very productive ponds are harmed by ingesting gas bubbles. Problems due to deficiencies in the supply of oxygen, however, are much more common.

With the exception of a few types of specialized microorganisms (cf. Section 4.3.10), heterotrophic organisms need at least some exposure to oxygen to provide electron acceptors for respiration. Aquatic organisms have many morphological, biochemical, and behavioral adaptations to deal with the problems of limited or changing concentrations of oxygen (Dejours 1975, Prosser 1986). Some organisms live in water but use atmospheric oxygen. Others can use anoxibiosis to survive periods between exposures to dissolved oxygen.

The cellular exchanges of oxygen and CO_2 require that the gases diffuse across a boundary surface. The rate of gas exchange, therefore, depends on the surface area available. The relationship of the surface area to volume of small organisms is larger, and hence their ability to take up oxygen is better, because the surface area increases as a square and the volume as a cube function. Thus, protists and small metazoans simply use their entire body surface area for gas exchange.

Such simple solutions are not adequate for most larger metazoans and those with a hardened outer integument; they require specialized exchange surfaces such as gills that have a large surface area and are thin. These are usually protuberances of the skin that are highly branched, thin, and filled with hemolymph. They are delicate and protected in a chamber, as in fish and crustaceans. Larvae of insects such as mayflies have external gills that are thin membrane extensions of their tracheal system, which in terrestrial insects is entirely internal.

An oxygen gradient with an oxygen-deficient zone develops in the water near the surface area for gas exchange. Diffusion through this boundary layer supplies adequate oxygen for small organisms. With gills, there must be constant renewal of the surrounding water. Either the gills are moved in the water such as with mayfly larvae, or water flows past the gills as with fish and crustaceans. Larvae of caddisflies and chironomid midges produce a flow of water through their cases by rhythmic body movements. It is somewhat easier for some insect larvae in flowing water, for they can also utilize the stream flow to facilitate gas exchange with the water (cf. Section 4.2.5).

Some insects come to the surface to breathe atmospheric oxygen. The "rat-tailed larvae" of the fly *Eristalomyia* have an extreme adaptation to this mode of breathing. They can live in decomposing organic sediments that are low in oxygen by getting oxygen through a breathing tube that extends up to 15 cm to the water surface. The water spider *Argyroneta* lays away a supply of air bubbles in its submerged web. Pulmonate aquatic snails *Planorbis* and *Lymnea* breathe either through their entire body surface underwater, or by means of a lung, when they come up to the water surface from time to time.

Some aquatic bugs and beetles can carry a bubble of air with them that serves as a "physical gill." As the insect breathes from this air supply via its tracheal system it reduces the oxygen concentration and increases the CO_2. The increased higher partial pressure of CO_2 in the bubble causes this highly soluble gas to diffuse into the surrounding water, and, similarly, oxygen diffuses into the bubble in response to lowered partial pressure there. Normally, such physical gills are unprotected and therefore compress as the animal dives under water, causing nitrogen gas to diffuse into the water (Henry's Law), thereby gradually reducing the size of the bubble. Thus physical gills function for only a limited period, after which the insect must return to the surface to get a new supply of air. Some insects place a small supply of air between the hydrophilic hairs on their underside (*plastron*). The air cannot pass between the fine hairs because of the water surface tension and the bubble does not compress, providing a physical gill that need not be frequently replaced.

[55]

Some aquatic animals have the blood pigment hemoglobin that has a high affinity for oxygen and enables the animals to live in habitats with extremely low oxygen concentrations. Some species of *Daphnia* synthesize hemoglobin when they are placed in low oxygen conditions. Animals that come from oxygen-poor conditions, such as the lower regions of the thermocline in eutrophic lakes in the summer (cf. Fig. 3.9), are easily recognized by their pink color. They can use this zone of the lake as a refuge from predators that are unable to tolerate these low oxygen concentrations.

How a species regulates its metabolic rate in response to changing oxygen concentration can affect its distribution. There are two basic ways animals react to changing oxygen concentration (Fig. 4.4). *Conformers* have respiratory rates that are dependent on the external oxygen concentration. Thus their metabolic rate decreases with decreasing oxygen concentration (Fig. 4.4; note the example of the cladoceran *Simocephalus vetulus*). *Regulators* are able to maintain a relatively constant respiration rate over a wide range of oxygen concentrations. Below a certain lower limit of oxygen, regulation becomes inadequate. The river limpet, *Ancylus fluviatilis,* illustrated in Fig. 4.4 compensates for decreasing oxygen levels by increasing its ventilation movements. This figure also indicates that there are some species with intermediate types of respiration regulation.

Anoxibiosis allows some organisms to live for periods without any oxygen, although this is energetically inefficient (cf. Section 4.3.10). End products of anaerobic metabolism, including lactic acid, amino acids, succinate, and ethanol, must be broken down to repay the oxygen "debt" after the animal returns to aerobic conditions. Inhabitants of the deep water of lakes are well known for their anoxibiosis (cf. Section 7.6.2). Two important examples are the red chironomid midge larvae (*Chironomus*) and tubificid oligochaete worms (*Tubifix*). These bottom-dwelling animals can survive for many weeks without oxygen. Even copepods bury themselves in anoxic sediments when they undergo diapause. Animals become inactive and metabolic rates are very reduced dur-

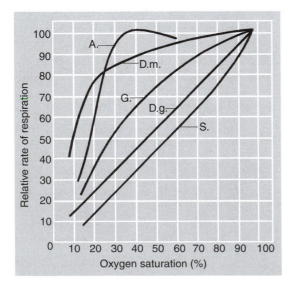

Figure 4.4 Regulation of the respiration rate of some aquatic animals with decreasing oxygen concentration. A, *Ancylus fluviatilis*; Dm, *Daphnia magna*; Dg, *Daphnia galeata mendotae*; G, *Gammarus fossarum*; S, *Simocephalus vetulus* (from Lampert 1984).

ing anoxibiosis. The phantom midge larva *Chaoborus* is an especially interesting example of switching between aerobic and anaerobic metabolism. They carry out a diel vertical migration in which they spend the daytime in the anoxic hypolimnion and bottom sediments and undergo anoxibiosis. At night they migrate into the oxic epilimnion to feed and "pay back" their respiration debt (Fig. 7.10a–e). This ability to use anaerobic metabolism allows *Chaoborus* to avoid predation by fish (cf. Section 6.8.4).

4.2.3 pH

It is difficult to distinguish between direct and indirect effects of pH because of its many influences on the water chemistry. Aquatic organisms have ranges of tolerance and optima for pH itself, in addition to the many indirect effects. Enzymes function within specific pH values and thus require that the pH of the cell plasma be held relatively constant. This process becomes more and more costly, the further the pH of the surrounding water deviates from the optimal range for the cell. The internal regulation of pH is not perfect. The cyanobacterium *Coccochloris peniocystis*, for example, is able to maintain maximum photosynthetic activity at external pH values of 7 to 10, even though the optimal pH range for the enzyme RuBP-carboxylase is in the narrow range of 7.5 to 7.8. When the external pH drops to 5.25, the cell can only adjust its internal pH to 6.6 and photosynthesis stops completely (Coleman and Coleman 1981).

The problem of *acidification* of weakly buffered waters by atmospheric pollutants (*acid rain*) has become very serious (cf. Section 8.6.3) and has led to many investigations of the effects of acid stress on aquatic organisms. A small decrease in the pH of the blood of fish, for example, results in a decrease in the ability of hemoglobin to transport oxygen. For the benthic amphipod *Gammarus lacustris* pH values of less than 5.5 cause death within a few days, and the cladoceran *Daphnia magna* has a striking decrease in survivorship below pH 4.5 (Fig. 4.5). Such physiological effects are probably due to the influence of pH on the ion transport through the cell membrane.

The most important indirect effects of pH are related to its effects on the calcium–carbonic acid equilibrium (cf. Section 3.1.6), the dissociation of ammonium ions, and the solubility of metallic ions, especially *aluminum*. Low pH has a strong effect on the solubility and speciation of metallic ions, many of which can be very toxic. The solubility of iron, copper, zinc, nickel, lead, and cadmium increases with decreasing pH, whereas vanadium and mercury tend to become less soluble in acidic conditions. Aluminum is an important aspect of the problem of acidification of natural waters. It is a ubiquitous component of siliceous rocks and one of the most common elements in the earth's crust. Thus it is often available in inexhaustible quantities in the watershed regions of lakes and streams. Below is a chemical description of the weathering of siliceous rock and the release of aluminum ions:

$$Al_2Si_2O_5(OH)_4 + 6\,H^+ \rightleftharpoons 2\,Al^{3+} + 2H_4SiO_4 + H_2O$$

$$Al(OH)_3 + H^+ \rightleftharpoons Al(OH)_2^+ + H_2O$$

$$Al(OH)_2^+ + H^+ \rightleftharpoons Al(OH)^{2+} + H_2O$$

$$Al(OH)^{2+} + H^+ \rightleftharpoons Al^{3+} + H_2O$$

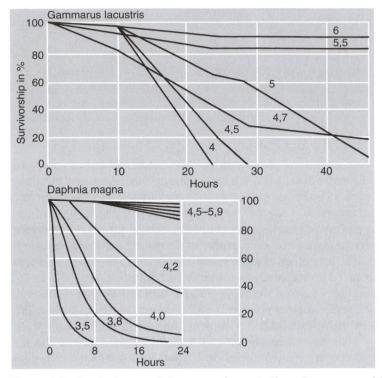

Figure 4.5 Survival rates of *Gammarus lacustris* (from Borgstrom and Hendrey 1976) and *Daphnia magna* (from Parent and Cheetham 1980) in relation to the pH of the water.

The Al^{3+} ion is toxic, but is relatively uncommon. Increase in the number of protons by acid rain enhances the weathering of aluminum and shifts the chemical equilibrium to toxic Al^{3+} ions (Overrein et al. 1980). The loss of species in acidified waters is often due to the combined effects of acid stress and aluminum toxicity.

The most important side effect of high pH for aquatic animals is related to changes between the harmless *ammonium* ion (NH_4^+) and the toxic undissociated **ammonia** (NH_3). The dissociation of ammonia is controlled by the pH. Below a pH of 8 there is almost exclusively ammonium, but above a pH of 10.5 it becomes almost entirely ammonia (Fig. 4.6). Conditions of high total ammonium concentration and high pH, caused by photosynthesis in poorly buffered and nutrient-enriched lakes, can lead to sudden fish die-offs when the critical pH is exceeded and concentrations of NH_3 develop.

4.2.4 Other Ions

The boundary between saltwater and freshwater represents one of the most critical limits for the distribution of aquatic organisms. A minimum number of species occur near the mouths of rivers and in other transition zones at salinities between 0.5 and 0.7%.

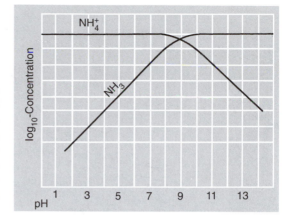

Figure 4.6 The effect of pH on the fractions of NH_4^+ and NH_3 at a constant concentration of total ammonium. The scale of the concentrations is relative, since the absolute values depend on amount of total ammonium available. Each line on the scale represents an order of magnitude (from Morel 1983).

The osmotic effect of the total concentration of dissolved salts is of primary importance. In the oceans, protists and most invertebrates are in *isotonic* equilibrium with the seawater, meaning the osmotic pressure is the same inside and outside of the cell. There is no need for *osmoregulation*, since the salt concentrations in the ocean changes little. The osmotic value of the protoplasma and the body fluids is adapted to the minor fluctuations in salinity of the surrounding water; that is, they are *poikilosmotic*. The absence of osmoregulation does not mean there is no *ion regulation*. Organisms can store certain ions and eliminate others, so that the total ionic strength remains constant.

Poikilosmotic organisms are rarely able to tolerate large fluctuations in salinity, such as are common in the brackish waters where rivers flow into the ocean. The osmotic value of freshwater is so low that "isotonic life" is not possible. Brackish and freshwater habitats can only be occupied by organisms capable of osmoregulation. There are few animals, such as the shrimp *Palaemonetes varians*, that are perfect *homeoosmotic* organisms able to maintain a constant osmotic pressure in their body fluids in both *hypotonic* and *hypertonic* surroundings. Hypertonic regulation, in which the osmotic pressure inside the body must be kept higher than pressure in the water, is most common in both brackish and fresh water organisms (Fig. 4.7a and b). The ability to osmoregulate varies greatly from species to species. The freshwater crayfish *Potamobius fluviatilis* keeps an almost perfect salt balance with its surrounding water, whereas the brackish water polychaete *Nereis diversicolor* is essentially poikilosmotic.

Species that have moved from freshwater into habitats with elevated salinities, such as salt lakes, have lower osmotic pressures inside their bodies compared to the surrounding water. One of the most efficient and most common hypotonic regulators in the salt lakes, including the Great Salt Lake in the United States, is *Artemia salina*. This salt water shrimp can live in salt pans that have salt concentrations at the saturation limit for sodium chloride.

Hypertonic regulation works against the invasion of excess water and loss of ions, whereas hypotonic organisms must minimize water loss and excrete ions. Organisms must have an active uptake and removal of water and ions, since surfaces that serve

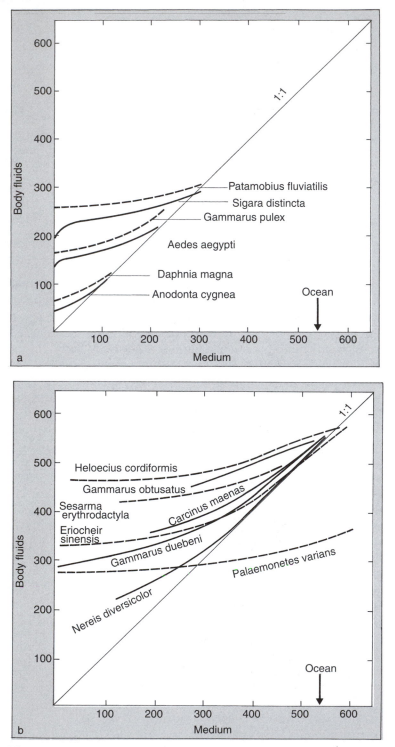

Figure 4.7 Relationship between the concentration of ions (mM l^{-1}) in the external medium and the body fluids of invertebrates from fresh water (a) and brackish water (b) (from Beadle 1943).

[60]

for exchange with the environment, such as cell membranes of osmotrophic protists, gills, and intestinal epithelium, cannot be impermeable.

Energy must be expended to regulate the concentration of ions against a concentration gradient. Since this energy would otherwise be used by the organism for other functions such as growth and reproduction, one can assume that a species that occurs outside its optimum range for ion concentrations will have a diminished fitness, even though it can survive there. The concept of tolerance is too simplistic to define the success of a species when its biotic interactions are included (cf. Chapter 6). Very small differences in reproduction and mortality can have a tremendous effect on relative fitness. Fitness parameters, such as population growth rates (cf. Section 5.2.3), are much more powerful predictive tools than tolerance curves that only deal with survivorship of organisms. Figure 4.8 illustrates the use of a fitness parameter to define a relationship referred to as a *standard response curve*.

Calcium is usually the dominant cation and carbonate or bicarbonate the dominant anion in freshwater. Researchers often relate the occurrence of species and higher taxa with the calcium concentration of the water. Although correlations are often found, the causal relationships remain largely unexplained. *Holopedium gibberum*, a cladoceran with a large gelatinous sheath, is a well-known example of a species that only occurs in lakes that are very low in calcium.

Calcium can be considered a "micronutrient" for phytoplankton, since they need only micromolar concentrations. The calcium concentrations in most lakes are much higher, in the range of 0.1 to 6 mM, and is thus not usually a limiting resource for phytoplankton (cf. Section 4.3.3). Indirect effects of low calcium concentration are more important, such as the decreased ability of water to take up CO_2 and the reduced buffering capacity.

Another important characteristic of calcium is its ability to form complexes with *humic substances*, which then precipitate (*coprecipitation*). Calcium-rich waters therefore tend to be relatively clear, even if they receive considerable dissolved humic sub-

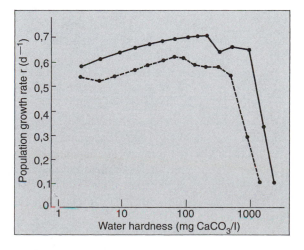

Figure 4.8 The effect of water hardness on the population growth rate of two cladocerans, *Daphnia magna* (solid line) and *Ceriodaphnia dubia* (dashed line) (from Cowgill and Milazzo 1991).

[61]

stances from outside sources such as bogs, whereas calcium-poor lakes that are rich in humic substances are typically "brown water." Coloration by humic substances alters the light environment by increasing the vertical attenuation (cf. Section 3.2.1) and shifts the spectral light distribution toward the long wavelengths. Humic lakes have shallower mixing depths than comparable clear water lakes, due to the greater absorption of the heat producing radiation. Dissolved humic substances also form complexes with other metallic ions, thereby influencing the availability of trace elements and the toxicity of poisonous ions.

One might expect that organisms most directly affected by calcium are those that have heavily calcified shells or carapaces, such as mollusks and crayfish. There are, in fact, more species of amphipods, isopods, decapods, mussels, and snails in calcium-rich than in calcium-poor lakes. After molting, the freshwater crayfish *Cambarus affinis* requires 32 days of exposure to constantly renewed calcium-rich water (1.7 μM Ca) to harden its new exoskeleton completely. In water with only 0.5 μM Ca, the exoskeleton stays paper-thin and soft and even the claws do not harden. The calcium supplied in food cannot substitute for the calcium that must be absorbed from the water to harden the exoskeleton (Mann and Pieplow 1938). Nevertheless, there are animals in soft (calcium-poor) water that are heavily calcified. The river pearl mussel, *Margaritifera margaritifera*, is normally found in soft water streams, but it has a very thick shell. This mussel, which lives for more than a century, grows very slowly, forming its shell over a long period. The difficulty *M. margaritifera* has in maintaining its calcium balance can be seen in corrosion of the older parts of the shell; which is, in fact, an identifying feature of this mussel.

4.2.5 Water Flow

Stream organisms are faced with the danger of being carried downstream, away from the habitat to which they are adapted. They have evolved morphological and behavioral adaptations to avoid this. The flow in a stream is not the same throughout (Lancaster and Hildrew 1993). Barriers on the stream bed such as stones and plants form "dead water regions" in which there is little water movement. Organisms, such as amphipods (*Gammarus*) and aquatic pill bugs (*Asellus aquaticus*), that lack specialized devices for attachment utilize these areas. Fish in fast-moving streams also use these dead water regions to reduce the energy required for swimming.

Stream algae form thin coverings on the rocks within the boundary layers (cf. Section 3.1.5), or they occur as flexible threads that swing in the water with little resistance. Such algae are either attached to the substrate or sometimes fused to the rock by a layer of calcium carbonate deposited by algal photosynthesis. In mountain streams with a flow rate of about 1 m/s, the boundary layer is millimeters thick. Animals that graze on the upper rock surfaces are flattened and streamlined (Fig. 4.9) and are rarely more than 4 mm high. They can press themselves against the substrate with varying degrees of pressure, depending on the flow rate (Ambühl 1959). Amphipods are laterally flattened and swim against the current while lying with their side against the substrate.

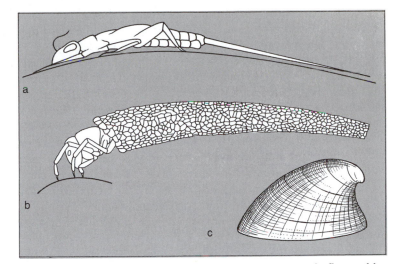

Figure 4.9   Stream animals that are adapted to flowing water: (a) The extremely flattened larvae of the mayfly *Rhithrogena* (ca. 10 mm). (b) Larvae of the caddisfly *Neothremma*, with a stream-lined case (ca. 10 mm). (c) Stream limpet *Ancylus fluviatilis*.

Recent investigations have demonstrated that the hydrodynamics involved in adaptations to stream flow are more complicated than would seem at first glance. One can measure very precisely the *flow field* around a stream organism with new laser- Doppler anometrics (Fig. 4.10), describing how the flow field varies with the flow conditions as well as the size of the animal (*Reynolds number*) and its position on the substrate. The dead water space that occurs behind the animal also determines the resistance of the animal to the flow. The stream limpet (*Ancylus fluviatilis*) gives the impression that it has an optimally streamlined silhouette (Fig. 4.9c). Flow measurements indicate, surprisingly, that the limpet has the same water resistance whether it is faces forward or backward into the stream flow (Statzner and Holm 1989). Such studies have also disproved the assumption that certain caddisfly larvae weigh down their cases with pebbles, for the increased weight achieved is counteracted by the increase in resistance.

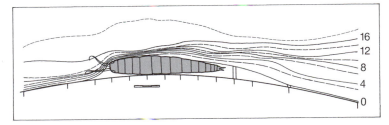

Figure 4.10   Lines of equal flow velocity (cm/s) around a laterally compressed amphipod *Gammarus* lying against a rock. The flow comes from the left. Note the "dead-water space" behind the animal (from Statzner and Holm 1989).

The shape of a stream organism is apparently a compromise between different factors. Water resistance, friction, and flotation must be minimized to prevent the organism's being swept away. On the other hand, being in a high-water-velocity and, therefore, smaller boundary layer, offers the advantage of better diffusion of oxygen and ions through the gills. Moving water can also bring food to an animal that does not move about. Some stream animals have thus evolved specialized feeding mechanisms that utilize the flowing water. Some have, for example, a special filtering apparatus that is held in the flow, and others spin submerged nets (cf. Fig. 4.25a and b).

As mentioned in Section 3.1.5, water moving over an object produces a lift that pulls the organism away from the substrate. It is therefore not enough to merely lie flattened; an organism must also have a means of attachment, such as hooks and suction cups. The foot of *Ancylus fluviatilis* acts like a suction cup. Also, the larvae of the net-winged midge (*Blepharoceridae*) has ventral suction discs on each segment, permitting it to live in fast riffle areas, and even waterfalls, where there are few competitors or predators.

Many small species and the young stages of larger species live below the stream bottom in the spaces between the sand and gravel. There is no sharp boundary as one goes downward in a stream. There is a system of openings within the stream sediments in which subterranean water can flow and is often connected to the groundwater, which, in turn, flows parallel to the stream. Such regions with underground water systems that extend downward from a few centimeters to a meter, depending on the structure of the stream, are called **hyporheic** or **hyporheic interstitial** systems. The water flow in hyporheic habitats depends on the size of the openings, with slowest flow rates in fine sand. Hyporheic species are thus protected from high stream velocity, but may still receive renewed oxygen supplies from the flowing water. Early stages of many species can penetrate so deep into the stream bottom that they are not threatened by moving sediments during floods. When flood waters retreat, these animals return to the surface to inhabit the stream bottom.

Even stream organisms that are adapted to living in fast-flowing water occasionally become dislodged and are carried short distances to a new location downstream. When organisms are found floating downstream, they are referred to as *stream drift*. By placing a net in the water, one can measure the drift rate as the number of organisms caught per unit area or volume over a given time interval. Stream drift has a pronounced diel periodicity, with the largest number of organisms generally caught during or shortly after sunset and the fewest number found in the drift nets during the day. The initiation of drift is regulated by both the light intensity and the relative change in light (Haney et al. 1983).

The evening increase in drift is related to the movement of animals from underneath the rocks to the rock surface, where they feed. Species that are found on the upper sides of the rocks become more active at night (Cowan and Peckarsky 1994). This heightened nocturnal activity results in a greater probability that animals get swept downstream. There is growing evidence that many species are very capable swimmers and actively release themselves from the substrate (*behavioral drift*) to enter the drift. Such drift movements can serve to distance prey from areas of high predation or help in the location of habitats with better food resources. This effect is compounded by the fact that many invertebrate predators in streams are more active at night, causing their prey to release from the rocks.

Since the flow of a stream is unidirectional and all stream organisms may at some time enter the drift, one might predict that a population in a stream should gradually move downstream and the headwaters should eventually become depleted of organisms. This does not happen, however, indicating that there must be some mechanism that compensates for the losses due to drift. First, many aquatic animals are positively *rheotactic*, meaning they move against the direction of the water flow, tending to move them in an upstream direction. Sometimes one can observe groups of amphipods (*Gammaridae*) moving upstream along the stream bank where the flow is not so strong.

Also, many stream organisms are insects that spend only part of their life cycle in the water. Many of these species have *compensatory flights* upstream, allowing them to lay their eggs in the upper reaches of the stream. Their young gradually drift downstream after hatching, occupying the appropriate habitat once again. It is not yet clear how such adult insects orient their flights in an upstream direction. This behavior is not necessary for a species that produces an excess of eggs and whose populations of larvae are regulated by density-dependent factors. In such cases, the flight direction of the adults might well be random, for enough eggs will be laid upstream to maintain the downstream population.

Drift should not be viewed as simply a disadvantage to stream organisms, for it is an important mechanism for colonization that allows species to find new and better habitats and may also reduce competition for space and food (cf. Section 5.7).

4.2.6 Density of Water

Sinking and floating
Gravity and buoyancy determine whether a particle in water is suspended, sinks, or floats. Since gravity does not change, the specific density of the particle is especially important. Most organisms are more dense than water. Although the fresh weight of all organisms is mainly water, other components are heavier, such as carbohydrates (approx. 1.5 g/ml), proteins (approx. 1.3 g/ml), and nucleic acids (approx. 1.7 g/ml). Minerals are even heavier, as, for example, polyphosphate granules (approx. 2.5 g/ml) and diatom frustules (2.6 g/ml). Lipids (approx. 0.86 g/ml) and gas vacuoles of cyanobacteria (approx. 0.12 g/ml) are unusual in that they are less dense that water. Most aquatic organisms that lack heavy mineral components have a density of 1.02 to 1.05 g/ml, whereas diatoms, which have much silica in their shells, may have densities up to 1.3 g/ml.

Most aquatic organisms tend to sink in water. Only a few have specializations that reduce their specific gravity to less than that of water. The flow around sinking particles is laminar when the Reynolds numbers are small (cf. Section 3.1.5). Under such conditions the sinking velocity of the particle can be calculated according to *Ostwald's* modification (Ostwald 1903) of Stoke's law for falling spheres, which adjusts sinking rates for the effect of form resistance of nonspherical objects:

$$V_s = (2/9)gr^2(q' - q)\,\mu^{-1}\phi^{-1}$$

where V_s is the sinking velocity (m/s); g, the earth's acceleration (9.8 m/s²); r, the radius of a sphere with equivalent volume; q', the density of the sinking particle (kg/m³); q, the density of the medium (kg/m³); μ, the dynamic viscosity of the medium (kg m⁻¹ s⁻¹); and ϕ, the form resistance (dimensionless; for a sphere it is 1; for most other bodies it is $>$1; only for vertically oriented, elongate bodies it is somewhat $<$1).

This formula is an excellent description of the sinking rate of phytoplankton, since they have such small Reynolds numbers. Even one of the largest diatoms found in freshwater, *Stephanodiscus astraea* ($r = 25\ \mu$m, $v = 0.1\ \mu$m/s) has a Reynolds number of about 0.001, at which there is negligible deviation from Stoke's formula. Deviations from Stoke's law become important at Re$>$1 (Reynolds 1984), such as for zooplankton. These relatively large plankton, however, depend more on active swimming to stay in suspension.

Sinking imposes a strong selective pressure on the ability of plankton to float. Photosynthetic organisms must maintain themselves in the euphotic zone, and heterotrophic organisms that feed on these phytoplankton must also stay where the food is located. Stoke's formula indicates which parameters can be modified through evolution to reduce the sinking speed, namely, the particle radius, the density of the suspended particle, and its form resistance. Particle size is the most sensitive parameter, since the effect of the radius is squared in the formula. This is one of the reasons that, as a rule, planktonic organisms are so small (phytoplankton are usually $<$1 mm, zooplankton $<$ 1 cm).

Small changes in the density also can have a dramatic effect, since the difference between the medium and the organism is small ("*excess density*," q'-q). A diatom with a density of 1.2 g/ml is only 15% heavier than a green algal cell (1.04 g/ml), but if they were both the same size and had the same form resistance, the diatom would sink 5 times as fast. Organisms can lessen their average body density by increasing the storage of lighter components such as lipids and gas bubbles and by excreting a *mucilaginous* or *gelatinous sheath*. Mucilage is a gel that consists mainly of water, held together by a network of hydrophilic polysaccharides. Since the sheaths consist of a much higher percentage of water than protoplasm they reduce the density of the entire organism. Phytoplankton that are especially large often possess mucilaginous sheaths. The cyanobacteria *Microcystis* is an extreme example. It forms colonies that consist of relatively small cells that are held together in a common mass of mucilage that can be several millimeters long. Note the sheath can reduce the sinking velocity only within limits, since the density of the mucilage is never less than the water. Also, sheaths increase the size of the organism, which in turn increases the sinking velocity disproportionately as a squared function. At some point, the accelerating effect of size will equal the braking effect of the sheath.

It seems unlikely that reduction of sinking rates is the ultimate cause for the development of sheaths by large organisms. Zooplankton such as the cladoceran *Holope-*

dium gibberum and the colonial rotifer *Concochilus unicornis* form gelatinous sheaths. The sinking rate of *Holopedium* is in fact reduced by about 50% by the presence of its sheath. Recent evidence concerning the costs of producing the gelatin (Stenson 1987) and its advantage in predator–prey interactions suggests the clear capsule is a protective mechanism (cf. Section 6.5.5). This is also probably true for algal colonies enclosed in mucilage.

Deviations from a spherical form result in a reduction in the sinking velocity by increasing the *form resistance*. For example, an elongate body with a length 4 times its width has a form resistance of ca. 1.3. The long, needlelike diatom *Synedra acus* has a length-to-width ratio of 15:1 and a form resistance of 4.

Complicated shapes of colonies further increase the form resistance. An individual rod-shaped cell of the diatom *Asterionella formosa* has a form resistance of 2.5, whereas the form resistance of a star-shaped colony consisting of 8 *Asterionella* cells is 3.9 (all values are from Reynolds 1984). Overlapping boundary layers may develop from the laminar flows around the cells resulting from the low Reynolds numbers (cf. Section 3.1.5). A colony can therefore act like an individual particle (parachute effect). With increasing size, colonies face the same problem as with increases in mucilage: The effective diameter of the particle increases, thereby increasing the V_s, so that eventually the size effect exceeds the form effect.

Thus the *selective advantage* of large colonies is therefore not just the increased form resistance. Scientists first suggested that the complicated morphologies of planktonic organisms were adaptations to reduce sinking by increasing form resistance. Spines were commonly referred to as "flotation appendages." More recently it has become clear that many of these examples are adaptations that protect against predation (cf. Section 6.4.2), and they are interpreted as *defense mechanisms*. This change is evidence of the increasing emphasis ecologists place on biotic factors.

Most of the sinking rates have been determined for relatively large algae. Sinking rates range from about 1 m/day for large diatoms, several dm/day for the desmid *Staurastrum cingulum*, and several cm/day for small diatoms. The swimming speed of flagellates is usually at least 10 times the sinking rate of similar-sized algae. Thus flagellated phytoplankton have little loss due to sinking.

Some cyanobacteria can control their density by means of *gas vacuoles* to the point that they become lighter than water and float. These cyanobacteria can migrate up and down in the water column by metabolically regulating the number of gas vacuoles (Walsby 1987). Photosynthetic production of carbohydrates allows for the regulation of excess weight in the cell. Cyanobacteria store these photosynthetic products as glycogen. Large amounts of glycogen (density 1.5) are formed under high light conditions in the surface waters, acting as ballast and causing the cell to become heavier. At the same time the pressure inside the cell increases because of the increased glycogen, causing the gas vacuoles to collapse. The cells sink, thereby moving into a depth where light is limiting. Here the glycogen is utilized for cell metabolism and the cells becomes lighter, causing them to move upward again. The light dependency of photosynthesis results in a daily rhythm of vertical migrations during windless periods (Reynolds et al. 1987). Some of the largest phytoplankton in freshwater include cyanobacteria species with gas vacuoles (large colonies, e.g., *Microcystis*). Size has the same effect on floating as on sinking. For example, large particles that are negatively buoyant due to gas vacuoles have a higher velocity of ascent than smaller particles. Thus, as buoyant cells aggregate, they form large colonies that move more rapidly toward the surface of a lake.

The best known example of a zooplankter that can control its buoyancy is the phantom midge *Chaoborus*. *Chaoborus* can finely regulate its density by means of two pairs of contractible swim bladders that are connected to its tracheal system. Normally these larvae float horizontally, waiting for prey. However, they can undergo pronounced diel vertical migrations (cf. Fig. 7.10a–e).

The sinking velocity (v_s) calculated from the Stoke equation is valid only for water without currents. In the epilimnion, however, there is much turbulence, which reduces sedimentation. This mixing allows particles to remain in suspension, despite the fact they are also sinking. The extent of the actual population losses due to sinking depends on the relationship between the depth of mixing (size of the epilimnion, z) and the sinking velocity (v_s) (Fig. 4.11). When there is no turbulence, within the time interval t' all individuals sink the distance v_s/t'. If all individuals N_o are homogeneously distributed within the epilimnion, at the end of the time interval, $N_o v_s t'/z$ individuals will sink out of the epilimnion and $N_o(1 - v_s t'/z)$ individuals will remain in the epilimnion. If the epilimnion is subjected to turbulent mixing, the individuals remaining will be redistributed within the epilimnion. The epilimnion will not be completely depleted as long as $v_s t'$ is smaller than z.

We can use following two extreme examples to illustrate how sinking losses over extended periods can be estimated:

1. During calm periods there is a time interval of about one day (t') between mixing events, the result of convective mixing due to cooling during the night. If there is no reproduction and no losses other than sinking, the number of individuals present at time t (N_t):

$$N_t = N_o(1^{-v_s t/z})^t$$

where t is defined as number of days.

2. When there is a wind effect and little daily warming there is continuous mixing; that is, the time interval t' tends toward 0 and the number of mixing events per day tends toward infinity. This case can be calculated according to Reynolds (1984)

$$\frac{dN}{dt} = N(v_s/z) \quad \text{or} \quad N_t = N_o \, e^{-v_s t/z}$$

In nature, actual sinking losses would lie somewhere between these two extremes. In both situations the losses increase with the increasing quotient v_s/z. Diatoms with a sinking velocity of 1 m per day and a mixing depth of 2 m have a loss rate of 50% per day of its original population according to the single mixing model and ca. 39.4% per day with the continuous turbulence model; with a 10 m mixing depth the losses are only 10% and 9.5%, respectively, per day.

[68]

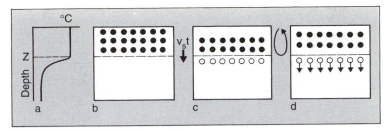

Figure 4.11 The effect of sinking velocity (v_s) and mixing depth (Z) on the losses due to sinking from a mixed epilimnion: (a) Thermal stratification. (b) Distribution of particles at the onset of the quiet water period t'. (c) Distribution of particles at the end of the quiet water period t'. All particles have traveled the distance $v_s\, t'$. A third of the particles (open circles) have settled below the lower boundary of mixing. (d) Distribution of the particles following a mixing event at the end of period t'. The particles below the mixing boundary are not resuspended and continue to sink.

Swimming

There is no clear division between floating and active swimming, for it is largely a matter of scale of observation. Even though many phytoplankton and zooplankton can actively swim, we consider them "plankton." Their swimming movements prevent them from sinking out of the water column, but they cannot avoid being transported horizontally by water currents. Nevertheless, zooplankton can undergo impressive vertical as well as horizontal migrations. Various groups of aquatic organisms have developed a variety of means of propelling themselves through the water, including the ciliary movements of flagellates and ciliates, paddling appendages of insects and crustaceans and highly specialized swimming movements of fish that cannot be explained by normal hydrodynamic models. The ability to move, relative to the water, depends largely on the Reynolds number (cf. Section 3.1.5), which in turn depends on the size and shape of the organism. Different laws apply to very small organisms that move in essentially a laminar environment than for large organisms whose environment is turbulent. This is summarized nicely by Vogel (1995).

Active swimming always costs *energy*, which is then not available for other uses, such as reproduction (Koch and Wieser 1983). The amount of energy used depends on the environmental conditions and the *swimming activity*. Many freshwater fish, such as the perch *Perca fluviatilis*, swim on the average at a speed of 0.5–0.8 m/s. An Atlantic salmon (*Salmo salar*) can swim as fast as 6 m/s during its spawning migration. Such swimming activity costs an enormous amount of energy. A fish swimming slowly (0.8–1.0 body lengths/s) doubles its metabolic rate (respiration) compared to its resting condition. This increases to an eightfold increase in respiration for a Pacific salmon (*Oncorhynchus nerka*) swimming at 4.1 body lengths/s. However, this rate can only be sustained for about 1 h (Brett 1965).

There are no comparable measurements for small organisms, such as zooplankton, that essentially swim only to compensate for sinking. Estimates based on hydrodynamics and an elegant experiment by Alcaraz and Strickler (1988) show, however, that the energetic costs for swimming are negligible (less than 1% of the total energy used for metabolism). This also seems to hold for small, motile planktonic protists (flagellates) that swim with flagella (Crawford 1992).

4.2.7 Surface Tension

The high surface tension of water (cf. Section 3.1.4) makes the *surface film* an important habitat for a specialized group of organisms called *neuston*. The surface film provides both a mechanical attachment point for neustonic organisms as well as a surface that collects organic substances. Organic liquids have a much lower surface tension than water, causing them to accumulate in the gas–water boundary layer. Similarly, this boundary layer acts as a trap for particles carried in the atmosphere and organic matter transported in aerosols. Substances caught in the surface film are exposed to higher light, especially higher UV radiation, than substances dissolved in the epilimnion. This results in chemical reactions that make the water surface a habitat that differs chemically from the epilimnion. Of special importance is the higher availability of organic substances (low molecular weight) that can be utilized by bacteria, which leads to higher densities of bacteria. This, in turn, creates better conditions for organisms that feed on bacteria.

The surface film acts as a mechanical barrier preventing the penetration of particles with hydrophobic surfaces. Small particles with high densities may stay on top of the water surface (*epineuston*). Similarly, this allows *hyponeuston* to attach to the underside of the surface film. To live on the surface film an organism must either have an entire body that is hydrophobic or at least certain *hydrophobic* structures with which it attaches.

Neustonic bacteria, algae, and protozoans usually have an entirely hydrophobic body. However, there are certain neustonic algae that hang from "swimming umbrellas" (*Nautococcus mammilatus*) or sit on mucilaginous stems that are hydrophobic (*Chromulina rosanoffii*). It has been determined that, just as in higher plants, these single-celled epineuston have an internal movement of water by transpiration. Rhizopodia located inside the mucilaginous stem penetrate into the surface film. These organelles collect water, which is then excreted by the cell body that is extended into the air.

Certain spiders and insects are the largest organisms that are supported by the surface film despite their high density. The most well known is the water strider *Gerris*, having leg hairs and the underside of its body that are hydrophobic.

4.3 RESOURCES

4.3.1 What are Resources?

All organisms must feed; that is, they must remove energy and matter from their environment in order to maintain their basal metabolism, form new body substance, and reproduce. In addition, many organisms require other consumable environmental factors (e.g., space for sessile organisms, nesting places). All these consumable environmental factors are referred to as "resources." A shortage of one or more of these resources leads to a reduction in the growth rate or even to death from starvation. Resources are consumed; this distinguishes them from other factors that limit physiological rates and growth (temperature, water current, toxic substances in the environment, etc.). *Consumption of resources* means a reduction in their abundance in the environment and thus involves interactions between organisms and their environment.

Organisms can be divided into fundamental trophic types according to the type of energy source, their electron donors, and their source of carbon. Light (*phototrophic* organisms) or the energy from exergonic chemical reactions (*chemotrophic* organisms) can serve as sources of energy; inorganic matter (*lithotrophic* organisms) or organic matter (*organotrophic* organisms) can be used as electron donors; CO_2 (*autotrophic* organisms) or organic matter (**heterotrophic** organisms) can serve as C sources. Within the heterotrophic organisms there are those that consume dissolved carbon compounds (bacteria, fungi), as well as those that consume particulate food (animals). *Mixotrophs*, which are capable of forming body substance from autotrophy or heterotrophy, are transition forms between auto- and heterotrophs. **Auxotrophic** organisms cover most of their C requirements from CO_2 but cannot synthesize a few organic substances (vitamins), which must thus be taken up by heterotrophy (Table 4.1).

TABLE 4.1
Nutritional types of organisms and the primary resources used for obtaining energy and synthesizing biomass (POC = particulate organic carbon, DOC = dissolved organic carbon).

	Energy source	C-source	e-donator	e-acceptor
Photoautotrophs				
Plants, Cyanobacteria	Light	CO_2	H_2O	CO_2
pigmented sulphur bacteria	Light	CO_2	H_2S	CO_2
sulphur-free purple bacteria	Light	CO_2	H_2O	CO_2
Chemolithoautotrophs				
colorless sulphur	S*	CO_2	S*	O_2
bacteria (*; H_2S, S, or S_2O_3)	S*	CO_2	S*	NO_3
nitrifying bacteria	NH_4	CO_2	NH_4	O_2
	NO_2	CO_2	NO_2	O_2
iron-oxidizing bacteria	Fe^{2+}	CO_2	Fe^{2+}	O_2
methane bacteria	H_2	CO_2	H_2	O_2
	H_2	CO_2	H_2	NO_3
Chemolithoheterotrophs				
Desulfovibrio	H_2	DOC	H_2	SO_4
Chemoorganoheterotrophs				
animals	POC	POC	POC	O_2
aerobic bacteria, fungi	DOC	DOC	DOC	O_2
denitrifying bacteria	DOC	DOC	DOC	NO_3
desulphurizing bacteria	DOC	DOC	DOC	SO_4

4.3.2 Consumption of Resources ("Functional Response")

The rate at which organisms consume their resources depends on the availability of the resources as well as the abilities of the organisms themselves. Even if a resource is very abundant, there are limits to the rate an organism can consume it. Animals require time to locate, hunt, handle, and devour their prey. An increase in prey density may reduce the time needed to locate and hunt the prey, but will not reduce the time required to handle and devour each prey. Microorganisms possess a limited number of transport systems, limiting the amount of dissolved resources that can be transported to the inside of the cell per unit time. When all transport systems are functioning at maximum speed so that further increase in resource concentration does not result in an increase in the consumption rate per individual, it is referred to as *saturation*. At this point, the resource will be consumed at a species-specific *maximum rate of consumption, v_{max}*, which may vary depending on the external boundary conditions (e.g., temperature). At resource concentrations below the saturation limit consumption cannot proceed at maximum speed, either because the animal does not encounter the prey frequently enough or because the enzymatic transport of dissolved substances cannot go faster because of the shortage of substance to be transported. The consumption rate thus becomes limited by the availability of the resource; that is, it depends on the concentration or availability of the resource as well as on the ability of the consumer to acquire a resource in short supply.

The dependency of the specific rate of consumption (v, consumption rate per individual or per unit biomass) on the resource availability (S) is described by a *saturation curve*. These curves typically have a linear or nearly linear increase at low concentrations. This increase defines the ability to use a limited resource ("affinity"). The major difference between the various models for resource limited processes is whether the upper limit (v_{max}) is approached asymptotically or whether it is reached by a sharp change from limitation to saturation. The most common asymptotic model is the right-angle hyperbole, which appears under different names in the literature. Its original version is the *Michaelis–Menten* equation, which was developed to describe the kinetics of enzyme use of a substrate. The most widely used model for an abrupt transition (rectilinear model) is the **Blackman** model, which assumes a linear relationship between consumption rate and resource concentration in the region of limitation (Box 4.1).

Michaelis–Menten and Blackman kinetics were developed for organisms with homogeneously distributed resources that occur as small units (dissolved molecules or small particles, relative to the size of the consumer). For this situation, the time needed to ingest an individual resource unit is negligible. This is not true for predatory animals, since the prey tend to be large, relative to the size of the predator. In this case, the consumption rate depends on the length of time it takes the predator to consume a prey item, as well as on how many prey it can capture per unit time. While the predator is devouring its prey it cannot look for other prey. Thus, in addition to search time, *handling time* is also important.

Holling (1959) described three types of *functional response* for predators (Fig. 4.12). His Type I curve applies to cases where the handling time is very short. It is similar to the Blackman model. Type II describes a predator with a relatively long handling time. It corresponds to the equation of Michaelis-Menten, where the half-saturation constant

BOX 4.1 **MODELS OF THE EFFECTS OF RESOURCE ON THE CONSUMPTION RATE (FUNCTIONAL RESPONSE)**

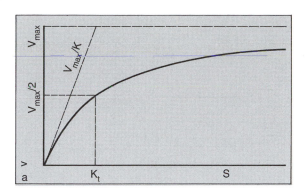

Michaelis–Menten Model

$$v = v_{max} * S/(S + k_t)$$

where v is the consumption rate (mass/time); S, the concentration or availability of a resource (mass/vol); v_{max}, the saturation value for the consumption rate; and k_t, the half-saturation constant, the resource concentration at which $v_{max}/2$ is reached. If resource uptake requires a minimum concentration, a threshold value (k_o) must be introduced. In this case, $S - k_o$ replaces S; one-half the maximum consumption rate is then reached at $S = k_o + k_t$. When resources are scarce, the important initial rise in the Michaelis–Menten curve is v_{max}/k_t.

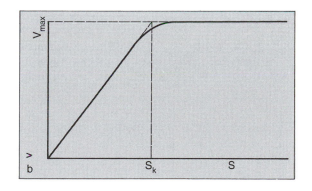

Blackman Model

$$v = S\alpha, \qquad \text{for } S < S_k$$

$$v = v_{max}, \qquad \text{for } S > S_k$$

$$\alpha = v_{max}/S_k$$

where S_k is the saturation concentration, and α is the initial increase. Threshold values can also be introduced into the Blackman model.

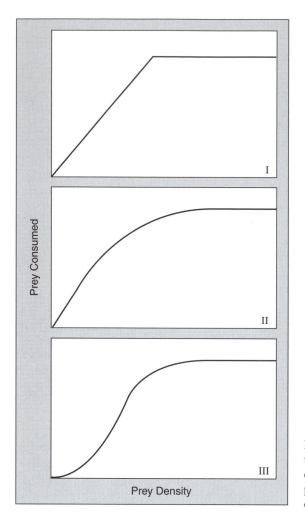

Figure 4.12 Models illustrating the relationship between feeding rate of a predator and the density of its prey. Three functional response curves of Holling (1959).

(k_t) is replaced by the maximum rate of consumption v_{max} and a specific consumption rate (F). The consumption rate (v) as a function of the prey concentration (S) is

$$v(S) = v_{max}\, S/(v_{max}/F + S)$$

F indicates, for example, the space a predator can search per unit time. Holling's Type II model can be used for many predator–prey relationships, such as for the rotifer *Asplanchna*, which feeds on other rotifers; for the phantom midge larva *Chaoborus*, which eats small zooplankton; and for the backswimmer *Notonecta*, which selectively feeds on *Daphnia*. Holling's Type III model has a sigmoid shape. The number of prey eaten per unit time first increases with the prey density, then reaches a plateau. This type of response might be expected when the predator learns over time how better to catch the prey or if the predator becomes increasingly motivated by the increasing density of a particular prey species and begins to feed on it selectively. Fish, which are very capable of learning, might show this type of behavioral change. However, there is no such case documented for fresh water.

Fish often eat relatively large, scarce items, such as benthic animals, and therefore have long search times. The handling time for fish is, however, relatively fast, since they simply swallow their prey. The classic saturation model developed for fish comes from Ivlev (1955): $v = v_{max}(1 - e^{-\delta s})$. The constant δ describes the steepness of the increase in the curve.

There are also many other ways to describe limitation-saturation models mathematically (cf. Jassby and Platt 1976). Actual data usually have such large variability that it is difficult to select the best model (Mullin et al. 1975). Nevertheless, some fairly well-accepted conventions have developed: The dependency of nutrient uptake on its concentration in water for bacteria and algae is usually described with the Michaelis–Menten equation; the Blackman model is most often used to describe the relationship between feeding rate of filter-feeding zooplankton and the food concentration (Fig. 4.13a–c). The importance of the parameters of these equations in the comparative physiology of potentially competing organisms is a strong case for following these conventions, even if a different model may provide a somewhat better description of the data.

To unify the models the consumption rate is defined as v (or v_{max}). Other symbols have been used in the literature, for example, I (I_{max}) for ingestion rate and P (P_{max}) for predation rate.

4.3.3 Regulation of Abundance and Growth by Resources (Numerical Response)

In the previous section we used the concept of *limitation* by a resource. The concept can have several meanings. Originally, J.v. Liebig meant the limitation of harvest from a field caused by a shortage of indispensable (*essential*) nutrient elements; the limited size is thus a static parameter, namely, the biomass or number of individuals per unit surface or space. It is instructive to note that essential elements can set an upper limit for the biomass or population density that can be achieved in a habitat, since the elemental composition of biomass can vary only within certain limits and no more than 100% of the amount of an element that is available in a habitat can be utilized. The elements in the biomass are in a stoichiometric relationship to one another that varies little. Thus the element that is least available relative to this stoichiometry (*minimum factor*) must determine the size of the harvest (*Liebig's "law of the minimum"*). When the availability of the minimum factor is increased, the biomass harvest will increase, up to that point where another resource becomes the minimum factor.

The law of the minimum assumes that resources are not *substitutable*. This is undoubtedly true for essential elements, which cannot replace one another in their biochemical functions (e.g., phosphorus in nucleic acids cannot be replaced by other elements). This is also true for chemical compounds that a particular organism cannot synthesize. Resources are substitutable if they contain the same essential elements or chemical compounds, even if in different proportions. We will explain this later.

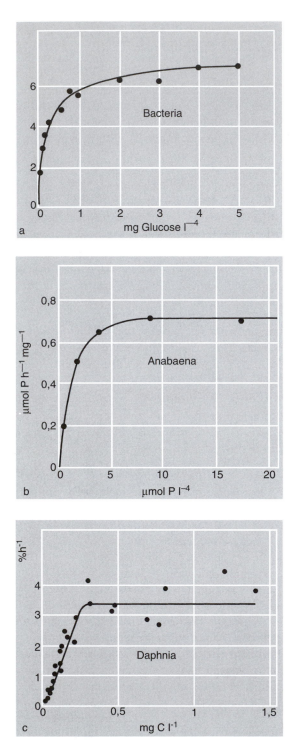

Figure 4.13 Specific rate of consumption as a function of the concentration of resources, for bacteria, phyto-, and zooplankton: (a) Uptake rate of radioactive labeled glucose by a natural assemblage of bacteria in Lake Erken (from Wright and Hobbie 1966). (b) Specific phosphorus uptake rate (in μmol P h^{-1} mg^{-1} dry weight) by the cyanobacteria *Anabaena variabilis* (from Healey 1973). (c) Specific feeding rate of *Daphnia* on *Scenedesmus* (expressed as % of its own biomass per hour; from Lampert and Muck 1985).

The process of growth lies between the act of consumption and the development of biomass. If both consumption rate and the maximum biomass are resource limited, it is likely that the intermediate process of growth is also resource-limited. Resources that are consumed may be first deposited in a *reserve pool*, which can later be used for growth processes. This can lead to a temporal decoupling between growth and the consumption of the resource, so that when there is temporal variability in the resource concentrations, the growth rate may reflect the earlier conditions.

It is necessary to distinguish between individual growth and population growth (reproduction) when analyzing the *kinetics* of resource-limited growth. Microorganisms generally divide after they have doubled their body size. Thus one can study their growth limitation using the increase in the number of individuals (*population growth*). At this point before the chapter on populations, it is thus necessary to introduce the concept of growth rate for microorganisms.

The reproductive activity of an individual per unit time is referred to as the *specific growth rate*. If each individual doubles within a time unit, after x doublings there will be 2^x individuals. The temporal changes in the number of individuals follows a geometric increase; the absolute increase in individuals per unit time gets larger and larger, although the reproductive activity of each individual remains constant. In a large population individuals do not all divide at the same time (synchronously), but rather divisions are distributed randomly through time. In such populations there is a continuous exponential increase in the number of individuals, rather than a stepwise geometric change:

$$N_2 = N_1 \, e^{\mu(t_2 - t_1)}$$

where N_1, N_2 are the number of individuals at time 1, 2; t_1, the beginning of the time interval; t_2, the end of the time interval; μ, the specific growth rate in d^{-1} or h^{-1}. A specific growth rate of 0 means there is no reproduction; a specific growth rate of ln 2 (ca. 0.69) d^{-1} means one doubling per day. Assuming there are no losses of organisms, μ can be estimated from the change in the number of individuals:

$$\mu = \frac{dN}{dt} \frac{1}{N} = \frac{\ln N_2 - \ln N_1}{t_2 - t_1}$$

Often a biomass is used instead of the number of individuals in the calculation.

The kinetics of resource-limited growth can be described with saturation curves similar to those used for the kinetics of resource consumption. If resource consumption and growth are not decoupled, the growth rate can be directly related to the concentration of available ("free") resources in the environment. The *Monod equation*, which is identical to the Michaelis–Menten equation, is usually applied to such growth (Box 4.2). If there is a temporal decoupling of consumption and growth (e.g., P-limited growth of algae), either the growth rate must be related to the amount of resource already in the organism (*cell quota*) or experimental provisions for a constant supply of resources must be made (chemostat cultures, Box 4.2).

[77]

BOX 4.2	STATIC AND CONTINUOUS CULTURES

A Static Culture (Batch Culture)

A static culture consists of a small number of organisms (inoculum) added to a known amount of medium, with no further additions of medium. Following a brief adaptation period (lag phase), the organisms increase exponentially and continue to reproduce at a maximum rate until a resource becomes limiting. The concentration of available resources declines as the increasing population of organisms consumes the resources, virtually eliminating the minimum factor. Uptake of the limiting resource no longer allows for further reproduction of the organisms. If there is continued reproduction, the amount of the resource contained in each organism decreases (called *cell quota* in microorganisms). The decline in the cell quota leads to a drop in the growth rate until the cell quota reaches its minimum value, at which there can be no further growth (*stationary phase*). The growth rate can be expressed as a function of the cell quota (Droop 1983):

$$\mu = \mu_{max} (1 - q_0/q)$$

where μ is the growth rate (d^{-1}); μ_{max}, the maximum (resource-saturated) growth rate (d^{-1}); q, the cell quota (e.g., pg/cell or μg/mg biomass); and q_0, the minimum cell quota for growth.

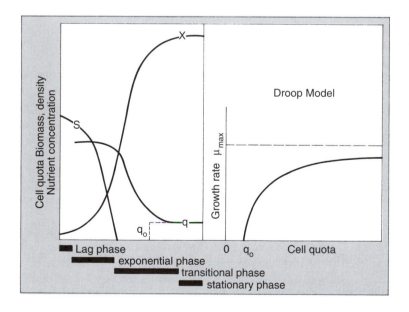

The more dense the organisms become, the more they must share the available resources. The decrease in the growth rate can therefore be viewed as a function of the population density. The more dense the population, the less the growth rate (*density-*

BOX 4.2   (*Continued*)

dependent control of growth). This relationship is described by the *logistics growth function* (cf. Section 5.2.4). The negative feedback between the growth rate and the population density is dependent on how close the population is to its maximum density (capacity, K), rather than on the absolute population density:

$$\mu = \mu_{max}\,(K - N)/K$$

Therefore, $\mu = 0$, if $N = K$, and $\mu = \mu_{max}$, if $N < K$.

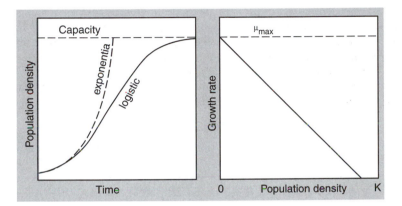

The logistic growth curve does not indicate the mechanism that limits the growth rate and requires no knowledge of the limiting factors. Factors other than resources, such as the excretion of harmful metabolites, can also create a density-dependent decrease in the growth rate. Once a limiting nutrient has been identified as a critical factor, the logistic and Droop equations can be combined:

$$K = S_{tot}/q_0$$

where S_{tot} is the total concentration of the limiting nutrient (dissolved and incorporated in the organisms).

Continuous Culture (Chemostat)

A chemostat is a container in which new nutrient medium is added through an inflow at a constant flow rate (F; volume/d) and, simultaneously, the same amount of suspension including the organisms is removed through an outflow. The *dilution rate (D; culture volume/flow rate; d^{-1})* equals the *elimination rate* ("mortality") of the organisms. The biomass of the organisms will increase as long as the available resources allow for a growth rate that is higher than D. As organisms increase they use up the limiting resource, causing a reduction in the growth rate. If the growth rate declines to the dilution rate, the system is in *steady state*. This is a self- regulating state. If the den-

---

BOX 4.2   (*Continued*)

---

sity of organisms exceeds the equilibrium value, the concentration of resources will decrease below the equilibrium value. This causes the growth rate to become smaller than the dilution rate and the population density declines. The self-regulating mechanism responds in the opposite direction if the density of the organisms is below the equilibrium.

In a steady state the resource limitation of growth can be described with the Monod (1950) equation, which is similar to the Michaelis–Menten equation:

$$\mu = \frac{\mu^{Max}S}{S + k_s}$$

where μ_{max} is the maximum (resource-saturated) growth rate (d^{-1}); k_s, the half saturation constant (mass/volume); and S, the concentration of limiting nutrient (mass/volume) remaining in solution. If there is a threshold value (k_0) for the resource concentration, S must be replaced with ($S - k_0$). Equilibrium conditions are: $\mu = D$. The equilibrium biomass or density of organisms (X) can be estimated from the amount of limiting resource consumed and the growth rate at the corresponding cell quota:

$$X = (S_0 - S)/q$$

where S_0 is the resource concentration in the medium before consumption by organisms (inflow).

Two culture procedures, representing idealized models for two extreme situations in natural systems (Box 4.2), are used to investigate the kinetics of resource-limited growth of organisms that can be cultured in suspension (bacteria, planktonic algae, some zooplankton). The static culture (*batch culture*) corresponds to the explosive settlement of free habitat (for planktonic algae, usually at the beginning of the growth period). The continuous culture (*chemostat*) is comparable to a *steady-state equilibrium* in which production and elimination of organisms, as well as consumption and supply of resources, are in balance with one another (approximates plankton communities during the summer).

4.3.4 Nonsubstitutable and Substitutable Resources

In the previous models we must assume that the rate of consumption, growth rate, and the maximum attainable biomass are limited by a single resource. Most organisms actually consume more than one kind of resource. This raises the question whether the extent of the limitation by one resource is affected by the availability of other resources. The critical question is whether resources can be substituted for one another. In this regard there are important differences between animals and autotrophic organisms. The nutrition of an animal has a "package character"; that is, a piece of food (a prey, a plant as food) serves both as an energy source as well as a source for carbohydrates, proteins, lipids, vitamins, essential elements, etc. Such "packages" can generally be substituted by others, even if the substitute package has a less favorable composition or is more difficult to capture. A cat that does not catch mice can also be fed canned cat food. It is possible that even animals are not limited by only the total amount of food, but rather by the individual components of food (e.g., carbohydrates, lipids, proteins, in extreme cases even individual amino acids).

Autotrophic organisms are in a different situation. Their resources (light, essential elements) occur separately and cannot be consumed as combined packages. Energy and the various essential elements are not substitutable. The biochemical functions of an element cannot be replaced by another element (cf. Section 4.3.3). Such resources are referred to as "essential resources." For autotrophic organisms, the only substitutable resources are those compounds that contain the same element, such as CO_2 and HCO_3^- as carbon sources or NO_3^- and NH_4^+ as nitrogen sources.

The difference in the growth response to mixtures of essential resources and mixtures of substitutable resources can be shown graphically for pairs of resources (Tilman 1982). The availability of each resource is plotted on the x and y axis, respectively. The growth response is shown by isoclines for growth rates (*growth isoclines*). Each point on a particular isocline is defined by a mixture of resources which permits the growth rate characteristic for it (Fig. 4.14).

The isoclines for *essential resources* are parallel to the axes, because no growth is possible if one resource is zero. If there is no interaction between both resources (as assumed in *Liebig's law of the minimum*), the isoclines form a right angle. If there is some interaction at low resource availability, the corners of the isoclines are rounded. The isoclines for *substitutable resources* cross both axes, because growth is possible if one resource is zero. If there is neither an advantage nor a disadvantage in a mixed diet, the isoclines are straight lines (*perfectly substitutable resources*). If a mixed diet confers an advantage in growth (*complementary resources*), the isoclines are bent toward the origin. If growth is better with a one-sided diet (*antagonistic resources*), the growth isoclines are bent away from the origin.

Bacteria have an intermediate position between plants and animals in this regard: On the one hand, numerous organic compounds can be substituted as energy and carbon sources, and some even have a particulate, package character, since they contain several nutrient elements (e.g., amino acids as sources for C and N). On the other hand, inorganic ions can partly cover the mineral nutrient requirements (e.g., phosphate). Essential organic substances that the consumer itself cannot synthesize (e.g., certain vitamins) cannot be substituted.

In addition to the two basic types of essential and completely substitutable resources, there are a number of other types of hypothetically possible interactions between resources, but for which there is little real evidence (Tilman 1982). The theory of resource limitation is an important prerequisite for understanding the mechanistic models of competition (cf. Section 6.1.3).

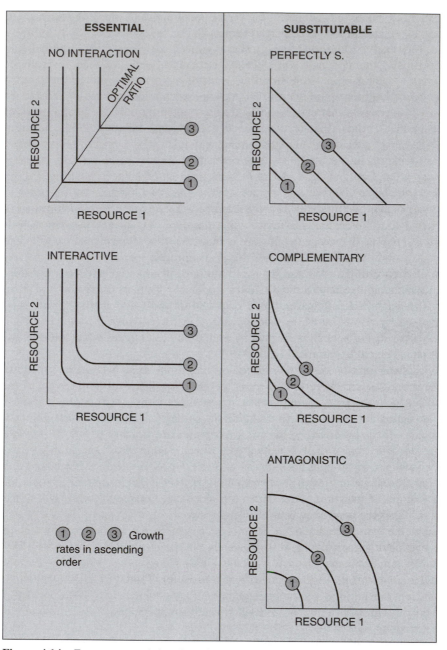

Figure 4.14 Zero-net-growth isoclines (ZNGIs) for a single species limited by two essential resources (left) or two perfectly substitutable resources (right). The species cannot exist in the shaded area.

4.3.5 Light

The use of light as an energy source in *photosynthesis* is by far the most important of processes by which inorganic matter is converted to organic matter (*primary production*). We will not go into the details of the biochemistry of photosynthesis, as it is covered well in all textbooks on plant physiology. Photosynthesis can be presented in a simplified form according to the formula:

$$6 \text{ CO}_2 + 6 \text{ H}_2\text{O} \rightarrow \text{C}_6\text{H}_{12}\text{O}_6 + 6 \text{ O}_2 - 2802 \text{ kJ}$$

Sulfur bacteria utilize H_2S for photosynthesis in place of water as an electron and hydrogen donor; accordingly, sulfur is formed in place of oxygen. This process is strictly anaerobic.

One can see from the formula that photosynthesis can be measured as either the release of oxygen or the use of CO_2. Oxygen changes in water can be measured with relative ease, while it is difficult to measure the small decrease in CO_2, for normally there is a large supply of bound CO_2 (bicarbonate–carbonate system) present in lake water. There are more sensitive methods for measuring the incorporation of CO_2 into organic matter (carbon fixation) (^{14}C, Box 4.3).

Two resources that could possibly limit the rate of photosynthesis are the energy source (light) and the inorganic substrate (CO_2, H_2S for purple-sulfur bacteria). Contrary to the widely held concept, light is not just a boundary condition. It is actually a consumable resource, as indicated in the discussion of the effect of organisms on the vertical attenuation of light (see Section 3.2.1). For example, Tilzer (1983) demonstrated that during the phytoplankton biomass maximum in spring and summer in Lake Constance up to 70% of the light attenuation was due to phytoplankton chlorophyll. At the time of the seasonal minimum of phytoplankton biomass (ca. 0.3 μg chlorophyll/l) the vertical light attenuation coefficient in this moderately plankton-rich lake was ca 0.25 m^{-1}, whereas during the biomass maximum (ca. 30 μg Chl/l) it was ca. 0.75 m^{-1} (cf. Fig. 3.4).

The light limitation of nonacclimated photosynthetic rates is usually described with a modified Blackman model (Box 4.1). In this model the photosynthetic rate decreases above a certain light intensity (Fig. 4.15). In the literature this is usually called a *P–I curve* (*P* for photosynthetic rate, *I* for light intensity). All the parameters of this curve are influenced by the species of phytoplankton as well as by their physiological condition (light adaptation). Planktonic algae have been most studied in this regard and will therefore be used here as examples. The basic principles also apply to benthic microalgae ("periphyton") and to higher aquatic plants (macrophytes).

The increasing portion of the *P–I* curve passes through the origin if one considers only the process of photosynthesis (*gross photosynthesis*). There are also, however, losses due to respiration which uses organic carbon and oxygen. When these losses are subtracted (*net photosynthesis*), the entire curve is shifted downward. Although it is not entirely correct to assume respiration is not affected by light (Reynolds 1984), this assumption is adequate for most practical purposes. The point intercepted by the net photosynthesis curve along the *x* axis defines the light intensity at which the gross pho-

BOX 4.3	MEASUREMENT OF PHOTOSYNTHETIC RATE

Oxygen method

With the oxygen method, samples (suspensions of planktonic algae, leaves of macro-phytes, etc.) are incubated in light and dark bottles and the oxygen concentration is measured at the beginning and at the end of the incubation period. It is assumed that the increase in dissolved oxygen in the light bottle corresponds to net photosynthesis (photosynthesis minus respiration) and that only respiration occurs in the dark bottle. The gross rate of photosynthesis is calculated as the difference in the final concentrations in the light and dark bottles. Two assumptions are sometimes not valid: The first assumption is that the respiration rate of heterotrophic organisms inadvertently enclosed in the bottle (e.g., zooplankton) is negligible. The second assumption is that respiration rates in the light and dark bottles are identical. The latter assumption is not valid for high light intensities where light-dependent photorespiration causes higher respiration rates than in the dark bottle. The greatest limitation of the oxygen method is its low sensitivity, which is dependent on the resolution of the usual methods of measuring dissolved oxygen (ca. 0.1 mg O_2 l^{-1}). The method is therefore only useful for nutrient-rich lakes.

^{14}C method

The ^{14}C method involves the addition of a small amount of radioactive $^{14}CO_2$ to the natural CO_2 available to organisms. The incorporation of the radioactivity in the biomass (operationally defined as the filterable particulate matter) is then measured. Trace quantities of ^{14}C (usually as bicarbonate) are added to light and dark bottles that are then incubated for 2–4 h in situ. After the incubation, the radioactivity of particles retained on a membrane filter is measured. The organisms photosynthesize the radioactive ^{14}C slightly more slowly than the ^{12}C (correction factor 1.05).

$$\frac{^{14}C_{takenup}}{^{14}C_{available}} \times 1.05 = \frac{^{12}C_{taken\ up}}{^{12}C_{available}}$$

Dark bottles used here do not measure respiration, but rather provide a background value for physical adsorption and dark fixation of ^{14}C. The amount of ^{12}C available must be determined chemically (usually from alkalinity and pH). Some of the photosynthetically formed organic matter that is also excreted (e.g., as glycolate) during the incubation can be determined by acidifying the water and bubbling off the CO_2. The radioactivity remaining in solution is a measure of the newly formed dissolved organic matter.

There is much debate over whether the ^{14}C method measures gross or net photosynthesis. This largely depends on how much of the newly formed photosynthate was respired during the incubation. Certainly at first the incorporation of ^{14}C is a measure of gross photosynthesis, since the algae have only respired ^{12}C. With time, however, as the ^{14}C content of the algae increases, an increasing percentage of the respired CO_2 is ^{14}C. The incorporation of ^{14}C measures the net rate of photosynthesis only after there is isotopic equilibrium (uniform labeling of all algal components). At the normal incubation times used, some undefined value between gross and net photosynthesis

BOX 4.3	(*Continued*)

is measured. Another important limitation is that the method cannot measure negative rates of photosynthesis in the complete darkness (respiration). Despite its limitations, the ^{14}C method is much more widely used than the oxygen method, since the sensitivity can be increased considerably simply by increasing the amount of radioactivity.

Measures of photosynthetic rate

In studies of production biology the photosynthetic rate, usually referred to as the rate of primary production, is related to the volume of the water sample or the surface area of the lake. The extrapolation of rates measured for short periods (2–4 h) to daily rates requires assumptions about the relationship between the rate of photosynthesis and light intensity. This requires measurements of a vertical profile of photosynthesis, the vertical light gradient, and changes in incident light during the course of the day. The various models for calculating the daily production (e.g., Talling 1957) assume a vertically homogeneous distribution of phytoplankton, which is only true if the water column mixes to the lower boundary of the measured profile. Wetzel and Likens (1991) provide a detailed description of the method for measuring primary production.

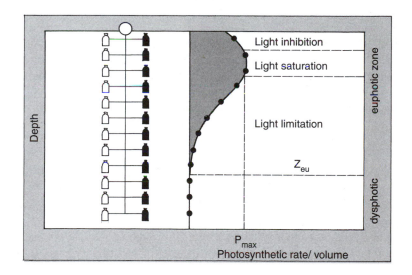

Above is a typical vertical profile of photosynthetic activity per unit volume of water, showing inhibition near the surface, maximum photosynthesis at shallow depths with decreasing rates deeper in the water column. The measurements of photosynthesis were made with pairs of light/dark bottles suspended at different depths. The zone in which there is detectable photosynthesis is the "euphotic zone." The photosynthetic rate per unit surface area of the lake, indicated by the shaded surface area, can be approximated as $P_{max}Z_{eu}/2$ (Z_{eu} is the depth of the euphotic zone).

BOX 4.3   (*Continued*)

Photosynthetic activity per unit biomass (specific rate of photosynthesis) can be related to the physiology of the organisms. Carbon or chlorophyll is usually selected as a measure of biomass. There is a relatively constant percentage of carbon in the total organic biomass (ca. 45–50% by weight), whereas the chlorophyll content is more variable and, for example, can increase as an adaptation to low light intensity. C-specific and chlorophyll- specific rates of photosynthesis are not always proportional to one another.

---

tosynthesis is sufficient to compensate for respiratory losses. At this "compensation point" there is neither growth nor loss. In a vertical light gradient in a lake this corresponds to a particular depth or *compensation level*. The layer above the compensation level is referred to as the *euphotic zone,* and the layer below the compensation level is called the *aphotic zone.* As a rule of thumb, the compensation level for phytoplankton lies at a depth where there is about 1% of the surface light intensity. Figure 4.15 shows how the vertical profile of photosynthesis changes with different surface light intensities and algal biomass, following the *P–I* curve response (cf. Box 4.3).

The beginning increase in the *P–I* curve is especially important for photosynthesis at low light availability. Measurements for planktonic algae range between 2 and 37 mg C (mg Chl)$^{-1}$ E^{-1} m^{-2}, most frequently between 6 and 18 (Reynolds 1984). In the light-limited range of photosynthesis the photosynthetic rate is determined exclusively by photochemical processes that are not temperature dependent (cf. Section 4.2.1). Thus the increase in the *P–I* curve is independent of temperature.

The region of transition from the light-limited (increasing) to the light-saturated (horizontal) region of the *P–I* curve is determined by the intensity of saturation (I_k). I_k values in the literature range from 20 to 300 μE m^{-2} s^{-1} of photosynthetically active radiation (PAR), most frequently between 60 and 100 μE m^{-2} s^{-1} (Harris 1978). On a clear summer day [$E_d(0) = 2000$ μE m^{-2} s^{-1}] in a lake with a moderate amount of plankton ($k_d = 0.5$ m^{-1}), this would be roughly equal to the light intensity at a depth of 6 m (ca. 100 μE m^{-2} s^{-1}).

The highest specific photosynthetic rates for light-saturated conditions measured in temperate zone lakes are about 7.5 mg C (mg Chl)$^{-1}$ h^{-1} and in tropical lakes approximately 12 mg C (mg Chl)$^{-1}$ h^{-1}. The magnitude of the light-saturated photosynthesis is temperature dependent. For temperatures below the optimum, the Q_{10} is ca. 1.8–2.5. Since the light-saturated photosynthetic rate is temperature dependent, but the initial region of the *P–I* curve is not, the I_k value is indirectly temperature dependent.

The photosynthetic rate does not follow the Blackman model at high light intensities (above ca. 200–1000 μE m^{-2} s^{-1}), where it begins to decrease (Fig. 4.15). This *light inhibition* is caused by photochemical damage of the chloroplasts by UV radiation as well as increased photorespiration.

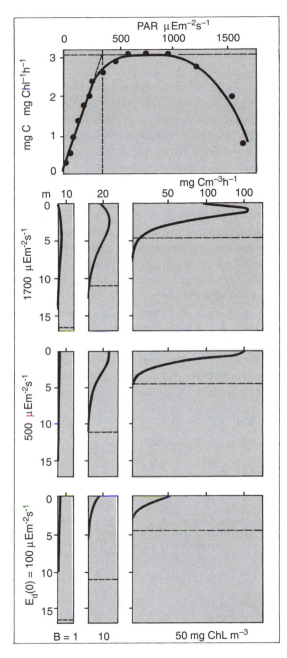

Figure 4.15 The effects of incident light intensity and algal biomass on the vertical profiles of photosynthesis: Above: extremely high-light-adapted *P-I* curve (l_k ca. 300 $\mu E\,m^{-2}\,s^{-1}$), measured in Lake Constance. Below: vertical profiles of volume-specific rates of photosynthesis ($mgC\,m^{-3}\,h^{-1}$) at three different surface light intensities $E_d(0)$ and three different algal concentrations (as chlorophyll concentration). - - -: the compensation depth, as conventionally defined [1% $E_d(0)$]. All assumption about light attenuation are derived from Lake Constance data (from Tilzer).

Plants can also adapt physiologically to low light intensities. Light adaptation can be tested by comparing the *P–I* curves of the same clones of phytoplankton that have been cultured under different light intensities. There are two different types of light adaptation (Jørgensen 1969). The chlorophyll content increases in the "*Chlorella*-type" light adaptation, resulting in higher rates of photosynthesis for cells adapted to low light. In the *Cyclotella*-type light adaptation there is a restructuring of the photosynthetic apparatus that only results in an elevation in the slope of the initial region of the *P–I* curve (Jørgensen 1969). The ability to adapt to low light intensities means that the long- term response to light, unlike the instantaneous response, does not show the linear increase of the Blackman model and can be better described with a Michaelis–Menten-type model (Fig. 4.16). The *P–I* curve therefore only describes the immediate reaction to light conditions, whereas the long-term correlations between the underwater light environment and the occurrence of different species are better explained by the light-adapted model.

When there is adequate surface light intensity, one might expect to find most species of phytoplankton at depths of optimal (i.e., saturated) light. It is not always possible, however, for the algae to remain stratified at a particular depth. Even a wind velocity of only ca. 3 m/s can destroy stratification patterns of flagellated algae. The phytoplankton then become exposed passively to a light intensity equal to the average light intensity of the mixing zone. During calm periods even closely related species can show clear vertical separation (cf. Fig. 3.8). For example, the flagellate *Rhodomonas minuta* stratifies at about 50% of surface light intensity, whereas *Rhodomonas lens* locates at ca. 10% (Sommer 1982). Studies of the effect of light on growth rates indicate these are close to the optimal light conditions for both species.

Low-light-adapted species of algae can develop very pronounced maxima in the metalimnion of lakes with a stable thermocline as long as the light intensity in these layers is not below the compensation point for that species. The best-known example of this are the red pigmented cyanobacteria *Planktothrix* and *Limnothrix* (prev. *Oscil-*

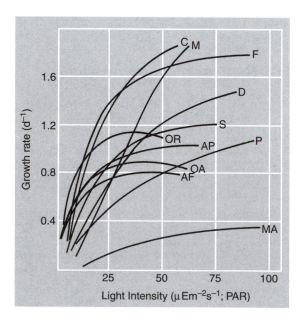

Figure 4.16 The effect of light intensity on the growth rate of planktonic algae, with constant light and at 20°C: C, *Coelastrum microporum*; M, *Monoraphidium minutum*; F, *Fragilaria bidens*; D, *Dictyosphaerium pulchellum*; S, *Scenedesmus quadricauda*; OR, *Oscillatoria redekii*; AP, *Aphanizomenon flow- aquae*; P, *Pediastrum boryanum*; OA, *Oscillatoria agardhii*; AF, *Anabaena flos-aquae*; MA, *Microcystis aeruginosa* (from Reynolds 1989).

latoria). *Planktothrix rubescens* stratifies in moderately enriched, thermally stratified lakes in Middle Europe (Fig. 4.17), and *P. agardhii* var. *isothrix* in similar lakes in Scandinavia. The "burgundy red alga" *Planktothrix rubescens* formed blooms in the early stages of eutrophication of Lake Zürich. These populations then disappeared with increasing eutrophication as the algal densities in the epilimnion increased, decreasing the water transparency. *P. rubescens* finally reappeared once again after sewage treatment plants succeeded in reducing the algal densities and increasing the transparency of the water. With an I_k of ca. 10 μE m^{-2} s^{-1} and light inhibition at about 130 μE m^{-2} s^{-1} (Mur and Bejsdorf 1978), these red pigmented *Planktothrix* and *Limnothrix* species can be considered true shade plants.

Photosynthetic bacteria (*Chlorobiaceae, Chromatiaceae, Rhodospirillaceae*) can develop below the zone of phytoplankton in lakes with an anaerobic hypolimnion and a buildup of H$_2$S, as long as light penetrates into the H$_2$S zone. These bacteria have light requirements that are comparable to low-light-type phytoplankton. For example, purple bacteria have I_k values of 25–70 μE m^{-2} s^{-1}, and sulfur bacteria, which are usually stratified below the *purple bacteria,* have I_k values of 20–25 μE m^{-2} s^{-1} (Pfennig 1978). These *sulfur bacteria,* have spectral optima for photosynthesis at wavelengths that are weakly absorbed by the phytoplankton that are stratified above them: 700–760 nm for the green sulfur bacteria and >800 nm for purple bacteria.

In contrast to phytoplankton, there is very little known about the P-I curves for the algae that grow on submersed surfaces (*periphyton, aufwuchs*). One of the few published such studies of periphyton deals with the reed region of Lake Maarsseveen in The Netherlands. The I_k in the spring was ca. 250 μE m^{-2} s^{-1} and in the summer ca. 100–200 μE m^{-2} s^{-1}. The greater degree of low light adaptation in the summer was related to the shading by the leaves of the reeds. There was no measurable light inhibition.

There is likewise little experimental information concerning the light requirements of macrophytes that grow on the bottom of lakes. It is generally recognized that the light conditions set the maximum limits of depth distribution for *submersed macrophytes.* For example, Maristo (cited in Hutchinson 1975) found a linear relationship between the extent of macrophyte growth into deep water and the Secchi disk trans-

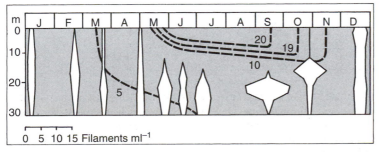

Figure 4.17   Depth distribution of *Planktothrix* (= *Oscillatoria rubescens*) in Lake Wöther (Austria) showing the isotherms for 5, 10, 19, and 20°C (from Findenegg 1943).

[89]

parency in 27 different lakes in Finland. In 17 of these lakes the aquatic mosses *Fontinalis antipyretica* and *Drapanocladus sendtneri* had populations at the greatest depth. The best-documented examples of deep macrophyte growth are reported for Lake Titicaca (Peru/Bolivia) and the extremely clear Lake Tahoe (U.S.) (Hutchinson 1975). In Lake Titicaca the flowering plant *Potamogeton strictus* occurs at depths to 11 m, the low-light alga *Chara spp.* to 14 m, and the moss *Hygrohypnum* to 29 m. In the macrophyte-poor Lake Tahoe, the flowering plants extend down to only 6.5 m, but *Chara globularis* and aquatic mosses have been found at 75 m. Flowering plants are generally not found at depths where there is on the average less than 2% of the surface light intensity.

Macrophytes show a characteristic *depth zonation* (cf. Hutchinson 1975). In addition to light, hydrostatic pressure and depth-related changes in the bottom sediments are also important determinants of macrophyte depth. As soon as the leaves reach the surface, the underwater light climate is unimportant for emergent macrophytes (e.g., *Phragmites, Typha*) and floating plants (e.g., *Nymphaea*). The maximum depth of these plants may be determined by the ability of the air-filled tissue (aerenchyma) to withstand the hydrostatic pressure at depth. Submerged species, however, are arranged according to their light requirements. Spence and Chrystal (1970) measured the relative photosynthetic activity of five *Potamogeton* species under low light conditions after being first adapted to high light intensities. The species that had the highest low-light photosynthetic rate were those that also were found deepest in the lake (Fig. 4.18).

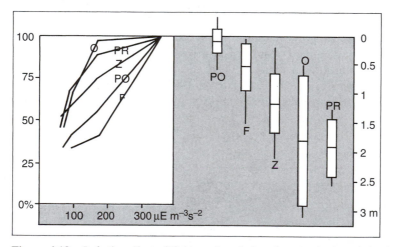

Figure 4.18 *Left*: the effect of light on the relative photosynthetic activity (as percent of photosynthesis at 350 μE m^{-2} s^{-1}) for high-light-adapted leaves of various species of *Potamogeton*: O, *P. obtusifolius*; PR, *P. praelongus*; Z, *P. zizii*; PO, *P. polygonifolius*; F, *P. filiformis*. *Right*: depth distribution of 5 *Potamogeton* species in Scottish lakes (from Spence and Chrystal 1970; light values are recalculated).

4.3.6 Inorganic Carbon

Dissolved organic carbon (DIC) occurs in three different forms, CO_2, HCO_3^-, and CO_3^{-2}. The abundance of each form depends on the pH of the water (cf. Section 3.3.2). All plants can utilize CO_2. When the supply of CO_2 has been depleted, the pH rises to 9 and only plants that produce the enzyme carbonic anhydrase can continue to photosynthesize, using HCO_3^-, essentially the only form of dissolved inorganic carbon present at that pH. If the supply of HCO_3^- also becomes exhausted, the pH will continue to rise to 11.

Consumption of CO_2 can become a problem, especially in stands of macrophytes, since CO_2 diffuses very slowly and there is little exchange of water within macrophyte beds. To deal with this aquatic plants have developed four different strategies (Bowers 1987):

1. Formation of aerial leaves (emergent macrophytes);
2. Utilization of CO_2 in pore water (interstitial) of the sediments (*Lobelia, Littorella*);
3. Temporal decoupling of the photosynthetic light reaction, similar to the CAM metabolism of some terrestrial plants. This makes it possible to fix CO_2 at night when it is released through respiration by other organisms (*Hydrilla, Isotes, Lobelia*);
4. Utilization of HCO_3^- (*Myriophyllum, Elodea*).

Plants that do not have such adaptations are usually especially efficient in using CO_2 at low concentrations in water. This can be recognized in such plants by the CO_2 compensation point, which is the minimum CO_2 concentration that is needed to compensate for the loss of carbon due to photosynthesis. Obligate CO_2 users that do not have access to alternate sources in the air and interstitial water have a CO_2 compensation point of 2–12 μM, compared to 60–110 μM for plants with alternate sources of CO_2 (Sand-Jensen 1987).

Even those plants that use HCO_3^- prefer to assimilate CO_2, and the presence of CO_2 will suppress the uptake of HCO_3^-. If the total amount of DIC remains constant, but the pH value changes, thereby altering the forms of DIC present, the photosynthetic rate of HCO_3^--using plants decreases as the pH exceeds a value of 7 (Sand-Jensen 1987; Fig. 4.19). On the other hand, HCO_3^--using plants can continue photosynthesizing to pH 11, but CO_2 users can do this only to about pH 9.

4.3.7 Mineral Nutrients

In addition to the elements C, O, and H that are part of the general photosynthetic equation, there are many other elements that are essential components of the biomass of living plants. Nutrients are usually divided according to the amounts required into the macro elements (N, P, S, K, Mg, Ca, Na, Cl), which usually make up >0.1% of the organic matter, and the trace elements (Fe, Mn, Cu, Zn, B, Si, Mo, V, Co, and possi-

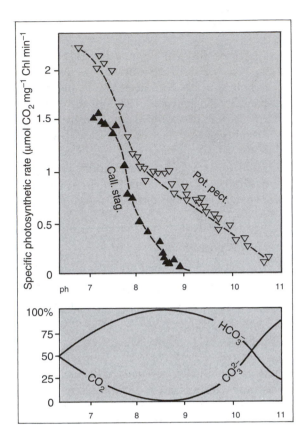

Figure 4.19 *Above*: the effect of pH on the specific rate of photosynthesis by *Potamogeton pectinatus*, which utilize HCO_3^-, and *Callitriche stagnalis*, which use only CO_2, at a DIC concentration of 5 mM (from Sand-Jensen 1987). *Below*: distribution of the species of DIC as a function of pH.

bly some other elements), which are needed in far smaller quantities. Silica is also a macro element for diatoms and some chrysophytes. All these elements must be taken up from the pool of nutrients dissolved in water. Theoretically, any of these elements could become an essential, limiting resource. In most lakes, however, many of them are almost always in excess (e.g., Mg, Ca, K, Na, S, Cl), so that the spectrum of limiting nutrients can be narrowed to N, P, some trace elements, and Si for the diatoms and certain chrysophytes. During the growth period the limiting nutrients dissolved in the water are reduced to concentrations that are often below detectable limits.

Before nutrients can become incorporated into the structure of organisms, they must first be removed from the environment. Only dissolved forms make up the pool of available nutrients. Usually, potentially limiting nutrients must be taken up at extremely low concentrations and transported across biological membranes against concentration gradients. This requires energetically costly active transport of ions, which is mediated by enzymes. The uptake rate is dependent on the concentration of nutrients and can be described by a Michaelis–Menten equation (Section 4.3.2). The parameters of this equation (V_{max}, k_t) are not constant and depend on the nutritional state of the organisms. Highly nutrient-limited organisms often have increased maximum uptake rates with which they can exploit short-term pulses of increased nutrient concentrations.

[92]

For many ecological questions it is more interesting to examine the effect of nutrient limitation on the growth rate than on the rate of nutrient uptake. The introduction of chemostat cultures for phytoplankton has provided valuable kinetics data for Monod-type models (cf. Box 4.2). The k_s values for dissolved phosphate at 20°C are between 0.003 μM for *Synedra filiformis* and 1.83 μM for *Volvox globator* (Tilman et al. 1982). The pennate diatoms and chrysophytes generally have low k_s values (<0.20 μM). Even in P-poor lakes at times of year when the algal biomass is sparse, there is usually more than 0.05 μM dissolved phosphate present. Thus phosphorus limitation only develops as a consequence of phosphate consumption by organisms. Figure 4.20 shows the P-limited growth kinetics for three extreme types of algae: *Synedra filiformis*, a most extreme case of an adaptation to low P concentrations; *Chlorella minutissima*, an example of adaptation to high concentrations; and *Volvox globator*, an example of an algal species that has low growth rates at all P concentrations.

k_s values for nitrogen-limited growth have been rarely determined for phytoplankton in freshwater. Because of the higher cellular requirements they usually are much higher than for P. The cyanobacterium *Planktothrix agardhii* has a k_s value at 20°C of 1.2 μM for NO_3^- and 1.1 μM for NH_4^+, in contrast to 0.03 μM for phosphate (Ahlgren 1978, Zevenboom 1980).

Certain cyanobacteria can utilize molecular nitrogen (N_2) as a source of nitrogen by *nitrogen fixation*, whereas eukaryotic algae and all higher plants are dependent on nitrate or ammonium (occasionally urea). Nitrogenase, the enzyme needed for nitrogen fixation, can only function in an oxygen-free environment. Pelagic cyanobacteria (Family Nostocaceae) create sites for nitrogen fixation by forming specialized cells called *heterocysts,* which release dissolved organic substances. Attached bacteria respire the dissolved organic matter, creating an oxygen-poor microenvironment around the heterocysts. Many species of Nostocaceae produce heterocysts in proportion to the amount

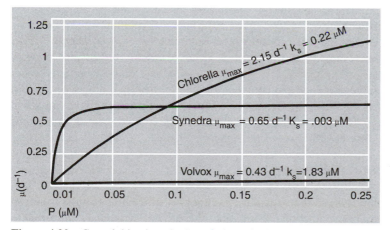

Figure 4.20 Growth kinetics of selected phytoplankton species under phosphate limitation and at 20°C: *Chlorella minutissima* (high μ_{max}, intermediate k_s; from Tilman et al. 1982); *Volvox globator* (low μ_{max}, high k_s; from Tilman et al. 1982).

of nitrogenase activity. Production and maintenance of heterocysts for nitrogen fixation are processes that require energy and are only beneficial when bound N sources are scarce. On the other hand, the atmosphere represents an almost inexhaustible supply of N_2 for nitrogen fixers (Paerl 1988).

The k_s values for silica-limited growth of diatoms at 20°C are between 0.88 μM (*Stephanodiscus minutus*) and 19.7 μM (*Synedra filiformis*). The last value is extremely high and indicates that silica can also become limiting during seasonal maxima (many lakes have no more than 70 μM Si). It is interesting that those diatoms that have a higher k_s for Si also have a lower k_s for P (many pennate diatoms, especially *Synedra* and *Asterionella*), whereas diatoms with low k_s values for Si have a relatively high k_s for P (*Cyclotella, Stephanodiscus*). Such differences in nutrient uptake ability represent "tradeoffs" that result in differing competitive outcomes and resulting phytoplankton species composition, depending on trophic condition of the lake (P concentration) as well as the availability of silica.

The use of Monod formulas for field data can lead to erroneous results. Many planktonic algae can store nutrients for short periods, enabling them to have growth rates that are higher than would be predicted by the Monod equation. In this case the growth rate is a function of the concentration of nutrients within the cell according to the Droop equation (cf. Section 4.3.3). Typical minimal *cell nutrient quotients* (q_o) for P are about 0.0014 mol P/mol C (Sommer 1988a); the extreme values are 0.02 mol P/mol C (*Microcystis aeruginosa*) and 0.0002 mol P/mol C (*Asterionella formosa*). Typical values for nitrogen are about 0.02 mol N/mol C. Phytoplankton that are neither N nor P limited often have a stoichometric C:N:P relationship of about 106:16:1 (*Redfield ratio*) (Goldman et al. 1979). The stoichometric composition of the plankton biomass is thus a reasonably reliable indicator of the nutrient status (Sommer 1990), even though some species are exceptions. If one of the two nutrients is present in the biomass in noticeably smaller concentrations than the Redfield relationship predicts, then there is a good probability that this nutrient is limiting, at least for the species that make up most of the biomass.

Most other *macronutrients* (Ca, Mg, K, Na, S, Cl) are not limiting, for they are present in excess. There are, however, examples of limitation by *micronutrients* ("trace elements"). One such example is iron, because of the low solubility of Fe^{3+} ions in neutral and alkaline water (approx. 10^{-8} to 10^{-14} mol l^{-1}). On the other hand, normally some iron can be released from complexes it has formed with dissolved organic substances (primarily humic acids). Some algae, especially cyanobacteria, even release complexing compounds ("siderochromes") that take up the iron complexes and cannot be utilized by other species (Simpson and Neilands 1976).

4.3.8 Inorganic Sources of Energy

There are autotrophic bacteria that obtain the energy needed for biosynthesis from exergonic chemical reactions rather than from light (Table 4.1). These *chemolithotrophic* organisms use chemical energy sources, inorganic electron donors, and CO_2 as a carbon source. Some of these organisms are extremely important in the nutrient cycles of lakes. Chemolithotrophic bacteria often live in the boundary between oxic and anoxic zones, such as in the sediment–water boundary and in the thermocline of lakes with anoxic deep water. The anoxic water provides a constant supply of reduced compounds into boundary layers.

In the case of bacterial nitrification, ammonium, which comes from the aerobic and anaerobic decomposition of N-rich organic compounds, is first oxidized to nitrite and then further oxidized to nitrate. Both of these reactions are carried out by two genera of chemolithoautotrophic bacteria:

Nitrosomonas:

$$NH_4^+ + \frac{3}{2} O_2 \rightarrow NO_2^- + 2H^+ + H_2O, \qquad \text{energy gain of 276 kJ/mol}$$

Nitrobacter:

$$NO_2^- + \frac{1}{2} O_2 \rightarrow NO_3^-, \qquad \text{energy gain of 75 kJ/mol}$$

Flagellated bacteria in the genus *Thiobacillus* get their energy by oxidizing reduced sulfur bonds. Most species can use many different sulfur bonds. Many are obligatory chemolithoautotrophs, but some can also utilize organic carbon sources. Examples of sulfur oxidizing reactions:

$$H_2S + \frac{1}{2} O_2 \rightarrow S + H_2O$$

$$S^2 + H_2O + \frac{3}{2} O_2 \rightarrow SO_4^{2-} + 2 H^+$$

$$S_2O_3 + H_2O + 2 O_2 \rightarrow 2 SO_4^{2-} + 2 H^+$$

Thiobacillus denitrificans, which occurs in anoxic environments, uses nitrate instead of oxygen as an oxidizing agent (cf. Section 4.3.9).

Iron oxidizing bacteria (*Ferrobacillus, Galionella, Leptothrix*) oxidize reduced *ferrous iron* (Fe^{2+}) to *ferric iron* (Fe^{3+}):

$$4 Fe^{2+} + 4H^+ + O_2 \rightarrow 4 Fe^{3+} + 2 H_2O$$

Most of the resulting oxidized iron precipitates out. There is disagreement as to whether bacteria are responsible for the similar oxidation of *manganese*, for it is not certain whether the structures referred to as *Metallogenium*, which appear during periods of Mn precipitation, are actually organisms.

Oxyhydrogen gas bacteria (e.g., *Alcaligenes eutrophus*) are facultative autotrophs. They obtain their energy from the oxidation of elemental hydrogen and are able to get their cellular carbon from CO_2:

$$6 H_2 + 2 O_2 + CO_2 \rightarrow CH_2O + 5 H_2O$$

These bacteria can grow just as well, and sometimes even better, with organic carbon sources.

4.3.9 Electron Acceptors in Anaerobic Respiration

Anaerobic, *heterotrophic* bacteria use DOCs as sources of carbon and energy, but cannot use oxygen as a terminal electron acceptor in respiration. Anaerobic respiration refers to the use of oxygen-rich compounds (nitrate, sulfate) instead of oxygen to oxidize organic matter. In such cases, the compounds used as terminal electron acceptors can take on the features of a limiting resource.

In *nitrate respiration*, nitrate is transformed in several reduction steps to either ammonium (nitrate ammonification) or to nitrogen (denitrification). Organic compounds are simultaneously oxidized to CO_2 and H_2O. The energy gain here is only about 10% less than for aerobic respiration. The steps in the reduction of nitrate are:

$$NO_3^- \rightarrow NO_2^- \rightarrow NO \begin{array}{l} \nearrow NH_2OH \rightarrow NH_3 \\ \searrow N_2O \rightarrow N_2 \end{array}$$

Denitrification is especially important for the nitrogen cycle of lakes, for it causes a loss of bound nitrogen that might otherwise be used by eucaryotic organisms.

Sulfur respiration (desulfurication) refers to the reduction of sulfate to hydrogen sulfide:

$$8 \, (H) + SO_4^{2-} \rightarrow H_2S + 2 \, H_2O + 2 \, OH^-$$

Here, unlike nitrate respiration, there is not a complete oxidation of the organic compounds and acetic acid is usually produced as the final product. If bacteria that respire sulfate also have hydrogenase (*Desulfovibrio*), they can follow the pathway of oxyhydrogen bacteria, oxidizing H_2 instead of organic substrates as a source of energy. Such bacteria are called *chemolithoheterotrophs*.

These specialized chemical reactions lead to characteristic vertical stratification of bacteria types in lakes with an anoxic hypolimnion (Fig. 4.21). The various nutritional types of bacteria actually depend upon one another, since they transform nitrogen and sulfur into the different forms.

4.3.10 Dissolved Organic Substances

Use by heterotrophic, aerobic bacteria

Most of the organic matter in lakes is in the dissolved form. In clear lakes with few humic acids the concentration of dissolved organic carbon (DOC) is from 2 to 25 mg/l, increasing with the trophic state of the lake. In these lakes the ratio of dissolved to particulate organic carbon (POC) is usually from 6:1 to 10:1 (Wetzel 1983). Humic lakes have DOC concentrations that are about an order of magnitude higher. Sources of DOC include *excretion* by living organisms, cell breakage (*autolysis*), microbial *decomposition* of dead organisms, as well as the contribution from allochthonous sources. Dissolved organic carbon is most important as sources of energy and carbon for aquatic bacteria, although it is also used by other protists and even pigmented mixotrophic algae. Dissolved organic substances utilized by organisms fall into the category of substitutable resources, but with differing nutritional value.

DOC is a mixture of various substances whose qualitative and quantitative composition cannot be completely explained. Often it is sufficient to give a very general characterization such as fractionation according to molecular size based on ultrafiltration or by gel chromatography. Heterotrophic bacteria prefer monomeric substances (monosaccharides, free amino acids, etc.). These always make up only a small fraction of the DOC. Amino acids in the open water as well as in interstitial water of the sediments are in very low concentrations (usually <10 $\mu g/l$ and up to 50 $\mu g/l$ for short periods in eutrophic lakes). The concentrations of monosaccharides, oligosaccharides, and simple organic acids are almost always less than 10 $\mu g/l$. Dissolved polysaccharides are in higher concentrations. Bacteria rapidly break down the most easily decomposed parts of the DOC. For this reason, the main component of DOC is "aquatic humus" that consists of a mixture of fulvic acids, humic acids, and humins, the remains of decomposition of plant material of both autochthonous and allochthonous origin. Such humic substances are difficult, if not impossible for bacteria to utilize (Geller 1985).

One can also describe the uptake kinetics of DOC with the Michaelis–Menten model (Fig. 4.13). Half-saturation constants of the uptake rate are in the same range as natural concentrations. Wright and Hobbie (1966) found that pelagic bacteria in Lake Erken had k_t values of 2–3 $\mu g/l$ for glucose and 8–15 $\mu g/l$ for acetate. In a more comprehensive study in Pluβsee, Overbeck (1975) found that glucose ranged from 3.8 to 47 $\mu g/l$. Amino acids in Lake Constance varied seasonally between 2.65 and 44 $\mu g/l$ (Simon 1985). The net uptake of a particular carbon source is essentially the same as

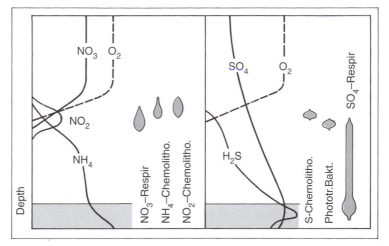

Figure 4.21 A diagram showing the relationship of the vertical distribution of O_2, NO_3, NO_2, NH_4, H_2S, and SO_4 to the depth distribution of nitrate-respiring heterotrophic bacteria, nitrite-using and ammonium-oxidizing chemolithoautotrophic bacteria, sulfur bacteria and sulfate-respiring bacteria in a lake with an anoxic hypolimnion.

new production of biomass. It is therefore possible to use the k_t value to predict lf-saturation constants for growth (k_s).

The comparison between k_t value for oligomeric substances and their concentrations in water may suggest at first glance that the growth rates of pelagic bacteria must be carbon limited. Oligomeric substances, on the other hand, are usually substitutable resources to which the law of the minimum does not apply. The degree to which aquatic bacteria are carbon limited cannot be estimated reliably, since neither the quantity nor exact composition of useable DOC is known and also the Michaelis–Menten kinetic has only been determined for a few substances.

Uptake of radioactively labeled oligomeres, mainly glucose, was introduced as a method of determining the *bacterial production* (Sorokin and Kadota 1972). This method is only superficially analogous to the method for measuring photosynthesis using ^{14}C-CO_2. ^{14}C-bicarbonate introduced into the water quickly changes into the fractions CO_2, HCO_3^-, and CO_3^{2-} according to the pH, whereas with ^{14}C-glucose only one of the many substitutable resources is labeled. The total uptake rate can be estimated only if one of two strict assumptions is met: first, if all other substitutable resources are taken up with the same efficiency and their total concentration can be determined or, second, if glucose is the only substrate taken up by the bacteria. Neither of these two assumptions is fulfilled in lakes. Glucose uptake rates are thus only a rough indication of bacterial activity and are sometimes referred to as *heterotrophic potential.*

To avoid these difficulties, a more recent approach is to use instead C sources that are required in only small amounts and whose uptake is directly related to the new formation of biomass. The most widely used substrate of this type is ^3H-labeled *thymidine*, an organic base of DNA (Fuhrman and Azam 1980). At this time, there is no method for determining bacterial production that has the recognized status of the ^{14}C and O_2 methods for measuring photosynthesis.

Specific growth rates from the thymidine method are rarely greater than $0.69 \ d^{-1}$ (Güde 1986). This is much lower than one would expect for maximum growth rates for organisms of such small size (cf. Section 5.2). One can conclude that natural populations of bacteria are substrate limited, assuming there are no major methodological errors. This conclusion can also be confirmed by another method. Lake water passed through a membrane filter with a $1 \ \mu m$ pore size allows bacteria, but not bacteria grazers (protists and zooplankton), to pass through. After incubation of this filtrate in situ the actual growth rate of the bacteria in the absence of grazers can be measured by counting the number of bacteria under an epifluorescence microscope. Using this method, Güde (1986) got bacteria growth rates that were in relatively good agreement with the thymidine method.

Use of DOC by anaerobic bacteria

Obligate and facultative anaerobic microorganisms can obtain energy through *fermentation* in the anoxic environment of the sediments, and hypolimnion of eutrophic lakes and in the monimolimnion of meromictic lakes. This process involves redox reactions in which polymers are hydrolyzed into monomeric organic molecules such as simple sugars, amino acids, and fatty acids. These monomers are then split, with one part being reduced and the other part being oxidized. The end product of the oxidation is CO_2 and the end products of the reduction are alcohols, organic acids, or very reduced gases (H_2, CH_4, H_2S). Ammonium is also the end product of amino acid fermentation. There is relatively little energy gained from fermentation, compared to aerobic respiration and nitrate respiration. When glucose is oxidatively respired, there is a gain of 2802 kJ/mol, whereas fermentation to ethanol releases only 67 kJ/mol and fermentation to lactic acid 111 kJ/mol. Thus the fermentation of organic substances is only important in anaerobic environments.

Methanogenic and methane oxidizing bacteria

Methane (CH_4) is an example of an organic substance that is both an end product of anaerobic decomposition as well as a resource for aerobic bacteria. *Methanogenic* bacteria are strict anaerobes that transform alcohol, organic acids, hydrogen, and CO_2 to methane. These transformations may even involve symbiotic interactions. For example, *Methanobacterium omelianskii* has been shown to consist of two different strains of bacteria, one of which transforms alcohol into acetic acid by splitting off H_2, and the other strain uses the split-off hydrogen to form methane:

$$CH_3 - CH_2OH + H_2O \rightarrow CH_3 - COOH + 2H_2$$
$$CO_2 + 4H_2 \rightarrow CH_4 + 2H_2O$$

The bacteria strain that splits off the hydrogen is also dependent on the methane- forming strain, since its growth is inhibited by H_2 and the methane-forming strain keeps the concentration of H_2 low. The methane-forming strain is a chemolithotroph, as it utilizes inorganic resources.

Methane oxidizing bacteria are aerobic bacteria that use the methane produced by methanogenic bacteria as a C source and electron acceptor. They are also heterotrophic bacteria that specialize in using C_1 compounds (methane, methanol, methylamine, formaldehyde, formic acid) and are therefore referred to as *methylotrophic* bacteria. The equation for methane oxidation is summarized as:

$$5\ CH_4 + 8\ O_2 \rightarrow 2\ (CH_2O) + 3\ CO_3 + 8\ H_2O$$

Methane oxidizing bacteria, like chemolithotrophic bacteria, occur mainly in the boundary between aerobic and anaerobic zones, since they depend on O_2 as an electron acceptor, but also require methane (Fig. 4.22).

Higher organisms

The use of dissolved organic substances by eucaryotes is limited to a few small organisms. The large quantity of DOC in lakes has lead many researchers to test whether higher organisms can directly take up dissolved organic substances. Much of the interest in this subject was stimulated by Pütter (1911), who discovered that *Daphnia* could live a long time in water that had been filtered free of particles. He postulated

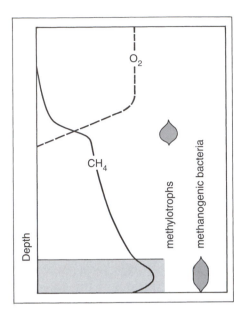

Figure 4.22 A diagram of the vertical distribution of methane bacteria compared to the profiles of O_2 and CH_4 in a lake with an anoxic hypolimnion.

that these animals could live from the direct uptake of dissolved organic substances. Later, careful testing demonstrated that this hypothesis was based on an error, for the diatomaceous filter he used was not fine enough to remove bacteria. We now know that *Daphnia* that are kept in water that has passed through a 0.2 μm membrane filter will starve in 5 days and lose about 50% of their weight. In Pütter's experiments the *Daphnia* probably indirectly used the DOC as it was actually first taken up by the bacteria, which in turn were filtered by the *Daphnia*.

Once ¹⁴C-labeled organic compounds were available, one could measure directly the incorporation of DOC into animals. This has been tested with radioactively labeled pure compounds (sugars, amino acids) as well as with algal cell hydrolysates. These experiments have demonstrated that the amount of DOC taken up by animals (primarily through the gut) is negligible compared to the amount of carbon they assimilate from particulate food. Larger quantities of DOC are taken up by soft-bodied worms that live in sediments where there are relatively high concentrations of DOC. Even here, however, these animals do not assimilate enough carbon from the DOC to cover a significant portion of their energy requirements. Direct uptake of DOC appears to be more important in marine than in fresh water environments.

4.3.11 Particulate Organic Substances

Particulate organic matter (*POM*) or particulate organic carbon (*POC*) occurs as either living or dead matter. The non-living particulate matter is called *detritus*. Detritus can be formed within the lake (*autochthonous*) or can be imported from outside the lake (*allochthonous*), as, for example, leaf litter. It is often not easy to distinguish between these two sources. Soon after an organism dies it is broken up and colonized by heterotrophic microorganisms (bacteria and fungi). These small particles of organic material are usually considered to be detritus, even though they consist, in part, of living matter. Detritus can also be formed from dissolved organic matter that flocculates under certain conditions. This is especially important in estuaries, where fresh and salt waters mix.

The amount of detritus in an aquatic system may be greater than the quantity of living material. In streams the POC can be almost exclusively detritus, whereas in lakes the proportion of living matter is usually larger, due to the high standing crop of algal biomass. The composition of POC in the pelagic zone of a lake can change very dramatically. During algal blooms large amounts of living matter are present, but this may be quickly changed to detritus after the algal population either dies or is eliminated by grazing zooplankton. In productive lakes the sediments are covered with a thin layer of detritus, and in shallow lakes this layer can be resuspended during storms.

It is difficult to quantify the importance of detritus as a source of food for organisms. Detritus usually has a low nutritional value, simply because it is a product of the feeding activities of higher organisms and microorganisms. On the other hand, the bacteria and fungi that live in the detritus may have a high nutritional content. It is not easy to discern whether organisms consume detritus for the detritus itself or to get the microorganisms that are attached to it. Many organisms cannot avoid eating detritus along with living food, because the detritus is so abundant. Sometimes it is the only source of nutrition for benthic organisms, especially for those that live in streams.

Size structure of particles

There is a great range of sizes of POC. The smallest particles that can be used by the smallest heterotrophic organisms are about 0.2–0.5 μm. Such small particles, often bacteria, can only be eaten by heterotrophic flagellates that themselves are only a few micrometers in size. The flagellates in turn are eaten by filter-feeding animals, thereby transforming the smallest particles into larger, edible particles. Some organisms specialize in producing small particles from large ones, making them available for other animals. *Shredders* are especially important in benthic communities. Gammarid amphipods, for example, transform the leaves that fall into the water into delicate skeletons. Shredding of leaves also increases the surface area of the particles so that they are more easily colonized and decomposed by microorganisms.

All *particulate feeders* are adapted to feeding on a specific range of sizes that correlates with their feeding structures. Many are very selective, having specialized filtering devices that detect or collect only specific types of food (cf. Sections 6.4 and 6.5). The size structure of POC is thus an important parameter for describing the resources of particulate feeders. Organisms are dependent on the occurrence of certain *particle sizes*. In streams, particulate feeders use a gradient of particle sizes shifting from large to small particles as one moves downstream (cf. Section 7.7.2). Fine filter feeders are often abundant in standing waters where small particles that sink slowly are more abundant (cf. Section 4.2.6).

Recently it has become more and more common to measure the size structure of parameters such as POC and chlorophyll. This is easily accomplished using membrane filters with defined pore sizes (e.g., Nuclepore filters, <10 μm) or nylon netting (e.g., Nitex >10 μm). One can also measure the fraction of particles that are in a particular size category (Fig. 4.23). Automatic *particle counters* provide a rapid analysis of particle sizes with very high resolution. These measurements are made by drawing water through a capillary tube containing an electrical field. Each particle that passes through the capillary tube alters the electrical field, causing a pulse to be recorded. Since the pulse size is dependent on the size of the particle, the particles counted can also be categorized according to size (Fig. 4.24). Such counters estimate the volume of a particle, but cannot distinguish its shape. Size classes of particles are expressed as equivalent spherical diameter (ESD), which is the diameter of a sphere with the same volume as the particle. Modern laser-based optical flow cytometers not only measure numbers and sizes of particles, but can also sort them into classes according to pigments and fluorescence. This provides additional information on the quality of the particles, making it possible, for example, to discriminate between living algae and detritus.

Microscopic analysis of particles has also become automated. Computers can analyze the number and size of particles using digitized video images. Computers can also recognize and count algal species, especially those that have simple shapes. Fluorescence methods can be used to distinguish cyanobacteria and algae. There is, however, no substitute for the human eye and a knowledge of the shapes of particles for detailed analysis and taxonomic identification.

Types of feeding

There are many mechanisms for the uptake of particulate organic matter. Protozoans encompass entire food particles in their cell vacuoles. Higher organisms usually concentrate or break up the particles before they ingest them.

The feeding categories *carnivore* (meat eaters), *herbivore* (plant eaters), and *detritivore* (detritus eaters) are based on the type of organic matter eaten. Strictly speaking,

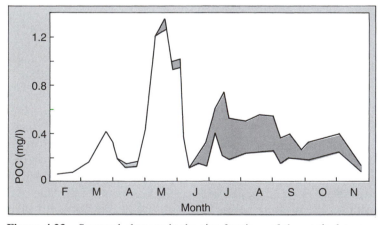

Figure 4.23 Seasonal changes in the size fractions of the particulate organic carbon in Lake Constance. The white region under the curve indicates the fraction <35 μm; the shaded area indicates the fraction between 35 and 250 μm (from Schober 1980).

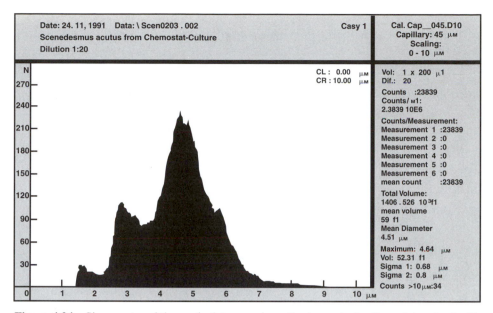

Figure 4.24 Size spectra of the particulate organic matter in an algal culture, determined with an automatic particle counter.

the term *herbivore* means "foliage eater" and cannot be correctly used for most aquatic animals. However, the expression has become well established in reference to algal-eating animals. There are no sharp boundaries between these feeding categories. Many organisms are **omnivores**; that is, they eat both plants and animals. Calanoid copepods, for example, can eat algae as well as small invertebrates. Cyclopoid copepods are herbivores in the juvenile stages and become carnivorous as adults, although even then they continue to eat some algae (Adrian and Frost 1993). Filter feeders often do not distinguish between living algae and detritus, making classification of their feeding difficult.

Clearer divisions can be made according to the mode of food collection. *Predators* must search for food and capture it (cf. Section 6.5.1). *Collectors* must also search, but their food does not try to escape or defend itself. There are many animals in the benthic community that "graze" on the algae of the aufwuchs and on microorganisms. These animals are called *scrapers*. Their mouthparts are highly specialized for scraping and collecting particles from substrates (Arens 1989). Many insect larvae and snails with rasping radula are scrapers. Also found exclusively in the benthic habitat are *shredders* that often feed on particles that are larger than themselves. Amphipods and isopods, as well as certain insect larvae such as caddisflies, live on or in leaf packs lodged in streams and are important in the initial breakdown of leaves.

Animals that live in the fine sediments are either collectors or *sediment feeders*. Chironomid larvae live in woven tubes in the mud and collect organic matter from the surface of the sediments immediately surrounding their tubes. Tubificids have their anterior end stuck in a tube, and their posterior end waves freely in the water. They ingest the sediment from several centimeters deep, assimilate the organic matter, and release their feces on the surface of the sediment. This results in a mixing of the sediments referred to as *bioturbation*.

If food particles are suspended in the water, they must be concentrated before they are eaten. There are many organisms in both the benthic and pelagic regions that are *filter feeders*. For filter feeding, a water current is used to bring the particles to the animal. Filter feeders either use existing water currents or create the current themselves.

Stream organisms can simply place their filter in the water current. There are several ways to accomplish this. Blackfly larvae (*Simulidae*) have their filtering surface on their head. They attach their abdomen to a substrate and hold their filtering antennae in the water current. The larvae close their filters from time to time, depending on the particle concentration, and ingest all of the particles. *Hydropsyche*, a caddisfly without a case, spins a fine net that it attaches between the rocks (Fig. 4.25a). The water current carries particles into the net, which is then eaten by the larva. In streams with a heavy load of organic matter, *Hydropsyche* can become so abundant that the entire stream bottom is covered with their nets. In still water filter-feeding animals must create their own water current. Mussels pump large amounts of water across their gills, where they retain food particles. Many zooplankton are filter feeders. The thoracic appendages of cladocerans form a complicated suction-pressure pump that forces the water through the filtering surfaces of the third and fourth appendages (Fig. 4.25b). Rotifers and copepods are also considered filter feeders, since they remove small particles from the water, although they capture individual particles in the water current and in this sense do not actually "filter."

The hydromechanics of different filters is very complicated, and the precise mechanisms of many are not understood. In most cases, it is probably not a case of pure "sieving" of the water, as this would be difficult with the very small Reynolds numbers that occur at the filtering structures (cf. Section 3.1.5). The "mesh" of the filter is often very small (a few micrometers), and some cladocerans have mesh widths of less than 1 μm. Due to the overlapping of the boundary layers, only water under sufficient pressure will be forced to pass through the filtering structures (Fig. 3.2). Cladocerans can create such an overpressure because they have a closed filtering chamber. Blackfly (*Simuliidae*) and other insect larvae that have open filtering mechanisms use the pressure of water moving against the feeding surfaces. Other mechanisms, such as electrical charges, may also be involved in attaching particles to the filter (Rubenstein and Koehl 1977).

The type of filtering mechanism determines the types of particles that are retained and, thus, the mortality of the different species of algae due to filter feeding (cf. Section 6.4.2). Figure 4.26 shows a graphic comparison of the size spectra of particles removed from the water by coexisting *Daphnia* and calanoid copepods. Differences in the spectra can be accounted for by the filtering mechanisms of these zooplankters. *Daphnia* uses a sieving method in which the lower size spectrum is determined by the mesh size of the filter. Particles smaller than the mesh simply pass through the filter (Brendelberger 1991). All larger particles are retained, except for those that are too large to enter the filtering chamber. The upper size limit is not so sharply defined, for

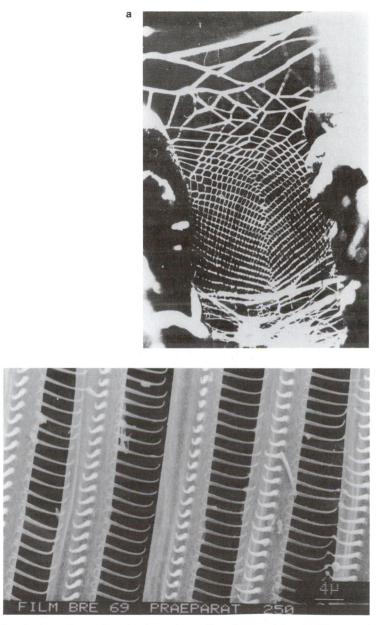

Figure 4.25 Filtering devices of aquatic animals used to collect fine particles. (a) The net of a caddisfly larva (*Hydropsyche*), which stretches between rocks and filters out particles as the water moves through the net. Height ca. 10 mm (photo by I. Schreiber). (b) The fine structure of the third pair of thoracic appendages in the filtering chamber of *Daphnia*. The filtering setules arising from the supporting structures (setae) are usually hooked at the end. Preparation for the scanning electron microscope has caused the adjacent setules to become unhooked. The scale in the photo is equal to 4 μm (photograph: H. Brendelberger).

[105]

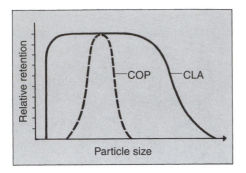

Figure 4.26 A model comparing particle selection by a cladoceran (CLA) that "sieves" water and a copepod (COP) that actively selects particles (from Gliwicz 1980).

it depends on both the shape and size of the particle. The copepod is much more selective. Its filter is smaller and is not enclosed in a chamber. The copepod uses the filter to enhance the capture of a particle, rather than as a passive sieve. The copepod's mouthparts create a water current. When an approaching algal cell is detected by the copepod, the water current is shifted to move the cell toward the second maxilla, which then captures it (Koehl and Strickler 1981). The particle spectrum depends on how the copepod handles each particle and not on the mesh size. Copepods can test the edibility of each particle and then either eat it or discard it (DeMott 1988).

Nutritional value and energy content

The types of particulate organic matter (POM) in water are too variable to be described with a single *nutritional value*. The nutritional value of a class of POM can only be defined for a specific consumer. Important nutritional features include ease of handling (size and shape), digestibility, energy content, and the content of essential nutrients. Since POMs are substitutable resources, a larger quantity can sometimes compensate for lower nutritional value. This is not always possible, and we can assume, therefore, that the energy and metabolism of animals can be limited by the quality as well as the quantity of food particles.

The *biochemical composition* and nutritional value of algae depends on environmental conditions. Zooplankton growth can be inhibited by feeding them with algae grown under nitrogen or phosphorus limitation that results in low N:C or P:C ratios. There are also strong diel variations in nutritional content of algae in the epilimnion. Daytime storage of food reserves results in a higher content of carbohydrates and lipids in the evening that by morning are lost through respiration. There is a recent debate concerning whether phosphorus can act as a limiting resource for zooplankton. Changes in the nutrient content of the water can cause considerable variation in the mineral content of algae (cell quota, cf. Section 4.3.3). For example, under P-limited conditions the algal cells may have a very low phosphorus content. On the other hand, zooplankton maintain a rather constant C:N:P ratio (Hessen and Lyche 1991), suggesting phosphorus availability might limit their growth as well. Phosphorus-starved algae could therefore have a low nutritional value for zooplankton, even though the algae provide an abundance of carbon and nitrogen. In fact, *Daphnia* grow more slowly when cultured with phosphorus-starved algae (Sterner 1993). Phosphorus deficiency may also have indirect effects on food quality. For example, P-deficient algae can be difficult to digest (Van Donk and Hessen 1993) or lack certain highly unsaturated fatty acids (HUFA) that the zooplankton cannot synthesize. Some phytoplankton considered to be poor food for zooplankton (e.g., cyanobacteria) also lack these HUFAs. The availability of HUFAs in the seston of a lake varies seasonally, depending on the nutrient conditions and the phytoplankton composition. The growth rate of *Daphnia* in the field is closely correlated with the amount of a specific HUFA contained in the seston (Müller- Navarra 1995).

Organic matter provides energy as well as essential elements for heterotrophic organisms. Thus *energy content* is also used to characterize food quality. Energy content can be measured as a combustion value in a calorimeter, but, of course, not all combusted heat is available to the animal. Chitin and cellulose, for example, can be utilized by only a few highly specialized organisms. The energy content of organic particles in water is quite variable. Values measured for algae are between 10 and 20 kJ/g dry wt. Much of the variation is caused by the different ash content. To measure the organic content of particulate matter one must first combust a preweighed sample for several hours at 550°C. The weight of the remaining ash is then subtracted from the dry weight to obtain the ash free dry weight (organic matter). Particles with a high mineral content, such as diatoms with siliceous shells, have a high proportion of ash (>30%), whereas flagellates have a very low ash content (<5%). This gives diatoms a correspondingly low energy content per unit dry weight and flagellates a high energy content.

One gets much more meaningful values if the energy content is related to the organic matter or carbon content. There is a strong correlation between the energy content and the organic or carbon content of particles, and particulate organic matter in lakes has on the average an energy content of about 23.5 kJ/g ash free dry weight. Variations are caused by differing biochemical composition of the organic matter. A more precise energy content can be estimated if one knows the percentage of carbohydrates, proteins and fat (lipids) in the organic matter. Typical values are:

carbohydrates	17.2 kJ/g
proteins	23.7 kJ/g
lipids	39.6 kJ/g

Since lipids have a relative high carbon content (80%) and carbohydrates and proteins have a low carbon content (40%), one can see from the above list that carbon content and energy content are highly correlated. Particles of animal origin often have a relatively high energy content because they contain more fat.

Energy content is only one aspect of the nutritional value of particulate organic matter. In addition to the possible importance of HUFAs discussed earlier, essential amino acids and vitamins can also be limiting factors. In nature, animals usually eat a variety of foods, thereby avoiding the deficits of any one food type. The food of detritus feeders, for example, includes bacteria and fungi that live on the dead organic matter.

Mixotrophy

Many flagellates can live both as autotrophs and heterotophs. They are able to photosynthesize as well as take up dissolved organic matter and ingest particles (*phagocytosis*). It is well known that some pigmented flagellates can grow on dissolved organic substances while in the dark, but photosynthesize when in the light. Recently, the ability of pigmented flagellates to ingest particles has received much attention, since these organisms, along with unpigmented flagellates and ciliates, can regulate the abundance of bacteria. This mixture of nutrition from photosynthesis and particulate matter is referred to as *mixotrophy* (Sanders and Porter 1988).

The uptake of particles has been observed in several groups of flagellates, Dinoflagellates, Cryptophytes, Chrysophytes, and flagellated green algae. Some of the best known examples are the Chrysophytes *Dinobryon, Ochromonas,* and *Chromulina*, the Dinophyceans *Gymnodinium* and *Peridinopsis*, and the Cryptophycean *Cryptomonas*. It is quite evident that mixotrophy occurs primarily in those groups that also have many representatives that are unpigmented and purely heterotrophic.

As with protozoans, many mixotrophic flagellates have a variety of mechanisms with which they can take particles into the cells, including the formation of pseudopodia. A flagellate can ingest up to 70 bacteria per hour, resulting in uptake rates of 5×10^3 to 10×10^4 bacteria per milliliter of lake water per hour. Mixotrophic feeding by flagellates can have a significant effect on the population dynamics of bacteria (Bird and Kalff 1993).

The contribution of phagotrophy to the growth of flagellates differs. Some species apparently cover a large part of their energy requirements from heterotrophic sources, and others are primarily autotrophic. Uptake of particles can be important as sources of nitrogen and phosphorus, especially when these are in short supply, as well as a source of vitamins. It appears that some flagellate species require the ingestion of food particles and therefore cannot be cultivated in the absence of bacteria (axenic).

4.4 ENERGY UTILIZATION

4.4.1 Net and Gross Production

Production is defined as the new formation of body substances from inorganic matter or organic raw materials (*anabolic processes*). The energy needed for synthesis comes from photo- or chemosynthesis or from the decomposition of organic compounds through oxidation and breakdown into small molecular substances (*catabolic processes*). Organisms need additional energy for mechanical work and the maintenance of body temperature, in the case of homeothermic animals. This energy requirement is also covered by catabolic reactions. Catabolism is a loss process in the metabolism of a organism, since it results in a decrease in mass. It is important to distinguish between this internal loss, which results from the organisms metabolic activities, and external losses, such as those due to feeding by other organisms and mechanical damage. Temporal changes in biomass are the net result of both simultaneously occurring anabolic and catabolic reactions, even if external losses have been excluded. The concept of *gross production* refers to a potential, but never realized, production that excludes catabolic losses. "Net production," in contrast, describes the difference between gross production and catabolic losses, which are the observed changes in mass, without correction for external losses. Although it is possible to exclude external losses in an experiment, this is rarely possible in nature.

4.4.2 Energetics of Photosynthesis

In photosynthesis, as in all energy transfers, only a portion of the energy is converted into a usable form (Second Law of Thermodynamics). The utilization of light energy can be described by the yield coefficient of quanta capture (ϕ), which is defined as the molar quantity of fixed carbon per mol of absorbed light quanta. The quantum yield coefficient is constant (ϕ_{max}) in the light-limited region of the *P-I* curve (cf. Section 4.3.5). In the light-saturated part of the curve the quantum yield coefficient decreases with the reciprocal of light intensity, since the rate of photosynthesis stays constant, despite increasing light intensity. Φ_{max} values of 0.03–0.09 have been found for phytoplankton (Tilzer 1984a). Assuming a caloric equivalent of 468 kJ for 1 mol C and 218 kJ for 1 mol of photosynthetically active radiation with a wavelength of 550 nm (mean value of the PAR spectrum), one gets an efficiency of energy use of 6.3–19.3%. The maximum possible quantum yield based on biophysical limits is 0.125, which corresponds to an energy efficiency of 26.8%.

One finds a much smaller utilization of light energy for the energy yield of the total photosynthesis per unit surface area of a lake (cf. Box 4.3). Much of the phytoplankton is located in water strata that have light saturation or even inhibitory light intensities. Thus a large portion of the total photosynthesis occurs under light conditions where ϕ_{max} is not reached. Nutrient limitation can also prevent a maximum use of available light energy. Also, the yield coefficient is related only to the light absorbed by living chlorophyll. In natural waters there are also light losses due to dissolved substances, suspended particles, and inactive breakdown products of the photosynthetic pigments. In Lake Constance, the planktonic chlorophyll absorbed between 6% (biomass minima) and 50% (biomass maxima) of the light energy penetrating the lake. The efficiency of light utilization varied seasonally from 0.16 to 1.65%, at least an order of magnitude below the maximum possible yield. These values are as high as those measured in most other lakes (Bannister 1974, Tilzer 1984a).

4.4.3 Energy Balance by Heterotrophic Organisms

Secondary production
The meaning of production is very different for heterotrophic and autotrophic organisms. Production by heterotrophs (secondary production) involves the transformation of organic matter, rather than its formation (primary production). Heterotrophs take up

organic substances (*assimilation*) and use part of it for metabolism. Only the remaining matter can be incorporated into the organism's own body. Simply written:

$$production = assimilation - metabolic\ losses$$

The above equation describes a change in mass or energy, but in this form can only be used to estimate production of individual organisms in the laboratory. One cannot follow each organism in populations of small species in nature. In this case, we measure the production of populations (cf. Section 5.2). Under field conditions, the observed changes in biomass do not necessarily equal the total production, since during the period of observation a portion of the production may have been eliminated, as, for example, by predators. If the rate at which the biomass is "harvested" exactly equals the rate of production, there will be no change in biomass, even if the production is high. Indirect methods must be used to determine secondary production in nature, for seldom can one directly measure the eliminated biomass (cf. Box 4.4).

Heterotrophic microorganisms

For microorganisms energy use is always measured for populations. Methods used to determine production of bacteria and protozoans are generally the same as used for measuring population dynamics. For cultures, one measures the reproduction of cells and multiplies this by the mass of each individual. It is also sometimes possible to give the microorganism a radioactively labeled substrate and measure its incorporation into the cells (cf. Section 4.3.10).

All parameters of the mass balance equation can be measured for cultures grown on defined media in the laboratory:

1. **Production** from growth of biomass;
2. **Assimilation** from uptake of the substrate or from incorporation of a tracer;
3. **Metabolic losses** from use of oxygen (respiration) or CO_2 buildup.

If one knows the production and respiration (metabolic losses), it is possible to calculate the *efficiency* (K_2) of transformation of organic matter into microbial biomass (as a percent):

$$K_2 = (P/A)100 = (P/(P + M))100$$

where P is production, A assimilation, and M metabolic losses. Efficiencies of approximately 25% have been measured for natural populations of bacteria (Sorokin and Kadota 1972), indicating that 25% of the energy consumed was transformed into biomass and 75% was lost through respiration.

Animals

The *energy balance equation* must be expanded to be used for animals that ingest particulate organic matter that is only partially useable:

$$P = I - F - R - E$$

BOX 4.4	ESTIMATING SECONDARY PRODUCTION

Secondary production is difficult to determine for field populations, since usually only the change in biomass is visible, not the proportion of production that was eliminated due to loss processes in the time interval. Indirect methods must therefore be used to include the eliminated production. Many methods for estimating secondary production were developed and tested as a result of the International Biological Program (IBP). These methods have been summarized in several different books (e.g., Edmondson and Winberg 1971, Winberg 1971, Downing and Rigler 1984). All these methods have fairly large errors due to the difficulties in obtaining quantitative samples and other uncertainties. It is thus probably better to refer to secondary production results as "estimates" than "measurements." Below are two typical examples:

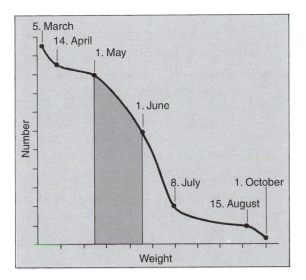

Allen curve

A graphic method (Allen curve) can be used for situations in the field where there is a class of animals of the same age, for example, a year class of fish (cohorts, cf. Section 5.2.6). It is based on the concept that animals grow over time, but their numbers continually decline as a result of mortality (cf. Fig. 5.5). By sampling the same cohort at different times, the number of animals at each sampling time can be compared to their weight. The surface area under the curve represents the production. In the hy-

BOX 4.4   (*Continued*)

pothetical example shown the sampling dates are plotted against the average weights. The shaded surface area shows the production for May and the area under the entire curve represents the yearly production.

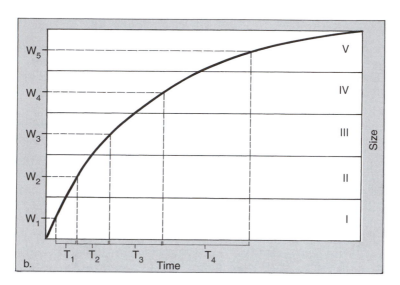

b.

Sum of the increments

Often generations overlap and it is not possible to follow cohorts. In this case the entire population can be divided into developmental stages or size classes, after it has been experimentally determined how long it takes the animals to grow from one class to the next. From these times and the increase in weight, one can then calculate the daily growth increment of an individual. This growth increment is then summed for all individuals. Below is an example of how this is done:

1. A growth curve is experimentally determined for an individual (thick line). The method strongly depends on the shape of this growth curve.
2. The field population is divided into five size classes.
3. The number (N) is determined for each size class.
4. The average weight (W) is determined for each size class
5. The time (T) in days that it takes to grow from one class to the next is determined from the curve.

The daily growth increment of class 1 is then:

$$P_1 = \frac{N_1(W_2 - W_1)}{T_1}$$

[112]

BOX 4.4	**(*Continued*)**

The total production can be estimated from the sum of all classes:

$$P = \frac{N_1(W_2 - W_1)}{T_1} + \frac{N_1(W_3 - W_2)}{T_2} + \cdots + \frac{N_1(W_{(i+1)} - W_i)}{T_i}$$

The specific production or mass produced per unit biomass and time (e.g., $\mu g \ mg^{-1} \ d^{-1}$) can be used as a measure of organism activity. Specific production of individuals is usually related to weight. For population production it is common to use P/B ratios (the ratio of production to biomass). It is a measure of *turnover* rate, since it describes what portion of the biomass is renewed per unit time.

---

where I is the *ingestion*, the amount of organic matter (energy, carbon) that is eaten; F, the *defecation* (egestion of unused food); R, the *respiratory losses*; and E, the *excretion losses*. R and E are as a rule combined as metabolic losses (M). *Assimilation* (A) is the difference between ingestion and defecation ($I - F$). Assimilation refers to the uptake of organic molecules through the gut wall. It is dependent on how much the animal has eaten and whether the ingested food can be digested. The *assimilation efficiency* can be used as a measure of digestibility:

$$AQ = (A/I)100 \ (\%)$$

If the feces can be collected, assimilation can be calculated as the difference between rates of ingestion and feces production. An example for this is the development of an energy balance for *Asellus aquaticus*. Small discs were cut from leaves, weighed, and offered to the animals. The leaf discs were weighed again at the end of the experiment to determine the quantity of leaves eaten. All feces were also collected and weighed. The energy content and the assimilation in energy units were calculated by combusting the leaf discs and the feces (Prus 1971).

It is not usually possible to collect all the feces of aquatic animals, since they may disperse or dissolve. Even if only a portion of the feces is found, one can still determine the percentage mineral ash content in both the food and the feces. From this one can estimate the assimilation quotient from the relative ash content of the food and feces, since only the organic portion can be used in digestion, leaving a higher percentage of ash in the feces. If no feces can be recovered, as for *Cladocera*, the assimilation rate can be measured as the rate of incorporation of ^{14}C from labeled food into body tissues. This is a very sensitive method, but requires a correction for [14]CO_2 respired (Peters 1984).

Metabolic losses are usually determined as *respiration* (O_2 consumption). It is difficult to measure the excretion of CO_2 in buffered water, since there is little change due to the large pool of bicarbonates and carbonates. The *respiratory quotient* (*RQ*), which expresses how many mol CO_2 are liberated per mol O_2 consumed, is used as a measure of conversion of oxygen into CO_2 production. The *RQ* depends on the type of substrate respired and ranges from 0.7 for respiration of fat to 1.1 for the respiration of carbohydrates and synthesis of fat (Lampert 1984).

Most animals are large enough to allow energy balance estimates to be made for individuals. Note that production is divided into body growth and reproduction. This *division of energy* between body and reproductive products is an important parameter that can be optimized through the evolution of life history strategies (cf. Section 6.8.2). Some animals grow their entire life and invest only a part of their production in reproduction. Others, such as the copepods, do not grow at all after becoming sexually mature, investing then their entire secondary production in egg mass.

The efficiency of energy use may be related to either the quantity of food eaten or the energy assimilated: The *gross efficiency (%)* is

$$K_1 = (P/I)100$$

This expresses the percentage of the energy ingested that flows into production. The *net efficiency (%)* (K_2; calculated as for microorganisms) is an expression of the percentage of assimilated energy that is available for production.

There are many published values of K_1 that deal with fish culture, since this tells how much food is needed to produce a given quantity of fish. K_1 depends heavily on how well the food is assimilated and digested and is thus quite variable. Typical values range from 10 to 15% and occasionally higher under laboratory conditions. K_2 also depends on the quality and quantity of the food. Aquatic animals can achieve K_2 values of over 70% with high-quality food, but in most cases are between 30 and 40% (Winberg 1971). Summaries are given by Winberg (1971), Grodzinski et al. (1975), and Zaika (1973). Figure 4.27 shows an example of the cumulative energy budget for *Simocephalus*, a littoral cladoceran. All parameters of the energy budget were summed

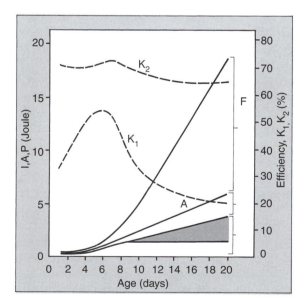

Figure 4.27 The cumulative energy balance for the first 20 days in the life of the cladoceran *Simocephalus vetulus*. I, ingestion; A, assimilation; F, defecation; R, respiration losses; P, production. The shaded area indicates the proportion of production that is invested into reproduction. The efficiencies K_1 and K_2 are shown as dashed lines (from Klekowski and Duncan 1975).

[114]

over time in this figure. It is clear that after reaching sexual maturity at 9 days the animals invest energy almost exclusively in reproduction. Some examples of energy balances for some different types of animals are given in Table 4.2. This comparison shows how energy utilization can differ. The cladoceran *Daphnia magna*, a filter feeder under optimal food and temperature conditions, had the highest efficiency. *Hyalella azteca*, an amphipod that eats detritus in the surface sediments, had the lowest efficiency.

Threshold concentrations

Feeding and assimilation rates are proportional to the amount of available food in the lower food concentrations. Maximum ingestion of food occurs above a *critical concentration* (functional response, cf. Section 4.3.2). The respiration rate is only slightly dependent on the amount of food available and never reaches zero, even for starving animals. This indicates that, below the critical concentration, the energy balance of an animal is dependent on the amount of food. At a certain food concentration, the animal assimilates the same amount of energy as it respires. At this point the production is zero. Figure 4.28 illustrates this for a filter feeder. Ingestion and assimilation rates increase with increasing food concentration. The point where the assimilation and respiration rate curves cross is where production equals zero. Above this concentration the animal can grow and produce offspring, whereas below this level it loses weight and eventually dies. The threshold concentration is thus an important parameter for the survivorship of an individual. There is also a minimal food concentration for the population growth. If a population is subject to mortality, the threshold concentration must be higher than for an individual, since the animals must not only maintain their weight, but also produce enough offspring to compensate for population losses. It is especially interesting to look at threshold concentrations for animals that compete for the same resources. The species with the lowest threshold concentration has the advantage under scarce food conditions (cf. Fig. 6.34).

TABLE 4.2
Energy balance parameters for *Daphnia magna* (food: green alga *Scenedesmus acutus*), river perch *Perca fluviatiles* (food: sludge worms *Tubifex*) (from Klekowski 1973) and the amphipod *Hyalella azteca* (food: detritus) (after Hargrave 1971).

Parameter	*Daphnia* (μg C ind^{-1} h^{-1})	Perch (J ind^{-1} h^{-1})	*Hyalella* (J ind^{-1} h^{-1})
Ingestion	0.90	392	0.220
Defecation	0.16	255	0.180
Assimilation	0.74	137	0.040
Metabolic losses	0.18	63.4	0.034
Production	0.56	73.6	0.006
AQ (%)	82	35.0	18.0
K_1 (%)	62	18.8	2.7
K_2 (%)	76	53.7	15.0

[115]

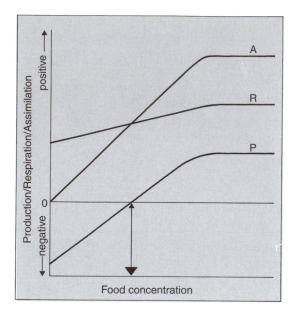

Figure 4.28 The effect of food concentration on the rates of assimilation (A), respiration (R), and production (P) of a filter feeder. The production rate is calculated as (A - R). The arrow indicates the threshold concentration at which there is zero production (from Lampert 1984).

Optimization of feeding

Foraging creates costs for search, handling, digestion, and biochemical transformations of the food. One could thus predict that behaviors should have evolved that keep the relative costs of foraging as low as possible. In other words, the evolved behavior has maximum profit from the available food (Pyke et al. 1977). Many behaviors have been described for terrestrial organisms, especially birds and insects, that optimize foraging (*optimal foraging*). Very few examples in freshwater can be interpreted in this way.

One opportunity to optimize energy gain is by selecting the best prey. A fish gets more energy from a large *Daphnia* than from a small one. As long as there are enough large prey, the fish should concentrate on the large ones and not waste its time hunting for small animals. Werner and Hall (1974) demonstrated this with sunfish. They offered the fish three size classes of *Daphnia*. The fish ate exclusively the largest size class as long as the time between encounters with these prey was less than 1/2 minute. With encounter intervals of 1/2 to 5 minutes they concentrated their feeding on the two largest size classes. When the large *Daphnia* became so rare that it took more than 5 minutes for a fish to see one of them, the fish ate animals from all size classes. It no longer paid to wait for large tidbits.

DeMott (1989) observed a similar behavior for the copepod *Eudiaptomus* when offered a mixture of algae of high and low food quality. When the concentration of food particles was high, the copepod selectively ate the high-quality algae. At low food concentrations, however, it abandoned the selective behavior and ate everything it could find.

Predators can optimize the time they must spend with an individual prey. The water scorpion (*Ranatra dispar*), an aquatic bug, is an interesting example. *Ranatra* is an ambush hunter that sits quietly on a plant and waits until a prey passes by. Then it

[116]

grabs the prey, injects a digestive fluid and sucks out the body contents. At the beginning of the sucking process the *Ranatra* gets much food per unit time, but the profit per unit time decreases as the prey becomes sucked empty. The water scorpion optimizes the relationship between the time it must wait for a prey and the time it spends eating the prey. If the density of prey is low, the time waiting for a prey to appear is long. Then the water scorpion completely empties the captured prey. If new prey appear frequently, it only partially sucks out the captured prey and catches a new prey before the old one is completely utilized. In this way it produces much waste, but still obtains more energy by spending half as long with each prey (Bailey 1986).

It may even be more profitable completely to cease searching for food when food is very rare. It costs a filter-feeding zooplankter, for example, energy to create water currents and the running of the "filtering pump." The filter feeder should really stop feeding when the particle concentration is so low that it costs more energy to pump water than is brought in with the food collected, checking only from time to time whether the situation has changed. If it is not filtering, it will starve, but at a slower rate than if it also were spending energy on pumping. Such behavior would alter the line of the functional response curve (Box 4.1) relating feeding rate to food concentration so that it would no longer pass through the origin (zero feeding at zero food) and would begin feeding above a food threshold value. Such a threshold value has been found for marine copepods, but not for lake zooplankton (Muck and Lampert 1980).

4.5 THE IMPORTANCE OF BODY SIZE

Body size is one of the most important parameters that determines the ecological and physiological characteristics of an organism. There are many examples found in this book, such as the flotation ability of plankton (cf. Section 4.2.6). Size differences between species can cause differential mortality through size-selective predation (cf. Section 6.5.6); they can also cause differences in competitive abilities (cf. Fig. 6.34). Size is important for the question of niche selection and coexistence of similar species. Peters (1983) has summarized the extensive literature about the physiological and ecological consequences of body size.

Often, the size dependency of each of the parameters of the energy balance equation (cf. Section 4.4.3) is determined, since body size has large effects on the energy balance of organisms. Some surprisingly constant relationships have been found. For example, using a linear dimension such as body length as a measure of size, both the mass as well as the physiological performance can be described according to an exponential function (allometric function)

$$F = aL^b \qquad \text{or} \qquad W = aL^b$$

where L is the length; W the weight; F the physiological activity; and a, b are constants). The exponent b is often between 2 and 3. An example is shown in Fig. 4.29.

One can obtain a straight line by plotting the data on double logarithmic axes as

$$\log F = \log a + b \log L$$

One can then determine the slopes (b) and intercepts (a) by linear regression. The equation for Fig. 4.29 is

$$I = 0.08 \, L^{2.19}$$

If one uses the weight of an individual instead of length, one gets an exponential function

$$F = aW^b$$

In this case the exponent is usually less than one. For the example given in Fig. 4.29, the corresponding equation is (W in μg)

$$I = 0.015W^{0.74}$$

The physiological rate can be expressed as a rate per unit body weight (specific rate)

$$F/W = aW^{(b-1)}$$

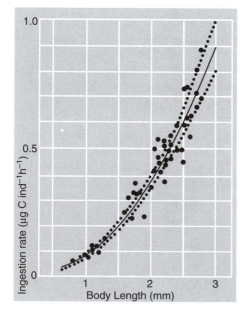

Figure 4.29 The effect of size on the feeding rate of *Daphnia pulicaria* at 15°C (food: *Scenedesmus acutus*, 0.17 mgC/l). Each data point represents an individual animal. Dotted lines are the 95% confidence limits for the regression line (from Geller 1975).

[118]

For our example we get

$$I/W = 0.015W^{-0.26}$$

These relationships hold for different species as well as for different sized individuals within a species. This principle can be demonstrated with an example. The respiration rates (R) at 20°C of different species of aquatic animals is plotted against body weight (Fig. 4.30). Weights range over four orders of magnitude, from rotifers (approx. 0.2 μg) to amphipods (approx. 1 mg). When plotted on a double logarithmic graph the points lie on a straight line with a slope of 0.794. Including all values up to 0.8 mg dry weight, we get the equation

$$R = 0.48W^{0.794}$$

Other physiological parameters also produce equations with exponents that are close to 0.75. Peters (1983) gives numerous examples of this. The ecological significance of this relationship is that, per unit biomass, large organisms have a slower metabolic rate than small organisms. They therefore use less resources, but grow more slowly.

It must be stressed, however, that this is only a general *trend*. Differences between organisms of similar size are not easily seen when the data over several orders of magnitude are plotted on double logarithmic graphs. The points in Fig. 4.30 appear to lie close to the regression line, but when converted to absolute values the deviations from the regression line are considerable. The measured respiration rates for a body weight of 3 μg dry wt, for example, lie between 0.02 and 0.07 μl O_2/ind/h. Within a narrow size range, there may be a negative relationship between metabolic rate and size for a particular species, but often not between species. In a comparison of species it is possible that the really important ecological adaptations are even reflected by the variability around the regression line.

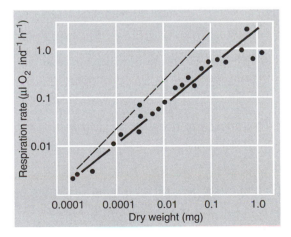

Figure 4.30 The effect of body weight on the respiration rates of various aquatic animals at 20°C. The slope of the line is 0.794. The dashed line has a slope of 1 (from Lampert 1984).

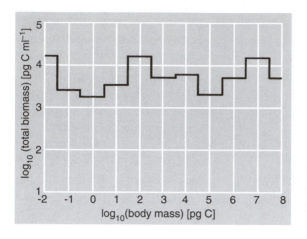

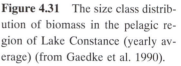

Figure 4.31 The size class distribution of biomass in the pelagic region of Lake Constance (yearly average) (from Gaedke et al. 1990).

Ecological factors also show trends with body size. Small species, for example, usually have higher densities per unit surface area than large species (other examples in Peters 1983). Sheldon et al. (1972) described an interesting rule for the marine pelagic region. If the organisms that live there are divided logarithmically according to their weight into classes and the biomass in each class is calculated, the biomass in each class is almost identical. This means that there is as much biomass per unit surface area in the size class of bacteria as there is in the size class of whales. The range is not as large in fresh water as in the ocean, but this phenomenon can be seen in the pelagic region of large lakes. Fig. 4.31 shows the example of Lake Constance, where the size differences from the smallest to the largest organisms nevertheless cover 10 orders of magnitude.

REVIEW QUESTIONS FOR CHAPTER 4

1. Why is tolerance to freezing less important for the distribution of aquatic organisms than for terrestrial organisms?
2. Chironomid midge larvae are important animals in the sediments of lakes. Members of the genus *Tanytarsus* are brownish, whereas members of the genus *Chironomus* are red because of their high hemoglobin content. Which chironomid genus would you expect in very eutrophic lakes and why would you expect them there?
3. Examine graphically Ostwald's law. Draw a graph showing the effect of the radius of a spherical particle on its sinking velocity. Assume that water has a density of 1000 kg m^{-3}, a viscosity of 1×10^{-3} Pa s. Construct graphs for a diatom with a density of 1.20 and a nonsiliceous alga with a density of 1.03. Calculate

the sinking velocity for diatoms with a radius of 5, 10, and 20 μm. How large would nonsiliceous algae have to be to have these same sinking velocities?

4. A phytoplankton species has a half-saturation constant of P uptake of 1 μmol l^{-1}. At what phosphorus concentrations does it achieve 10%, 30%, and 90% of its maximal uptake velocity?

5. The minimal cell quota of P-limited growth of many phytoplankton species is strongly size dependent. This variability can be reduced if the cell quota is expressed as a P content per unit biomass. For example, using carbon as a measure of biomass, the cell quota is about 0.001 atoms P per atom C. At what cell quota do such algae achieve 90% of μ_{max}?

6. A population of phytoplankton is growing in a chemostat until it reaches a steady state. Its μ_{max} is 1.4 d^{-1} and its k$_s$ is 0.1 μmol l^{-1}. The P concentration in the inflow medium is 3 μmol. How high is the remaining concentration of dissolved P in the culture vessel at dilution rates of 0.1, 0.3, 0.7, and 1.0 d^{-1}, respectively?

7. A phytoplankton species has a *P-I* curve characterized by the following cardinal points: compensation point at 10, saturation coefficient a 100, onset of light inhibition at 500 μE m^{-2} s^{-1}. The surface light intensity is 1000 μE m^{-2} s^{-1}. Calculate the compensation depth and the depth range of optimal photosynthesis for the following lakes: (a) an oligotrophic lake, extinction coefficient 0.2 m^{-1}; (b) a mesotrophic lake, extinction coefficient 0.5 m^{-1}; (c) a eutrophic lake, extinction coefficient 1.5 m^{-1}.

8. Why do purple bacteria and green sulfur bacteria grow at the lower margin of the euphotic zone of eutrophic and meromictic lakes, even though their photosynthetic rates might be higher at higher light intensities?

9. The chemolithoautotrophic sulfur bacterium *Beggiatoa* frequently forms thin layers of dense mats at the sediment surface. Under what chemical conditions would you expect such mats? Why are they so thin?

10. The "heterotrophic potential" of bacteria is estimated by measuring the uptake of radioactively labeled glucose. Why is this not identical with the actual production of heterotrophic bacteria?

11. In a feeding experiment, *Daphnia* are fed radioactively labeled algae with a specific activity of 4500 dpm/mg C (dpm = disintegrations of radioactive decay per minute, a measure of radioactivity). Each individual *Daphnia* can filter radioactive food for 10 min. In a series of food concentrations, the following results are obtained:

Algal conc. (mg C/L)	dpm/individual *Daphnia*
0.1	145
0.2	300
0.5	400
1.0	390

Calculate the *Daphnia* filtering rates (ml ind^{-1} h^{-1}) at each food concentration. Construct a graph showing the effect of algal concentration on filtering rate. What is the maximum filtering rate for *Daphnia*? What is the incipient limiting concentration (mg C/L) for *Daphnia* filtering rate?

12. For daphniids, the feeding or ingestion rate increases as a power function of body length. The ingestion rate of a 2.5-mm-long *Daphnia* has been measured as 3 μg C ind^{-1} h^{-1}. Calculate the ingestion rate of a 2-mm-long *Daphnia*, assuming an exponent of 2.3 in the length-ingestion rate relationship.

13. The respiration rate of an amphipod of 0.7 mg dry mass has been measured as 0.09 μmol O_2/h. Estimate the respiration rate of an immature animal of 0.1 mg dry mass.

14. A fish pond is stocked with 2,500 one-year-old carp (20 g each) on the first of May. When the pond is drained on 31 October, 1600 fish with an average weight of 250 g are recovered. Calculate the individual growth rate, the yield, and the total production. Estimate the amount of food consumed by the fishes during the season.

CHAPTER **5**

Populations

5.1 FEATURES OF POPULATIONS

Natural selection works on the phenotype that is carried by the individual. Clearly, evolution does not, however, take place at the level of the individual. For selection to occur, there must be a group of individuals that differ in their characteristics. Such a group of individuals of the same species that occupies a particular location at a given time is called a *population*.

The definition of a population is not a trivial point, for it can vary with the objective of the observer. In theory, all members of a species belong to one population; there is good reason, however, to view them as a collection of local populations. Groups of individuals that live in different geographical regions may very likely have very different selection pressures acting upon them. Thus it is useful to define a population as a group of individuals that form a *reproductive system* or *gene pool*. Most important is whether they actually interbreed, not whether it is theoretically possible. Individuals of a fish species in two different lakes, for example, can be viewed as two different populations, even though they belong to the same species and could easily reproduce if they were exchanged. Normally, however, there is no opportunity for the two populations to exchange. The fish in each lake form a reproductive system and can be seen as independent populations. This example makes two important points:

1. That a population is rarely completely closed. There is usually some degree of gene flow between populations;
2. That the space taken up by a population depends on the size and mobility of the species.

For example, in a collection of small ponds each pond may have its own population of ostracods. Frogs may return to their home pond for breeding, but there may still be some exchange between ponds. Such a population that is composed of local groups of individuals is sometimes called a *metapopulation*. For more mobile species such as flying insects (e.g., giant water beetle), all the ponds would make up a single population; ducks that visit these ponds would be considered only part of a population that occupies a much larger region. In limnology the distinction is often clear, for the boundary between water and air effectively isolates many species, enabling distinct populations to develop in each water body.

The concept of the population as reproductive system is not perfect. In fresh water there are many species, probably the majority, that rarely, if ever, reproduce bisexually. They reproduce by dividing or by parthenogenesis. Except for the effect of mutations, such gene pools consist of a multitude of genetically uniform clones, rather than genetic recombinations. For practical reasons, we usually view these clones as a single species, even though they cannot cross. Under such conditions *subpopulations* can arise that differ in certain features, such as their ability to undergo diel vertical migrations.

Since populations consist of many individuals, they have characteristics that result from the sum of the individuals:

1. They have a size or density that can change;
2. They exhibit phenotypic or genotypic variation;
3. They can have an age structure;
4. They have a specific spatial distribution pattern.

5.2 CONTROL OF POPULATION SIZE

5.2.1 Fluctuations in Abundance

Ecologists usually describe a population in terms of the density of organisms in a defined surface area or volume of water (*abundance*), whereas a population geneticist would be interested in the total number of individuals in a population. With microorganisms one must sometimes assume that all individuals of the population are identical and substitute simpler measurements, such as units of biomass (dry weight, carbon, cell volume), or other surrogate parameters, such as chlorophyll, for phytoplankton biomass.

The abundance of populations changes temporally and spatially. Typical temporal patterns include irregular fluctuations around a more or less constant level, increases over long periods, decreases over long periods, cyclic *oscillations,* and occasionally explosive increases by populations that normally exist at a low level. Frequent and long-term observations are needed to depict the actual development of populations and are an important aspect of ecology (Edmondson 1991).

[124]

Phytoplankton densities in the temperate zone may fluctuate over four orders of magnitude with the seasons. Yearly maxima or averages may be relatively constant over a long period, or they may show well-defined trends (e.g., *Rhodomonas minuta* in Lake Constance after 1981, Fig. 5.1). Zooplankton, such as cladocerans and rotifers, also have similar yearly changes in population density, whereas planktonic bacteria generally have noticeably smaller fluctuations (about one-tenth as great).

Longer-lived species (higher plants, fish, mussels) usually fluctuate much less and over a time scale of years. This is because these long-lived species consist of several year classes that compensate for "good" and "bad" years. These species may also show long- term trends. Reeds, for example, were only found in a few isolated locations in the Neusiedler Lake (Austria/Hungary) in 1872. Gradually they spread, and today a dense population of reeds occupies half of the lake.

5.2.2 Mechanisms of Change in Abundance

It is not easy to identify the causes for population changes simply from the pattern and rates of fluctuations. Constant population density within the period of observations may actually indicate that the changes are very slow. If the reeds in Neusiedler Lake had been observed for only a few years, no changes would have been seen. It is also possible that a population is very *dynamic*, but appears static because the processes of population increase and decrease are in balance. This is likely the reason for the relatively small fluctuations of the bacterioplankton.

The most important process of increase is *reproduction*, that is, the birth of new individuals of multicellular species or the cell division of protists. A population can also increase through *import* of individuals from outside. This can occur by active immigration or by passive transport. Import for one population means *export* for the original population.

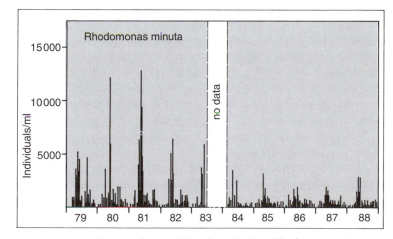

Figure 5.1   Long-term changes in the abundance of the phytoplankter *Rhodomonas minuta* in Lake Constance. Concentrations (no./ml) are averages for the upper 20 m.

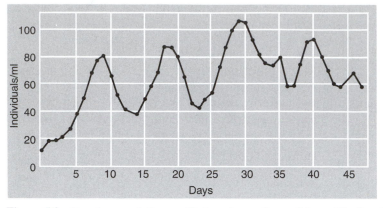

Figure 5.2 Regular population oscillations in a culture of the rotifer *Brachionus calyciflorus* with daily renewal of the culture medium and a constant amount of food (from Halbach 1969).

Individuals are not only born, imported, and exported; they also die. There are various causes for *mortality*, such as being eaten by predators and death due to disease, hunger, or lethal chemical and physical conditions. Multicellular organisms may die from old age not related to environmental conditions, but this does not occur in protists.

5.2.3 Population Growth Rate

We introduced growth rate in Section 4.3.3 as a way of characterizing the activity of organisms under natural conditions. The concept of *growth rate* is paramount for describing population dynamics and therefore must be clearly defined. For multicellular organisms, it is used to describe *numerical* (increase in numbers of individuals) as well as *somatic* growth of the individual (biomass gain per unit time). The meaning of growth rate is usually clear from the context of its use; so most authors refrain from referring more precisely to "body" or "population growth."

Zoologists generally use the term *growth rate* to mean the net change in abundance that result from additions and losses. For microbiologists and phytoplanktologists working with cultures, on the other hand, growth rate means the *rate of reproduction* (corresponding to μ; cf. Section 4.3.3), since there is negligible mortality in cultures lacking predators. This is equivalent to *birth rate* in zoological nomenclature. Phytoplanktologists who work in the field usually use "gross growth rate" for the reproduction rate and "net growth rate" to describe the observed changes in abundance.

Today, population growth rate is generally understood to mean *relative* (synonyms: *specific, per capita*) growth rate. This use refers to the temporal changes in a population expressed as numbers of individuals produced per individual and not the change in numbers per unit surface area or volume. An increase from 1 to 2 has the same growth rate as a change from 1000 to 2000. The (net) growth rate (r) is defined as:

$$r = \frac{dN}{dt}\frac{1}{N}$$

where t is time and N the number of individuals per surface area or volume). This has the dimension t^{-1}.

There are two different types of increase and loss: Multicellular organisms produce a series of offspring and then die, whereas unicellular organisms reproduce by division and are therefore potentially immortal. This has led to the use of two different sets of terms in reference to population growth rate. Ignoring import and export, the (net) growth rate is defined as:

1. In zoological terminology

$$r = b - d$$

where b is the *birth rate* and d the *death rate* (dimension: t^{-1})

2. In microbiological–phytoplanktological terminology

$$r = \mu - \lambda$$

where μ is the *gross growth rate* and λ symbolizes the *loss rate*.
The resulting *net growth rate* can be either positive (increasing abundance) or negative (decreasing abundance).

If a population grows at a constant growth rate, its size can be calculated for a future time (N_t) based on its initial size (N_1), the time interval, and the growth rate. If the population grows in discrete steps, the growth can be expressed geometrically:

$$N_t = N_1(1 + R)^{(t-1)} = N_1(N_2/N_1)^{(t-1)}$$

This formula can be used, for example, if the births occur at a particular time of the year or are in some other way synchronized at regular intervals. In each of the time intervals the population increases by a certain fraction (R) of the existing population. The time interval in the exponents must be an integer multiple of the time interval between the two reproductive periods. The formula can only be applied when the organisms develop in discrete generations (*cohorts*) that can be clearly distinguished. Examples of this include fish and insects that have one generation per year or copepods with long development times and few nonoverlapping generations per year (cf. Section 5.2.6).

Often births and deaths occur randomly through time, and there are thus no discrete pulses in population growth. Population changes then follow a smooth exponential curve rather than the stepwise curve of the geometric series. This can be expressed as an extreme case of the geometric growth with an infinite number of small time intervals:

$$N_2 = N_1 e^{r(t_2 - t_1)}$$

An exponential population curve can be changed to a straight line with a log transformation of the abundance. One can then estimate the population growth rate from the logarithms of the abundance:

$$r = (\ln N_2 - \ln N_1)/(t_2 - t_1)$$

Doubling time, the time it takes for a given population density to double in size, is frequently used as a graphic measure of the intensity of reproduction. Since the natural log of 2 is 0.69, an r of 0.69 d^{-1} is the same as one doubling per day and an r of -0.69 d^{-1} describes a population decreasing by 50% per day. The *doubling time* of an increasing population (t_d) is $\ln 2/r$.

5.2.4 Exponential and Logistic Growth

A population cannot continue to exist if its net growth rate is always negative. It follows then that all species that exist today have to be capable of positive growth rate under certain conditions; that is, they produce more offspring than are lost through mortality and export. When the growth rate is constant, the abundance follows an increasingly steep exponential curve (cf. Box 4.2). Obviously, even a slow-growing species would eventually completely overpopulate its habitat. It follows then that it is not possible to have unlimited exponential growth. Exponential growth that is sustained for some time only occurs in special situations, such as the new colonization of an environment free of competitors, the recolonization of a habitat following a catastrophic event, or the inoculation of batch cultures of microorganisms. Normally, there is an upper limit for the population density due to a limit in available resources. This is referred to as the *carrying capacity* of a population.

There are two ways a population is kept below the carrying capacity:

1. *Density-independent limitation*: Density-independent limitation occurs when the exponential growth is interrupted by external factors that are unaffected by the density of the population. These factors represent "catastrophes" for the population. These could be sudden changes in physical conditions (temperature, conditions during lake circulation, washout due to flooding, drying out of the system) or rapid chemical changes (e.g., poisons). Population control exclusively by population-independent factors is only possible if the catastrophic events occur frequently enough to keep the population from approaching its carrying capacity. This control is most important in disturbed habitats. Population-independent regulation does not, however, lead to nearly constant population levels, since catastrophes occur randomly.

2. *Density-dependent regulation*: Regulatory mechanisms that reduce the net growth rate of increasing populations are called density dependent. They can affect either the reproductive rate or the rate of mortality. Increasingly scarce resources with increasing population level can reduce the reproductive rate (resource limitation) as well as increase the mortality (lower life expectancy, starvation). Mortality is the critical factor when high population density leads to higher risks of infection by parasites and disease or to a shift in predator preference to the most frequent prey species. The effect of parasites and predators is usually dependent on the absolute density, whereas the effect of resource limitation usually depends on the relationship between population abundance and carrying capacity. Density regulation for mobile animals may also involve emigration.

Density-dependent factors can be viewed as genuine regulators, for they keep the abundance of a population near the carrying capacity. If the population exceeds the carrying capacity, the growth rate becomes negative and the population decreases. Below the carrying capacity, the growth rate becomes positive and leads to a population increase. The size of the positive or negative growth rate depends on the distance of the population density from the carrying capacity. The simplest mathematical expression for density-dependent regulation is the logistic growth curve that we already introduced in Section 4.3.3:

$$\frac{dN}{dt} = rN(K - N)/K \qquad \text{or} \qquad N_t = \frac{K}{(1 + [(K - N_0)/N_0]e^{-rt})}$$

The negative feedback between the population density and the net growth rate sometimes involves *time lags*. For example, the number of eggs produced may depend on the amount of resources available, but the offspring first begin using the resource after a certain period of development. A population may produce more offspring than the food resources can support. The food requirements of the population then exceed the carrying capacity, and the growth rate declines proportionately. Such time lags lead to regular population oscillations around the carrying capacity (Fig. 5.2). The effect of time lags can be expressed as:

$$\frac{dN_t}{dt} = rN_t(K - N_{t-T})/K$$

where T is the lag period.

Oscillations around the carrying capacity may or may not eventually dampen, depending on the length of the time lag (Fig. 5.3). In this way population densities may undergo many types of fluctuations independent of outside forces. The product of r and T is critical in determining the shape of the curve. Stable oscillations with a constant amplitude develop if rT is larger than $\pi/2$. Curves with steep peaks of short duration and long periods between them arise when there are long time lags ($rT \approx 2$). There is the danger to a population that during these long periods of very low density random events may lead to its extinction ("random extinction").

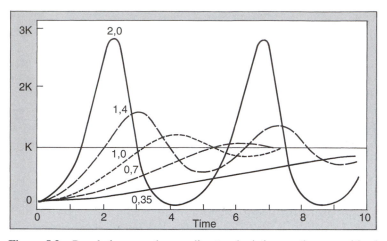

Figure 5.3 Population growth according to a logistic growth curve with a lag period T. Curves represent various products of r_{max} T. K, carrying capacity (from Hutchinson 1978).

5.2.5 The Generation of Deterministic Chaos by Density-Dependent Regulation

The classic form of the logistic growth curve assumes continuous reproduction, just as the exponential growth curve in the absence of density regulation. For populations where reproductive events occur in synchronized pulses, a stepwise version of the growth curve might be more realistic. In this case, the logistic growth equation reads:

$$\frac{\Delta N}{\Delta t} = \frac{RNK - N}{K}$$

where R is the relative rate of increase from step to step or $(1 - N_2/N_1)$ at infinitely small values of N. For many freshwater organisms such as fish and aquatic insects, the length of the time interval (T) is 1 year.

The behavior of the growth curve depends on the product of RT. At $RT < 1$ the abundance gradually approaches the carrying capacity (Fig. 5.4). Between $RT = 1$ and 2 there are damped oscillations between one value above K and one value below K. If $RT > 2$, there is no longer a dampening and there will be stable oscillations between two "attractors" (one $<K$, one $>K$). If RT exceeds 2.4495, the time series becomes more complicated. The number of attractors between which N oscillates increases to four ("period doubling"). Further period doubling occurs at $RT = 2.56$ (8 attractors) and at 2.569 (16 attractors).

If RT exceeds 2.5699, the growth curve loses its periodicity and becomes completely irregular. Because it is produced from a completely deterministic equation, the resulting pattern is called *deterministic chaos* (May 1974). Before the discovery of deterministic chaos irregular fluctuations were usually explained as the effect of random factors. Irregularity is no longer a feature that can be used to distinguish between determinism and randomness.

[130]

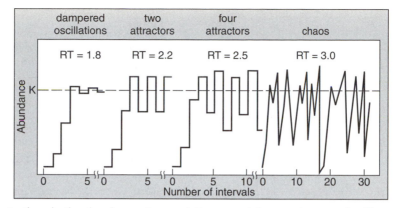

Figure 5.4 Fluctuations in the abundance of a population around the carrying capacity (*k*) at various values of *RT* (*R*, relative rate of population increase; *T*, time lag). Note that at *RT* = 3.0 the population changes around the carrying capacity become irregular.

According to the concept of deterministic chaos, even small differences in the initial conditions of an experiment can result in major differences after some time has passed in the experiment. This phenomenon, known as *weak causality*, leads to the question of whether one has indeed used true replicates in an experiment. Strictly speaking, experimental replicates must have identical initial conditions that, in turn, lead to identical results. In most sciences, and especially in ecology, where there is much variability, this concept of replication cannot be strictly adhered to.

The discussion of deterministic chaos has relatively recently entered theoretical ecology (May 1974). It has been found that other relatively simple models besides the stepwise version of the logistic growth equation (e.g., predator–prey models with three populations) produce chaotic behavior if parameter values are chosen appropriately. At present the question of the importance of deterministic chaos in actual ecological processes is still unresolved (Scheffer 1993). Some scientists consider it the new paradigm of complex systems analysis, while others simply view it as a game mathematical modelers play.

5.2.6 The Concept of Dynamic Equilibrium or Steady State

The fact that a population keeps its density at a constant level near the carrying capacity does not mean that the processes of population dynamics have stopped. It only means that reproduction (plus import) and mortality (plus export) are in balance. There is constant replacement of the population without changing its density. One can imagine this type of dynamic equilibrium or "steady state" occurring at the level of abiotic resources. Chemical substances can be simultaneously used and regenerated by import (solubilization, excretion by organisms). A population of autotrophs can be in steady state only when there is also a steady state of the limiting abiotic resources.

Although a perfect steady state is rarely if ever found in nature, this concept has become important in ecological modeling. Even only moderately complex models often can be solved analytically only by assuming equilibrium conditions ($dN/dt = 0$). Often transition conditions can only be calculated by numerical simulation. The most perfect experimental approach to the concept of steady state is the chemostat culture of microorganisms (cf. Section 4.3.3). Resource regeneration is provided by import of medium into the culture vessels and mortality occurs as export of organisms through the outflow (cf. Box 4.2).

Conditions similar to the steady state of a chemostat are sometimes found during the summer in the epilimnion of stratified lakes. Despite high rates of photosynthesis, there may be little change in the biomass of the phytoplankton. The newly formed organic matter is eliminated at about the same rate as it is formed and therefore does not accumulate. This also applies to individual populations of phytoplankton: The actual increases remain well behind the rate of cell division. While new phytoplankton are being formed by cell divisions, they are also being eaten, killed by parasites, sinking out of the epilimnion or dying from stress conditions at roughly the same rate. Zooplankton populations are also very similar. Even chemical substances that cannot reproduce, in contrast to organisms, can approach steady-state conditions: Changes in concentration are the net result of consumption processes (uptake by phytoplankton) and supply processes (excretion by zooplankton, addition from inflow, mixing from the hypolimnion).

Zooplankton have a role similar to that of the flow-through in a chemostat. The more algae they eat, the more nutrients they excrete (P as orthophosphate, N as ammonium and urea). This is analogous to the chemostat, in which increased flow rate causes increased removal of algae and a proportionate increase in the supply of nutrients.

5.2.7 Estimating the Parameters of Population Dynamics

Phytoplankton and other protists

The gross growth rate (μ) can be determined most easily in cultures in which population losses are either excluded or experimentally controlled. When there are no population losses, μ can be estimated simply from changes in the population density (N_1, N_2) between two observation times (t_1, t_2):

$$\mu = (\ln N_2 - \ln N_1)/(t_2 - t_1)$$

μ is a function of the limiting resource (cf. Box 4.2). The resource-saturated ("maximum") growth rate (μ_{max}) is a species-specific parameter that is also dependent on the temperature. The maximum growth rates of planktonic algae at 20°C vary from approx. 0.25 d^{-1} (large dinoflagellates) to 2.5 d^{-1} (very small green algae and cyanobacteria). The maximum growth rate generally decreases as the cell or colony size increases. This trend only holds, however, over a wide range of sizes of many orders of magnitude in cell volume, and not for a comparison of two species whose volumes differ by less than two orders of magnitude (Banse 1982).

Population losses can never be excluded under natural conditions. Thus only the net growth rate (r) can be observed in the field (cf. Section 5.2.3). It consists of both the reproduction rate (μ) and the loss rate (λ) and can also be negative:

$$r = \mu - \lambda$$

The loss rate can be divided into various parts, all of possible importance:

$$\lambda = \gamma + \sigma + \delta + \pi + \omega$$

where γ is the grazing rate (loss due to feeding by zooplankton); σ, the sedimentation rate; δ, the physiological mortality; π, the mortality from parasites; and ω, the washout loss to the outflow.

It is possible to determine r directly from counts, but in most cases it is extremely difficult to include all components of the equation. Rarely is it possible to measure the gross growth rate in the field directly. One of the most reliable field methods involves determining the frequency of dividing cells (mitotic index; Braunwarth and Sommer 1985).

The sedimentation rate (σ) can be estimated from the rate of sinking (v_s) and the average depth of mixing (z) (cf. Section 4.2.6). If there are no data available for the sinking rates one can directly measure the losses due to sedimentation by suspending sediment traps just beneath the euphotic zone (Sommer 1984). Cells that collected in open tubes over a period of days are then counted. For example, a sedimentation rate (V_s/z) of 0.5 d^{-1} would be in the same range as nutrient- or light-limited growth rates for phytoplankton (cf. Section 4.2.6). Even with relatively high levels of resources, it is difficult for heavy algae such as diatoms and desmids to stay in the epilimnion, especially if the mixing depth is shallow. If, on the other hand, the mixing depth were ten times as large as the daily sinking distance, sedimentation could cause a collapse of the population only under extreme resource limitation (i.e., low growth rates).

Grazing rates (γ) are estimated from the clearance rate (filtering rate) of the total zooplankton community (G) and the coefficient of selectivity (w) for the specific alga that is being considered:

$$\gamma = Gw$$

The total zooplankton clearance rate describes the proportion of the water volume that is cleared of particles, assuming that the food particles are eaten optimally ($w = 1$). It is the sum of all the individual clearance rates (individuals/volume) and has the dimension t^{-1}. The term *filtering* is used to mean the removal of algae, even though many zooplankton do not actually filter their food (cf. Section 6.4.1). Thus, clearance rate is a preferable term, since it does not imply a specific feeding mechanism. The coefficient of selectivity corrects for the differences in the edibility of different algal species. The grazing rate on species i is calculated relative to the grazing rate on the most edible algal species:

$$w_i = \gamma_i/\gamma_{opt}$$

[133]

It is difficult to make accurate estimates of grazing rates, since the coefficient of selectivity varies for different species of phytoplankton as well as for each species and growth stage of the zooplankton. This is least problematic for those species of phytoplankton that can be eaten optimally or nearly optimally by most zooplankton, as, for example, flagellates and thin-walled coccal algae in the size range of 3–30 μm. These phytoplankton may experience grazing rates of up to 2.5 d^{-1} or more during periods of maximum zooplankton biomass (Haney 1973, Lampert 1988a). This is equivalent to a daily population loss of 91.8% due to grazing alone ($N_1 = N_0 \, e^{-2.5} = N_0 \, 0.082$). Most algal species cannot reproduce rapidly enough to compensate for such high losses. Thus grazing can eliminate phytoplankton populations that have optimal nutrient and light conditions.

Physiological mortality (δ) can be determined only if one finds recognizable remains of dead plankton, such as empty shells of diatoms. The bodies or "corpses" sinking out of the epilimnion can be captured in sedimentation traps. Physiological mortality is an undefinable mixture of processes. *Mortality due to parasitism* (π) is also included in δ, for at present there is no way to distinguish those cells that have died because of parasites.

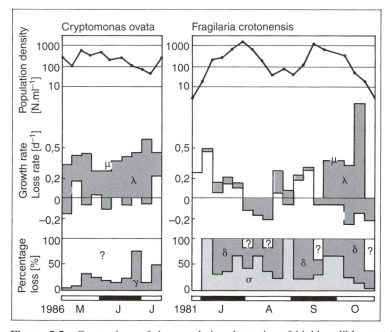

Figure 5.5 Comparison of the population dynamics of highly edible, nonsinking flagellates (*Cryptomonas ovata*; from Braunwarth 1988) with an almost completely inedible, rapidly sinking diatom (*Fragilaria crotonensis*; from Sommer 1984) in Lake Constance. (*Above*): Population density in the euphotic zone. (*Middle*): Gross growth rate (μ); net growth rate (r); loss rate (γ, difference between μ and r). (*Below*): Percent contribution of each component to the loss rate: γ, grazing by daphniids; σ, sedimentation; δ, physiological death, in this case caused mainly by parasites; ?, unexplained losses.

Losses due to washout (ω) can be estimated from the ratio of outflow to volume of the epilimnion. In most lakes this is a comparatively unimportant process, although it can become significant during periods of flooding, such as following a snow melt.

There are major species differences in the importance of the different loss factors (Fig. 5.5). Flagellates <30 μm, for example, are especially susceptible to grazing, but sink very slowly; large colonial diatoms are hardly grazed, but sink rapidly and are often parasitized. The loss factors for large, motile dinoflagellates that are resistant to both grazing and sedimentation are not yet understood.

It is rarely possible to determine all the components of the population dynamics of phytoplankton in nature. There are also large errors in many of the measurable parameters (counting errors, sampling error due to patchiness, errors in measuring sinking rates, coefficients of selectivity, etc.). Because of these problems, often other approaches are used. With "enclosures" (Fig. 2.1b) one can hold a particular parameter constant and compare the changes to that in the control. For example, grazing effects can be eliminated by removing the zooplankton; fungicides can be used to suppress parasitism by fungi; the enclosure can be mixed artificially to avoid the effects of sedimentation. Care must be taken, since any reduction in loss factors will lead to an increase in biomass. This, in turn, may cause an increase in the effect of resource limitation of μ, so that the gross growth rates in the control and manipulated treatment are no longer identical.

Animals

Changes in the animal populations are also the result of processes of reproduction and loss (cf. Section 5.2.3). The growth rate of a population (r) results from the difference in birth rates (b) and death rates (d):

$$r = b - d$$

In contrast to bacteria, protozoans, and algae, one can distinguish the juvenile, reproductive, and adult stages of most aquatic animals. Thus often one can often measure birth rate directly, although usually there is no means of getting reliable estimates of death rate and its various components (predation, natural death). Death rate in such instances can be estimated from the difference between birth rate and the growth rate of the population.

It is important to consider two different modes of reproduction:

1. Some animals reproduce in discrete pulses producing individual age classes or *cohorts* that can be tracked. The simplest situation is if each animal reproduces only once during its life and the reproduction of all the individuals is synchronous. Many insects, some copepods, and the Pacific salmon (*Oncorhynchus*) follow this pattern. By following each cohort, from the hatching of the young to the dying of the adults, one can estimate directly the parameters of the population changes. Theoretically, it is enough to get two measurements of the population density during the time the young are produced. If one can follow, however, what happens to a cohort, one can obtain valuable additional information, such as the mortality rates of the different stages. Figure 5.6 shows an example of a cohort of copepods that reproduce only

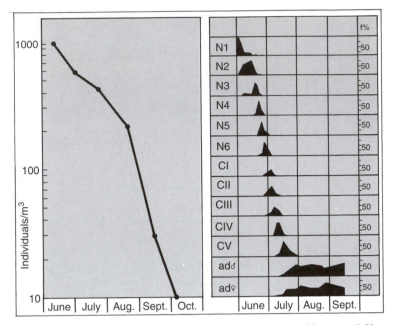

Figure 5.6 Typical development of a cohort. The calanoid copepod *Heterocope saliens* has only one generation per year in the Norwegian Lake Øvre Heimdalsvatn. The nauplii hatch in June. (*Left*) Decrease in the number of copepods throughout the summer. (*Right*) Temporal appearance of the various stages (from Larsson 1978).

once per year. It is easy to see how the population density decreases as the cohort of the copepod gets older. Species with several synchronized generations per year show a similar pattern.

The situation is more complex for long-lived animals that produce several times in their life, but in discrete pulses. An example of this type of reproduction is a fish species in the temperate zone that spawns once a year. Here the population consists of many year classes that form cohorts. It is possible to determine the exact cohort age by capturing a fish and counting the growth rings on its scales, otoliths (ear bones), or bones. Using this information one can estimate the average death rate of the population, even without precise estimates of the total population size. As long as one has a sample that is representative of all the age classes, the relative age distribution is sufficient to permit an estimate of the death rate, averaged over several years. In an example shown in Fig. 5.7, whitefish (*Coregonus* spec.) were examined in an unfished population in Schluchsee, a lake in the Black Forest (Germany). Relative age classes were determined based on samples of fish captured in a gill net with variable size meshes. For fish older than 3 years, there is a linear decrease in the logarithm of the proportion of each age class in the population with age class. The first two year classes are underestimated by the use of gill nets, which do not capture small fish efficiently.

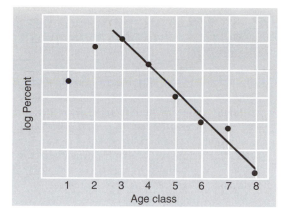

Figure 5.7 The composition of age classes of a population of whitefish (*Coregonus* sp.) in Schluchsee (Black Forest, Germany). The annual loss rates were determined by regression of the age classes 3–8. The age classes 0–2 are not quantitatively represented (from Lampert 1971).

Since we are dealing with cohorts, any newly hatched fish will belong to the age class 0. Thus all other age classes can only decrease with time. In this case $r = d$ and the negative slope of the regression line provides an estimate of the death rate. In the example above, the death rate was $0.87a^{-1}$, which corresponds to a loss of about 58% of the population per year. The variability of data around the regression line is caused by measurement errors as well as fluctuations in the size of each of the age classes, brought about by variable reproductive success.

2. Animals with frequent reproduction and overlapping generations lack cohorts and have *continuous reproduction*. It is impossible to determine from a field sample which young came from which adults. Both birth rates, in response to food, and death rates, in response to predators, can change rapidly in such populations.

Many zooplankton with continuous reproduction also carry their eggs with them, which can be easily enumerated. Since the mothers do not provide the eggs with nutrition during their incubation, the time to hatching is determined by the temperature at which the mothers were living. One can determine experimentally the relationship between temperature and egg development time. Egg development time decreases with increasing temperature (cf. Section 4.2.1). By knowing the number of eggs per animal and the temperature at which the animals resided, we can calculate the *instantaneous birth rate* (Box 5.1). The development of this concept has been discussed by Edmondson (1993). Although several models have been proposed for the estimation of the instantaneous birth rate, the formula from Paloheimo (1974)

$$b = \ln(E + 1)/D$$

has proved to be the most robust against violations of the underlying assumptions (e.g., uniform age distribution) and is easy to use (Gabriel et al. 1987). To estimate the in-

BOX 5.1	CALCULATION OF THE BIRTH RATE FOR CONTINUOUS REPRODUCTION BY THE EGG-RATIO METHOD

The instantaneous birth rate for continuous reproduction can be estimated by the *egg-ratio method*. The principle can be most easily explained by the procedure introduced by Edmondson (1968). It assumes a uniform age distribution of eggs, that is, that there are equal proportions of eggs from each age class, freshly laid eggs as well as nearly developed embryos. In that case, a certain proportion of eggs will hatch in each time interval. This proportion equals the ratio of the time interval to the time needed by an egg for complete development (egg development time D). For example, if the egg development time is 3 days, then at the end of 1 day all eggs will have hatched that had been older than 2 days at the beginning of the day, that is, one-third ($1/D$). The number of eggs in the population does not decrease during this period because hatched eggs will be replaced by freshly laid ones. If the average number of eggs carried per animal (egg ratio) is E, we can predict the proportional increase of the population as :

$$B = E/D$$

This proportion is called the *finite birth rate*. In fact, eggs hatch at any moment, not in daily pulses. Hence, we must consider infinitely small time intervals to estimate the *instantaneous birth rate* (b). If mortality is zero, the instantaneous birth rate will equal the rate of population increase [i.e., $b = r$ or $b = (\ln N_2 - \ln N_1)/t$].

Edmondson used an approximation to relate B to b. If $N_1 = 1$, then 1 day later $N_2 = 1 + B$. The upper formula then will read:

$$b = [\ln (1 + B) - \ln 1]/1 = \ln (1 + B)$$

or

$$b = \ln (1 + E/D).$$

This is only an approximation, as it is not a mathematically correct integration because the age distribution of eggs is not uniform in growing or declining populations. The correct formula, which is now widely used to calculate instantaneous birth rates, has been given by Paloheimo (1974):

$$b = \ln (E + 1)/D, \qquad \text{dimension d}^{-1}$$

stantaneous birth rate, one must count the number of eggs and embryos in the population and the number of female animals. D can be obtained from the literature (e.g., Bottrell et al. 1976) as long as the ambient temperature experienced by the eggs is known. The average temperature may be difficult to determine if the ambient temperature changes in a diel cycle in very shallow ponds or because egg-carrying females migrate vertically in a temperature gradient (cf. Section 6.8.4).

For continuously reproducing species, we can use two sequential measurements of population density to determine the population growth rate (r), as discussed earlier for the algae:

$$r = (\ln N_2 - \ln N_1)/(t_2 - t_1)$$

where N_1 and N_2 are the population densities at the times t_1 and t_2.

There is no direct method for measuring death rates, and they must be estimated from the difference: $d = b - r$. To calculate birth and death rates of animals with continuous reproduction, one must make population measurements over short time intervals (a few days). It is therefore possible to observe seasonal changes in the population parameters and attempt to relate them to changes in the environment. *Daphnia* are good examples of populations with continuous reproduction. The birth rate of *Daphnia longispina* (Fig. 5.8) is high in the early spring, since the number of eggs per animal is high during the spring pulse of algae. As the clear water phase develops (cf. Section 6.4) and the food conditions for the *Daphnia* worsen, the egg number and therefore the birth rate decline. Birth rate increases in the summer, but as a response to high temperatures and shortened development times and not because of high egg numbers. The death rate is low at first, but becomes as high as the birth rate later in the summer when there are many predators. This results in population growth rates (the difference between birth and death rates) that are high in the spring and low in the summer.

Figure 5.8 illustrates several problems with the estimation of death rates. Negative death rates appear in the spring, although technically these are not possible. The likely reason is that the death rate itself was not actually measured, but calculated from ($b - r$). Thus all errors that were made in estimating b and r accumulate in d. Hatching of young *Daphnia* from resting eggs deposited in the lake sediments the previous season and not included in the egg counts of the population would contribute to the negative death rate, for the population would grow more rapidly than the birth rate predicts. Other reasons include nonuniform egg age distribution, incorrect estimates of egg development time, or incorrect estimates of r due to low abundance and patchiness.

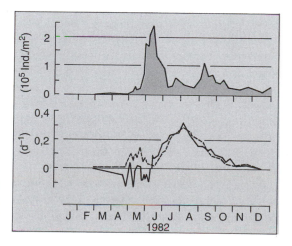

Figure 5.8 The population dynamics of *Daphnia longispina* in Schöhsee (Holstein). (*Above*): Seasonal changes in abundance. (*Below*): Birth rates (dashed line) and death rates (solid line) (from Lampert 1988).

5.3 PHENOTYPIC AND GENOTYPIC VARIABILITY

Phenotypic and genotypic variability are important features of populations. Individuals in a population are not all identical. Variability is also caused by the fact that although animals respond to changing environmental conditions (*reaction norm*), not all animals have been living under the same environmental conditions. Such environmental *modifications* do not alter the genome and are therefore are not hereditary. Another part of the variability is caused by individual genetic differences. These lead to different reactions (*differential reaction norms*) under the same environmental conditions. These differences are inheritable.

If we accept natural selection as the driving force of evolution, then we must be concerned with the selection of inheritable traits. Clearly, selection works upon a population of phenotypes that respond to environmental factors, but these phenotypes are an expression of the corresponding genotypes. Thus evolution can take place only when there is genetic variability. Ecological physiologists may be interested solely in the phenotypic reactions and adaptations of organisms to environmental conditions, but the ecological geneticist must be interested in the effects of environmental factors on the gene pool of a population, that is, natural selection.

The difference can be easily illustrated: There are many highly productive farm ponds in Illinois (U.S.). The high nutrient levels lead to extremely high densities of algae in the summer. In the day there is much primary production in the surface water, but in the deep water and at night including the surface water there is severe oxygen depletion. The cladoceran *Daphnia pulex* is a common inhabitant in these ponds. Under low oxygen conditions they have the ability to form hemoglobin, which allows them to extract oxygen out of the water, even at extremely low concentrations. It is easy to recognize the poor oxygen conditions in the summer by the pink color of the normally pale *Daphnia*. This is a phenotypic reaction.

A population of parthenogenetically reproducing *Daphnia* in a pond consists of many clones. Each clone consists of genetically identical individuals and differs from the other clones. Some clones are able to produce significantly more hemoglobin than others, under the same conditions. Population genetic methods such as allozyme electrophoresis can be used to demonstrate that throughout the summer there is an increase in frequency of clones that produce much hemoglobin and a decrease in the other clones (Weider and Lampert 1985). This is evidence that there has been a shift in the genetic composition of the population. We have observed the process of natural selection. Note, however, that no genotype has disappeared, for there was only a shift in the relative composition of the population. Although genotypes that do not produce much hemoglobin are rare, they survive in habitats that still have tolerable conditions, such as close to the surface of the pond. There appear to be costs for maintaining high hemoglobin (energy, higher visibility to predators). Thus the other genotypes will have an advantage when the oxygen conditions improve, and from fall to spring there will be a shift in genotype distribution back to the low hemoglobin producers.

Recently, many ecologists have begun to investigate the *genetic structure* of populations to get a better understanding of the selective factors and the mechanisms regulating the distribution and gene exchange between populations. Genetic variability means that certain characteristics of the organism can be expressed to differing degrees. For example, if one gene appears in several alleles, it is referred to as *polymorphic*. If a locus appears as two alleles (a and b), a diploid individual can have any one of the three possible genotypes aa, ab, or bb. The relative occurrence of an allele in a population is called its *frequency*. For example, if allele a occurs in 10% of all of the individuals investigated, it would have a frequency of p = 0.1. Allele b must have a frequency of $q = 0.9$, for there are only two alleles, and $p + q$ must equal 1. The relative frequency of certain genotypes (*gene frequency*) depends on the frequency of the alleles. Under equilibrium conditions genotype frequencies follow the Hardy–Weinberg law; that is, the frequency of genotypes follow the equation:

$$1 = p^2 + 2pq + q^2$$

This holds only for ideal conditions, in which:

1. The population is large;
2. All individuals have equal opportunity to reproduce;
3. There is no mutation;
4. There is no natural selection; and
5. There is no immigration or emigration.

Note that there will be no evolution in a population in which all these conditions are found. Certainly natural conditions in freshwater systems rarely meet these conditions. If the conditions are not fulfilled, identification of the causes for the deviation from the Hardy–Weinberg equilibrium can suggest which selective forces are at work.

Most morphological characteristics of organisms are controlled by several genes. It is rarely possible to distinguish clearly between alleles, since there is also phenotypic variability. We can gain some insight in the genetic structure of populations by examining proteins, the primary product of genes. The study of allozymes is especially well developed. *Allozymes* are enzymes that have the same function, but differ somewhat in their structure. When placed on an appropriate carrier gel in an electrical field, they migrate with differing speed (*electrophoresis*). After a certain time the migration is stopped and the enzymes are made visible as bands with a dye. The specific type of enzyme one has isolated can be determined by the distance it has traveled (Fig. 5.9). Only if an enzyme is polymorphic in a population will there be different bands for the different individuals. It is possible for an enzyme to be polymorphic in one population and monomorphic in another population; that is, only one band will be found with electrophoresis. There are often two alleles of a polymorphic enzyme in one population, but rarely more. A diploid organism can have either only one of the two possible bands, meaning the organism is homozygous, or both bands, meaning it is heterozygous. If polymorphic enzymes are examined in a large number of individuals, one can estimate the frequency of both alleles. One can then determine whether the population is at or near the Hardy–Weinberg equilibrium. Recently, molecular methods have been applied

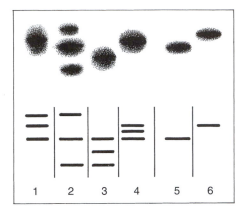

Figure 5.9 The allozyme pattern of glucose phosphate isomerase for six individual *Daphnia cucullata*. (*Above*): Bands, visible after electrophoresis on a starch gel. (*Below*): Diagrammatic representation of the individual bands.

to fish and small aquatic invertebrates to study genetic variation and relatedness of individuals by "genetic fingerprints." Small quantities of DNA (often mitochondrial DNA is used for this purpose) can be amplified by the polymerase chain reaction and the DNA can be analyzed by restriction fragment analysis or sequencing (Bermingham 1986, Taylor and Hebert 1993).

There have been many recent studies of aquatic organisms using electrophoretic techniques. The genetic structure of a population can indicate whether there is strong natural selection or whether different populations are exchanging genetic material. However, one point is especially important to note: Temporal and spatial differences in allele frequencies and the genotype composition indicate only that changes have taken place in the gene pool; they say nothing about the causes or types of selection factors. Allozymes are only *markers* for the genetic makeup of a population.

Let us examine once again the example of *Daphnia* under oxygen stress: The allozyme analysis tells us that there is a shift in the gene pool during the summer, but it cannot tell us why, since we have not investigated the gene that controls the synthesis of hemoglobin. The isozyme data allow us to identify specific genotypes that we can in turn examine physiologically. This would enable us to show that the adaptation to oxygen deficiency is not only a phenotypic reaction, for the best adapted clones have been selected for from the spectrum of genotypes that were present.

The prerequisites that we listed earlier for the Hardy–Weinberg equilibrium provide us with a series of questions that allozyme analysis can help address.

Chance plays an important role in small populations. Rock pools that fill with rainwater along the rocky ocean coast are a well-known example. These are very unstable habitats. The small pools can dry out and or they may become saline from the ocean spray. As rainwater fills them they gradually freshen again. These changes represent catastrophes that can cause populations to go extinct. Then repopulation becomes a dominant force. Chance determines which genotypes arrive first. Due to this *founder effect* there are great differences in the genetic structure of neighboring pools.

The harpacticoid copepod *Tigriopus californicus* lives in rock pools on the California coast. Individual populations often die out during catastrophes, since this species does not produce a resting stage. The rock pools are often found in small groups that are separated from one another by short distances of beach. Allozyme analysis indicates that these groups differ genetically, but there is little difference in genetic structure within a group of pools. Shortly after a catastrophe founder effects lead to major genetic changes in the pools. However, this is quickly countered by new genotypes that come from the adjacent rock pools. Not all pools are hit by a single catastrophic event, since the pools differ in their height above the sea level (Burton and Swisher 1985). There is much gene flow between pools within a group, but very little between groups of pools. It is therefore reasonable to consider the animals within a group of pools as a population, whereas the animals in the next group could be considered as a different population.

Founder effects can be overlaid by *selective factors*. The rock pools are also a good example of this effect. Rock pools have variable salt content. Pools that are near the high water level are saltier than pools that are farther inland that receive some salt from spray carried by the wind. The pools have a clear gradient of salt content, moving from the water line toward the interior. The genetic structure of a population of the cladoceran *Daphnia pulex* and *Daphnia magna* has been investigated in detail in the coastal region of Alaska (Weider and Hebert 1987). There was a clear effect of the salt content. The ponds nearer the coast had a higher frequency of salt-tolerant genotypes. Accidental colonization is also the important first step in these salty ponds following a catastrophe. Those genotypes that enter the pool but are not salt tolerant will not survive. This continues until by accident a salt-tolerant genotype arrives. The result is that there is greater genetic similarity of pools along the coast line than between these pools and the ones farther inland. There is a gradient of genotypes, and it appears that the salt content is the selective force responsible for this.

Populations in neighboring large lakes are usually more similar genetically than populations from small ponds. In large lakes selection is more important than the founder effect or chance genetic drift. Nevertheless, one can compare the genetic structure of populations of lakes by using allele frequencies as a statistical measure of genetic distance. Figure 5.10 illustrates an example with *Daphnia galeata* in lakes in Holstein, Germany. Lakes that are connected by streams are genetically more similar than those that are isolated. The fourth lake of the chain of lakes (Großer Plöner See) differs considerably. The allele F^+ (one that moves rapidly in the electrophoresis), which is rare in the other lakes, is frequent in the Großer Plöner See. Apparently, the genetic exchange between populations through the stream is not enough to outweigh the selective effects.

Even within a lake there can be genetic differences if there is limited water circulation throughout. For example, Jacobs (1990) was able to demonstrate that populations of *Daphnia cucullata* in the small Klostersee at Seeon, Bavaria (Germany), were genetically different in two basins connected by a zone of shallow water.

There can also be strong anthropogenic selective forces in aquatic systems. Aquatic organisms can develop resistance to harmful substances, much as terrestrial organisms. Maltby (1991) reported on an English stream that contained the aquatic pillbug *Asellus aquaticus* and that received mine drainage through a canal. The population of *Asellus* living above the mine drainage was sensitive to the waste water, but could produce

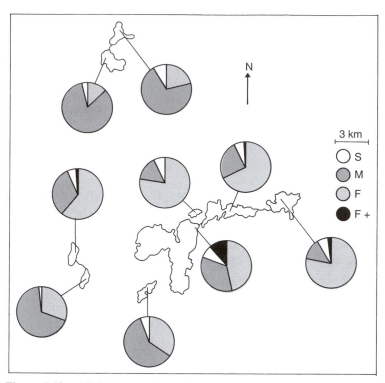

Figure 5.10 Allele frequencies of phosphoglucose mutase of *Daphnia galeata* in neighboring lakes in the "Holstein Switzerland." The segments of each pie represent the relative proportion of the four possible alleles. Lakes in the central group are connected by the small Schwentine River, which flows from east to west (from Mort and Wolf 1986).

a large number of offspring in clean water conditions. Another population of *Asellus* living only a few meters downstream and below the discharge could produce few young in clean water, but was resistant to the pollution. Surprisingly, the two populations were genetically distinct, despite the short distance separating them and the continuous drifting of *Asellus* downstream. It appears that selection by acidity and heavy metals was strong enough to maintain a stable genetic separation.

The use of pesticides to control aquatic larvae has produced similar results. For example, in Clear Lake, California, the concentration of methylparathion had to be increased after 13 years of applications to control the phantom midge larvae, since the larvae had developed a resistance to the insecticide. Between 1962 and 1975 the midge larvae became 10 times less sensitive to the poison (Apperson et al. 1978).

The first step to evolutionary change of a population is a change in the composition of the gene pool. Selection from the existing genetic variation is called *microevolution*. One might assume that under constant selective pressure there should be a reduction in the variability of characteristics that are not selection neutral. This raises the question involving both ecology and population genetics: Why is there so much genetic variability? Possibly the most important reason is that selective pressures are not constant; environmental conditions are constantly changing, giving different genotypes advantages at different times and in different locations. Also, there are newer models in population genetics, such as *frequency-dependent selection*, that explain how rare genotypes can gain an advantage, by not being selected by a predator that specializes in feeding on the most frequent genotypes.

This is a turbulent time in the development of ecological genetics. It is of great importance that ecologists bridge the gap between descriptive population genetics, which describes the temporal and spatial changes in genetic structure of populations, and ecophysiology, which attempts to quantify fitness.

5.4 DEMOGRAPHY

In considering population dynamics (Section 5.2.6) we considered all members of the population to be equal. We were concerned with rates of reproduction and loss from the entire population, looking only at the numerical response. This suffices for singlecelled organisms that reproduce by dividing and that are potentially immortal. For populations of multicellular organisms, however, individuals are not all equal. Reproduction always produces individuals of the youngest age class and different age classes have different probabilities of dying.

A useful feature of a population is therefore its age structure. Of course, it is not possible to predict exactly how long any individual will live. By looking at the entire population, we can assign to each individual a probability of its living after a given time. By examining Fig. 5.6, it can be clearly seen that one could estimate the probability that a copepod in larval stage (x) will reach stage $(x + 1)$ from the decreasing population size of each of the larval stages. If we refer to the *survival probability* from nauplius 1 to nauplius 2 as p_1, and the subsequent probabilities as p_2, etc., the probability (L_x) that the copepod will reach stage x is:

$$L_x = p_1 p_2 \cdots p_{x-1} p_x$$

The survival probability must get smaller and smaller with age, since the p of each older stage is smaller than 1.

The changes in the effect of age on survival probability differs for different species. These can be categorized into three general *survival curves*. In the first type, there is a brief phase of high mortality for juveniles, followed by a long period of relatively constant high survivorship; near the end, all remaining individuals die within a relatively short period. This type of curve applies to practically all mammals that care for their young, and when they pass the juvenile phase, have few predators. There are few aquatic organisms that fit this pattern.

In the second curve, the survival probability is the same rate for each age class, and the curve therefore declines as a negative exponential. This means that the losses occur randomly and affect all age classes equally. It is likewise difficult to find examples of freshwater metazoans that exhibit this type of survivorship for their entire life cycle, except, perhaps for some rotifers. Parts of the *life cycle*, however, may have this type of curve, as, for example, fish after the second year, as shown in Fig. 5.7.

The third model of survivorship is most commonly found in freshwater, where first the juveniles have high mortality rates, but after they reach a critical age, the probability of surviving becomes high and mortality occurs randomly amongst the remaining age groups. Copepods (Fig. 5.6) fit this type, for the nauplii have a higher mortality than either the copepodids or adults. Animals that do not care for their young, but produce large numbers of offspring, follow roughly this survivorship curve, such as mussels, insects, and even fish.

Rarely in nature is any one of these patterns strictly followed, and usually organisms have some combination of all three curves. The fish population in Fig. 5.7 is an example of this, for after the third year there is an exponential decrease (type 2), but in the first 2 years that are not quantitatively represented in this study, the mortality is much higher, typical of type 3.

Not all individuals contribute equally to the maintenance of the population. Juveniles do not contribute any offspring. Later, the number of young produced depends on the age of the individual. Organisms that reproduce only once and then die, such as many insects and Pacific salmon (*Oncorhynchus*), are exceptions, since their reproduction occurs in a single pulse. Age-specific fecundity curves can be drawn for organisms that reproduce frequently (Fig. 5.11). These curves show the average number of female offspring produced by an individual of a specific age class. Until sexual maturity, the age specific fecundity (designated as m_x) is zero. After that it often in-

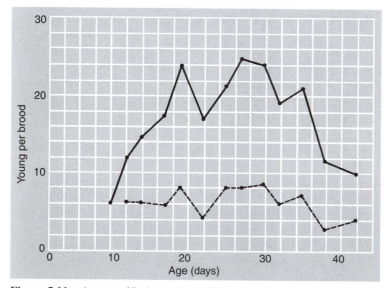

Figure 5.11 Age-specific fecundity of *Daphnia pulex* at high (solid line) and low (dashed line) food concentrations (from Richman 1958).

creases, and as the females grow larger they are able to produce more and more young. Finally, as the females become old, m_x decreases. The sum of the female offspring of an individual is simply the sum of all m_x values.

As a population colonizes a new habitat, its age structure depends on random factors. As the first young are born, the age distribution shifts toward the young animals. After sexual maturity, there follows a new pulse of juveniles. The age distribution and mean age of the population is therefore constantly changing. Since there are overlapping generations, the population can approach an equilibrium condition after several generations, as long as the environmental conditions do not change. The population approximates a "stable" age distribution that corresponds to the survivorship curve (cf. Fig. 5.7). Box 5.2 illustrates this stabilization process with a graphical model.

A high age class does not necessarily make a large contribution to the population, even if its age specific fecundity is high. Depending on the specific type of survivorship curve (Fig 5.6), only a small fraction of the population will reach the older age classes. Even though each individual of the older class may produce more offspring, the total number of offspring produced by the age class may be small. The total number of female offspring produced by an individual in the population (R_o) depends, therefore, on the age- specific fecundity (m_x) as well as on the probability of surviving to that particular age class (L_x):

$$R_o = \sum_{x=0}^{n} L_x m_x$$

In field populations this value must be smaller than an individual could attain if it were in a situation where it would die a natural death (e.g., laboratory culture).

R_o is the average number of female offspring each individual in a population produces, but we cannot use this directly to estimate the growth rate of the population (r), which depends also on the temporal distribution of m_x. When young animals reach a high m_x value, the population growth rate is larger than if most of the reproduction were by the older age classes, but with the same R_o. Offspring that were born earlier can contribute to the population while their siblings born later are still not reproductive. Rapidly growing populations therefore have a relatively young average age.

This fact is incorporated in the Euler formula, which is used to estimate population growth rate from demographic parameters:

$$1 = \sum_{x=0}^{n} L_x m_x e^{-rx}$$

This equation cannot be solved analytically, but an iterative solution is easy using modern computers.

Demographic parameters are very important in the population ecology of animals in freshwater. Parameters such as the time of the first reproduction and species-specific fecundity curves greatly affect the growth rate of the population and are therefore under strong selection pressure. To understand the evolution of the life histories of organisms, one must also understand the effects of changes in the demographic parameters (Stearns 1992). For example, one can estimate stepwise the contribution of each of the broods of *Daphnia* to the total population growth rate (r) using the fecundity

BOX 5.2	DEVELOPMENT OF A STABLE AGE DISTRIBUTION

Day		Juvenile	Adults	Total	% Juvenile
1	* 1	1	0	1	100
2	*	1	0	1	100
3	*	1	0	1	100
4	○ 1	0	1	1	0.0
5	● *	1	1	2	50.0
6	○ * 1	1	1	2	50.0
7	● * *	2	1	3	66.7
8	○ ○ * 2	1	2	3	33.3
9	● ● * *	3	2	5	60.0
10	○ ○ * 2	2	2	4	50.0
11	● ● * *	4	2	6	66.7
12	○ ○ ○ * 4	2	4	6	33.3
13	● ● ● * *	6	4	10	60.0
14	○ ○ ○ * 5	4	5	9	44.4
15	● ● ● * *	9	5	14	64.3
16	○ ○ ○ * 8	5	8	13	38.5
17	● ● ● * *	13	8	21	61.9
18	○ ○ ○ * 11	8	11	19	42.1
19	● ● ● * *	19	11	30	63.3
20	○ ○ ○ * 17	11	17	28	39.3
21	● ● ● * *	28	17	45	62.2
22	○ ○ ○ ● 24	17	24	41	41.5
23	● ● ● * *	41	24	65	63.0
24	○ ○ ○ * 36	24	36	60	40.0
25	● ● ● * *	60	36	96	62.5
26	○ ○ ○ * 52	36	52	88	40.9
27	● ● ● * *	88	52	140	62.9

BOX 5.2	(*Continued*)

The graphical model (Edmondson 1968) describes the fate of any individual in a population. The vertically connected symbols represent overlapping cohorts of hypothetical animals. A juvenile (asterisk) matures on day 4 (open circle) and gives birth (dark circle) to one offspring on each of the days 5, 7, and 9. It then dies and disappears from the population. All the offspring added to the population are indicated by the numbers placed above each cohort. The number of juvenile and adults in the population is the sum of all individuals presented in a horizontal row. For example, on day 11, two adults give birth to two juveniles, which are added to two 3-day-old juveniles already in the population. At that point in time the population consists of two adults and four juveniles. Sums of juveniles, adults, and the percentage of juveniles in the population for each day are presented in the right-hand columns. When the proportions are plotted against time (Fig. 5.2.2), it becomes evident that the fluctuations quickly become regular and slowly approach an equilibrium (stable age distribution). The time required to reach the stable age distribution depends on the life history of the individuals.

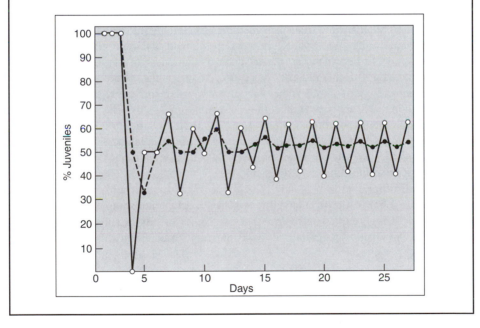

curve illustrated in Fig. 5.11. By the third brood, *Daphnia* already had reached over 90% of the maximum *r*. The fourth and all subsequent broods contribute very little to (*r*). Actually, few daphniids ever experience more than three broods in nature, since older (and therefore larger) animals are under heavy predation pressure. Thus it is adaptive that the first broods are the most important for population growth.

[149]

5.5 DISTRIBUTION

If one were to toss a square frame into a stream and count all the organisms of one species within the frame, a different number of organisms would be counted with each toss. One would find the same number only if the organisms were sufficiently small and evenly distributed.

There are three general patterns to describe the spatial distribution of organisms (Fig. 5.12):

1. The distribution may be *random*. Individuals have different distances from one another. If one measured the distance of adjacent organisms, rarely would two organisms be very close to one another or very far apart. The frequency of the distances follows a random distribution (Poisson distribution).

2. Organisms may be *evenly distributed*. In the most extreme case, all individuals would be exactly the same distance apart.

3. The distribution may be *clumped*. Such organisms tend to live in aggregations.

Random distributions can be expected for species that do not control their movements, such as small algae and free-suspended bacteria in the epilimnion of a lake. This also can be true for very small spatial scales of up to about one meter. For distances of even a few meters this becomes less likely. Water movements can cause even passively transported plankton to form "clouds." For example, water masses moving in opposite directions and passing one another, they can create vortices that aggregate plankton.

A distribution that is more even than random usually means that the organisms have some negative influence on one another, as, for example, competition for space. This distribution can be seen in streams, but it is not so common in lakes as in oceans. Net-spinning caddisflies are an example (cf. Section 6.2.2) or chironomid midge larvae (Chironomidae) that build tubes in the sand.

Clumped distributions occur where the habitat of the organisms is very heterogeneous, providing many microhabitats. Streams are an excellent example of this type of habitat. Zones of high and low stream velocity create a variety of substrate types (cf. Section 3.4). Organisms accumulate at those sites that offer the best habitats. Aggre-

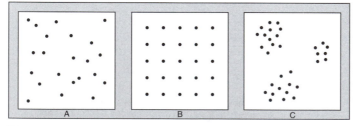

Figure 5.12   General types of distributions of organisms in a habitat. (A) Random; (B) even (infra- or underdispersed); (C) clumped (super- or overdispersed).

gations can also be formed purely by physical forces. In strong currents animals that are washed away may accumulate in protected areas. Such patches are very dynamic. While currents continuously accumulate new organisms, interference by competing organisms and the presence of predators may force some of the animals to leave the patch (Lancaster 1990). The distribution of food can strongly affect the distribution of stream invertebrates (Hildrew and Townsend 1982). For example, Shannon et al. (1994) studied the distribution of *Gammarus lacustris* in a dammed portion of the Colorado River. These amphipods showed a strong preference for patches of the filamentous alga *Cladophora*, but only if the filaments were colonized by epiphytic diatoms that they actually fed on. Clumped distributions are also common in the littoral zone of lakes, where biotic (e.g., macrophytes) as well as abiotic factors (e.g., wave action) create heterogeneous habitats.

Clumped distributions also occur in the relatively homogeneous pelagic region of lakes. Here plankton are seldom randomly distributed vertically, and often even the small organisms form pronounced layers (cf. Fig. 3.8). Such layers can be formed by active swimming or by vertical differences in growth rates or loss rates. Zooplankton can swim well enough to select their depth, even in a well-mixed epilimnion (cf. Section 6.8.4). Algae that move with flagella (flagellates) or gas vacuoles (certain cyanobacteria) may also stratify sharply at certain depths. Often organisms accumulate at the thermocline, where their sinking rate is slowed as they encounter rapid changes in water density.

Zooplankton often occur in clouds or swarms (*patchiness*). They can actively form swarms in addition to being passively aggregated by water movement. Cladocerans sometimes form dense collections in quiet water, where they orient to submerged structures, such as light patches. The significance of these swarms is not clear. It is likely that, just as for birds or fish, swarms offer planktonic organisms protection from predators. Predators tend to attack individuals that are outside a swarm, for they cannot concentrate on a single organism within the mass. Inexperienced sticklebacks have difficulty catching daphniids when they are together in a swarm (Milinski 1977). The zooplankton within the swarm are safer, since the fish always feed on the outside of the swarm. The swarm is maintained by the constant attempts of animals to move to its center.

Jacobsen and Johnsen (1987) suggested that dense swarms (over 9000 individuals per liter) of the cladoceran *Bosmina longispina* in the littoral of a Norwegian lake served as protection against stickleback fish that were abundant. There must be a considerable protective effect of the swarm, for there are also disadvantages in being inside the swarm. *Bosmina* is a filter feeder that depends on the concentration of particles for food. The concentration of food particles is significantly lower in the center of a swarm of *Bosmina* as a result of the collective filtration by so many animals. This is another example of the conflict between maximization of reproduction (requiring much food) and the reduction of mortality (protection). It is likely that the swarm breaks up in the dark, when fish can no longer orient optically. Swarms of fish that serve as protection against predatory fish and birds disperse at night so that the individual fish can search for food.

Fish swarms are also a good example of how scale can influence the observed distribution. Within a swarm, one would measure a remarkably even distribution with a constant distance between individuals. From the perspective of the entire lake, however, the fish would be extremely clumped. The choice of which scale should be used in a study depends on the question being posed.

These examples of uneven distributions point out the need to discriminate between the proximate causes (how do the animals orient to produce a swarm?) and ultimate causes (what is the fitness gained by having a clumped distribution?). "Shoreline avoidance" is a well-studied phenomenon of the pelagic zooplankton. Pelagic crustaceans such as *Daphnia* and *Diaptomus* are not found in the littoral during the day, although they are there at night. One hypothesis proposes that zooplankton actively avoid the littoral, but can do so only when it is light. Siebeck (1980) designed elegant physiological experiments that demonstrated that zooplankton can actively swim away from the shore in the early morning. The optical cue for this behavior is the light distribution field seen above by the zooplankton. Near the shore high objects such as trees and hills produce dark regions in the light field. By orienting toward the brighter sections of the light field the animals automatically direct their swimming path to the open water. At some distance from the shore even high objects no longer darken the light field; so that the animals can no longer orient to the light field. Some littoral zooplankton, such as the predatory cladoceran *Polyphemus* orient toward the dark part of the light field instead, causing them to form narrow bands along the shore. The Siebeck hypothesis states that the uneven distribution is the result of continuing selection against individuals in an unfavorable habitat. The "ultimate" factor causing this behavior might be predation by small fish that often concentrate near the shore during the day (cf. Section 6.5.3). If this is correct, then the shoreline avoidance behavior is the mechanism (proximate, light orientation) that the animals have evolved to avoid this predictable danger. Another possible explanation for the changing abundance in the shore areas of the lake has been proposed by Gliwicz and Rykowska (1992). They suggest that the zooplankton are simply decimated by young fish in the littoral during the day and replaced from the pelagic zone at night. An important implication of this hypothesis is that the zooplankton have not yet evolved a protective response to this predation pressure by fish along the shore.

5.6 *r* AND *K* STRATEGIES

Very different selection factors operate in a sparsely settled habitat that is well below its carrying capacity and in a habitat that is densely populated and near its capacity. The winner of a contest of populations in an undercolonized habitat will be the species that has the highest maximum net population growth (r_{max}) and a superior ability to disperse. On the other hand, in a habitat that is already fully colonized (near K), efficient use of resources, ability to compete, and avoidance of mortality are important.

MacArthur and Wilson (1967) viewed these two extremes as fundamental types of adaptive strategies and named them r and K strategies, based on the corresponding parameters of the logistic growth curve (Section 5.2.4). To avoid the teleological idea that the strategy is a planned adaptation, the concept should be thought of as a metaphor that refers to the package of features and abilities of a species that have been selected

[152]

for a certain set of conditions. It would be more correct to refer to *r*- and *K*-selected characteristics, combinations of features or organisms.

Species cannot be simultaneously adapted to the selective conditions of both sparse and dense colonization ("*r* and *K* selection"), for each unit of energy and material can be invested in only one place at a time. Energy invested in the production of offspring cannot be also invested in structures that protect against predation. This *problem of allocation* is the main reason that there is no "super species" that is superior over all other species and under all conditions.

Originally, there was an attempt to categorize all species and even major groups as either *r* or *K* strategists. Grasses and insects were given as examples of *r* strategists and trees and mammals as typical *K* strategists. It is more common today to consider a continuum of types. Generally, species and genotypes are no longer characterized as either *r* or *K* selected and are more likely to be described as "more or less" compared to one another, especially within a group of functionally similar organisms, such as phytoplankton, zooplankton, and fish.

The fundamental feature of *r*-selected organisms is a high maximum net growth rate (r_{max}). This may be achieved by high specific metabolic rates, short generation times, and a large allocation of energy to the production of offspring. For protists there is no separation between germ and somatic growth. For these unicellular organisms a high allocation in reproduction is also a high allocation to productive biomass components (e.g., chlorophyll in phytoplankton). The costs of a strategy aimed at maximizing reproduction is that the organism can invest little in reserves, structures, enzymes, and behaviors suited to obtaining scarce resources and in defensive structures. The disadvantages of this strategy appear when the organism must survive periods of starvation, compete for scarce resources, or resist predation pressure. Most *r* strategists are small, since small organisms are most capable of attaining a high specific rate of metabolism.

K-selected organisms must avoid a negative net growth rate, at least over long periods, even if their habitat is filled to capacity. These species can do this by minimizing mortality. Defensive structures are helpful if the mortality is due to predators; often an increase in body size is sufficient to protect against predation. Both adaptations are "paid for" by a reduction in the specific metabolic rate and reproductive rate. Storage of energy reserves protect against hunger-related mortality. Even such *reserve storage* has its costs, since the stored materials are metabolically inactive and must be removed from immediate investment in reproduction. Another method of avoiding a negative net growth rate is to gather and utilize resources in short supply more efficiently. Microorganisms achieve this with efficient uptake systems and mobility to localize at specific depths when the resources are distributed heterogeneously (e.g., vertical gradients). Predators invest in speed and a larger action radius and the ability to produce as much of their own biomass as possible for each unit of consumed resource (*high growth efficiency*).

The problem of allocation not only excludes the possibility of simultaneous adaptation to *r* and *K* selection; it may also create conflicts between different adaptations to *K* selection. An example of this was mentioned in Section 4.3.7: Diatom species that have a low half-saturation constant (k_s) for phosphorus have a high k_s for silica and vice versa. Apparently, there is a problem of allocation between the uptake systems for both limiting nutrients. Both large size and high nutrient uptake efficiencies (low K_s) are typical adaptations of K-selected organisms, but it may be difficult to have both adaptations. The uptake of nutrients depends not only on having the necessary enzymes, but also on the surface-area-to-volume ratio. Cells must be small to have a large surface-area-to-volume ratio, but such cells have the disadvantage of being more vulnerable to grazing. The largest phytoplankton (dinoflagellates, colony-forming green algae, and cyanobacteria) are thus inefficient at utilizing nutrients that are scarce, but they are protected against being eaten.

It is easy to imagine the ideal r strategist within a functional group of organisms, but there is no one ideal K strategist. It is impossible to maximize all components of the K strategy at the same time. The result is that within each group of organisms there is a continuum from low to high r_{max}. The species with a low r_{max} must have compensatory mechanisms that provide an advantage only when population densities are high. Species with maximum growth rates dominate during the early phase of the population growth. They return each time there is a disturbance that causes a decrease in population density. These are called *opportunistic species*, since they are the first to use unoccupied habitats. As the population density increases, they become suppressed by species with lower maximum growth rates.

Figure 5.13 shows phytoplankton that are examples of extreme r and K strategists. *Rhodomonas minuta* is an r strategist with a maximal growth rate of 0.95 d^{-1} at 20°C and a cell volume of approx. 80 μm^{-3}. This flagellate lacks a cellulose wall and is defenseless against grazing by zooplankton. In Lake Constance, which has considerable wind exposure that allows storms partially to destroy thermal stratification, this alga has several pulses of high densities following the start of the vegetative period ("spring bloom," April/May). At the other extreme in Plußsee, a sheltered lake with little disturbance of stratification during the summer, *Rhodomonas* has its high densities in the early spring. The dinoflagellate *Ceratium hirundinella* represents the K strategist. It has a maximum growth rate of only 0.26 d^{-1} at 20°C, but a cell volume of approx. 50,000 μm^3 and a thick cellulose wall that protects it from being eaten by most zooplankton. In both lakes *Ceratium* has its seasonal maximum at the time of the maximum total phytoplankton biomass, but it becomes a dominant species (>90% of algal biomass) only in the stable Plußsee and not in the frequently disturbed Lake Constance.

5.7 DISTRIBUTION AND COLONIZATION

The ability of a population to persist over a long term depends on the balance of reproduction and mortality. The first representatives of a species first of all must be able to populate a particular habitat. This colonization can result from active movement or from passive transport. Many organisms have special dispersal stages that are especially well suited for being transported.

One can view *lakes as islands* that are separated from one another by a completely different habitat. Even when lakes are connected by streams, the flowing water creates an effective boundary. Many lake organisms are not able to survive the turbulent wa-

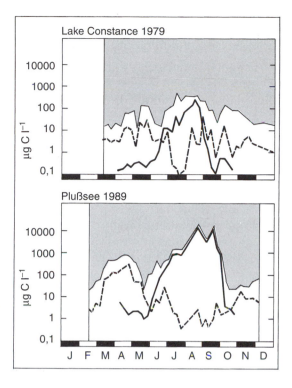

Figure 5.13 *r* and *K* strategies of phytoplankton: Seasonal changes in the biomass of the *r* strategist *Rhodomonas minuta* (- - -) and the *K* strategist *Ceratium hirundinella* (——), as well as the total biomass of the phytoplankton (clear area) in the physically stabile Plußsee and in the less stable Lake Constance.

ter or move upstream against the water current, and waterfalls may even interfere with the migrations of fish. For this reason, originally most of the high mountain lakes in the European Alps lacked fish populations before stocking was begun (mostly already in the Middle Ages).

Streams are *continuous systems*, in contrast to lakes. Physical transport is obviously unidirectional, resulting in a dominance of colonization from upstream. Thus the *drift* (cf. Section 4.2.5) of organisms in streams is an effective means of dispersal that is not simply an accident. Open areas of streams are colonized very rapidly. For example, a new stream bed was dug during the relocation of a stream in Canada. Only 1 day following the flooding of the new section of the stream 22 taxa of benthic animals were already present. Of these, 16 taxa had drifted in from upstream. Further colonization progressed more slowly; after 7 days there were 26 species, and after one year 41, 18 of which did not come from the drift (Williams and Hynes 1977).

Fortunately, such rapid recolonization makes streams considerably more resilient to environmental catastrophes than lakes. One of the most spectacular events of this sort was caused by a fire at the Sandoz Corporation in Basel, Switzerland, in November, 1986. This catastrophe resulted in the release of 10–30 tons of toxic chemicals, including insecticides, fungicides, and herbicides into the Rhine River. As a consequence, thousands of fish, especially eels, were killed as far as 400 km downstream from Basel, and the communities of macrozoobenthos were also adversely affected. Regular sampling demonstrated that the bottom-dwelling animals returned rapidly. Almost certainly, they migrated in from the Upper Rhine and tributaries. Another dramatic case that was less well publicized was the accidental poisoning of the Breitenbach stream in the town of Schlitz (Germany), where a few grams of cypermethrin, a pesticide used against bark beetles, was introduced into the brook by foresters. A single exposure to the poison killed all the insect larvae and amphipods downstream for a distance of 2 km. The effect on the stream was striking. Green filamentous algae soon covered most of the rocks in the stream, due to the absence of grazing larvae. The stream quickly recolonized, however, and within a year it once again had its "natural" communities.

Fish that migrate between the ocean and streams (eel, salmon) can act as connectors between different stream systems. However, even areas of the oceans may be impassable. Eels, for example, cannot migrate through the meromictic Black Sea, because of its anoxic deep water. Without artificial stocking there would be no eels in the Danube River system.

Opportunistic species rapidly colonize newly created water bodies. This can be easily observed in garden ponds. Flying insects, such as midges, water bugs, and beetles, are often the first colonizers. Other species may be dispersed from pond to pond by passive transport. Microscopic organisms (algae, protozoans) can disperse in aerosols. Many species have resistant resting stages that can be transported directly by the wind, attached to the feathers of waterfowl, or even in the digestive tract of birds.

Pelagic rotifers and cladocerans are very easily dispersed. Following many generations of parthenogenic reproduction, they can produce a bisexual generation that results in resting eggs. Cladocerans often encase these eggs in a thickened brood pouch called the ephippium, which is expelled as the animal molts (Fig. 5.14). Ephippia are extremely resistant to unfavorable conditions. They can survive drying out, freezing, and lying for several years in the anaerobic sediments. Because the surface of the ephippium is hydrophobic, they often float in the water surface and attach to the feathers of waterfowl, allowing them to be transported. Females hatch from the resting eggs under favorable conditions and begin a new population through parthenogenesis.

Humans, as they become more mobile, are also becoming dispersal mechanisms for aquatic organisms. Planktonic organisms are very often introduced during the stocking of fish. A spectacular case occurred during the 1980s, when the predatory clado-

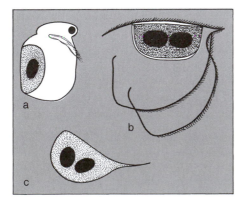

Figure 5.14 Ephippia, the dispersal stage of cladocera. (a) An ephippial female *Ceriodaphnia* sp.; (b) an ephippium of *Daphnia magna* that has retained part of the carapace; (c) an ephippium of *Daphnia longispina*.

ceran *Bythotrephes cederstroemi* was accidentally introduced into the North American Great Lakes (Sprules et al. 1990). Until then, *Bythotrephes* was the only European species of plankton, for which there was no counterpart in North America (Lehman 1987). *Bythotrephes* was quickly distributed throughout all these large lakes and, because its preferred prey was *Daphnia*, caused a major change in the plankton community. It appears that *Bythotrephes* crossed the Atlantic Ocean in the ballast water of ships from Eastern Europe.

The Zebra mussel (*Dreissena polymorpha*) is a case where rapid dispersal is made possible by its free-swimming veliger larvae, albeit only downstream. Although this species had been long established in the lower Rhine River and its tributaries, its sudden appearance in the 1960s in Lake Constance was a surprise, since this was upstream from a waterfall on the Rhine at Schaffhausen that should have prevented the larvae from passing. The mussels first appeared in a harbor in Lake Constance, suggesting that they had attached themselves with their byssal threads to the underside of boats and had been transported over land to the lake. Zebra mussels also appeared in the Great Lakes of North America in the 1980s where their populations exploded because of the veliger larvae. It is very likely that here also ships were the means of transport across the ocean (Hebert et al. 1989). They are now widespread all over the Great Lakes and its tributaries. Just as with many other species that invade a new territory, the zebra mussel developed enormously high population densities, probably well above their carrying capacity. Their impact on lake ecosystems is presently a matter of great concern (Leach 1992).

Passive transport can be especially effective for small organisms. As a consequence, *geographical distribution boundaries* are much less important for phytoplankton than for many animals. There is little difference in the phytoplankton in comparable lakes of the northern and southern temperate zones (e.g., southern South America; Thomasson 1963). There are only a few species (and not a single genus) that are found exclusively in the southern or northern hemisphere. Zooplankton are more distinct biogeographically than phytoplankton. There are generally fewer species of zooplankton in the southern temperate zones than in the north temperate lakes. One copepod genus (*Boeckella*) is exclusively found in the southern hemisphere, except for one species in the region of Mongolia and East Siberia.

Endemism is the extreme opposite of the cosmopolitan distribution seen for the phytoplankton. This concept is used to describe a taxon that is restricted to a particular location, even though habitats with similar conditions are found elsewhere. Endemic species are frequently found in the deep graben (fault block depression) lakes of the Tertiary Period (Lake Baikal, Lake Tanganyika, Lake Malawi, Lake Ohrid) that are much older (about 5 to 20 million years) than most other lakes (10,000 to 20,000 years) (Martens et al. 1994). It is easy to imagine that some species that used to be widely distributed now are found in a few lakes, since only a few lakes persisted through the ice age.

At the same time, the continuous existence of these lakes over long periods allowed for much speciation within each lake. Cichlid fishes are a particularly good example in the large East African lakes. Although it is less than one million years old, Lake Victoria alone has about 200 species of these "haplochromine" fish. There has been a

[157]

debate over whether these fishes all evolved in this one lake or in separate localities. Recent molecular studies using mitochondrial DNA have shown that this "species flock" is probably monophyletic. Some Lake Victoria species are morphologically very similar to species in Lake Tanganyika and Lake Malawi, but genetically they are more closely related to other species in Lake Victoria that look very different than to the morphologically similar "foreign" species (Meyer et al. 1990).

Lake Baikal in southern Siberia is the oldest lake (probably 20 million years old) with the highest degree of endemism. According to Kozhov (1963), 708 of 1219 species (58%) of the animals found within the lake are endemic, compared with 12 of the 150 phytoplankton species (8%). Varying degrees of endemism can be demonstrated within the animals. Rotifers (5 of 48 species; 10.4%), cladocerans (none of the 10 species; 0%), and calanoid copepods (1 of 5 species; 20%) have a noticeably low degree of endemism. Animals that can move actively over large distances have an intermediate degree of endemism: flying insects make up 24 of the 98 species (24%), and fish make up 23 of the 50 species (46%). Organisms with a strong association with the bottom sediments have a stronger tendency toward endemism than free-swimming species. For example, all endemic fish found in Lake Baikal belong to the benthic dwelling suborder of the Cottoidei, and none of the coregonids, salmonids, cyprinids, accipenserids, and other families has endemics. Extreme endemism is found in the turbellarians (all 90 species), harpacticoid copepods (38 of 43 species; 88%), ostracods (31 of 33 species; 94%), gammarid amphipods (239 of 240 species; 99.6%), and mollusks (56 of 84 species; 67%), all of which are animals associated with bottom sediments.

REVIEW QUESTIONS FOR CHAPTER 5

1. Two phytoplankton species are growing exponentially at net growth rates of 0.7 and 0.8 d^{-1}, respectively. The starting abundance of both species is 1 ind ml^{-1}. Calculate the abundance of both species after (a) 1; (b) 2; (c) 5; and (d) 10 days.

2. A phytoplankton species grows for 1 week at the exponential rate of 1.0 d^{-1}. Because of counting and sampling errors, there are 95% confidence intervals of +20% around both abundance estimates. Calculate the net growth rate for each of the following conditions: (a) the starting abundance (N_o) is 100 ind ml^{-1} and the final abundance (N_7) is 1000 ind ml^{-1}; (b) use the lower margin of the confidence interval for N_o and the lower margin for N_7; (c) use the upper margin of the confidence interval for N_o and the lower margin for N_7.

3. A population is growing according to the logistic growth equation with a carrying capacity of 1000 ind m^{-2}. The relative growth rate, $dN/(dtN)$, is the highest when the abundance is as close as possible to zero. At which abundance is the absolute growth rate (dN/dt) highest?

4. Two phytoplankton species have the same reproductive rate ($\mu = 0.9$ d^{-1}). One species is a diatom that sinks quickly ($v_s = 2$ m d^{-1}), but resists grazing ($w_i = 0.1$). The other species is a green alga that sinks slowly ($v_s = 0.1$ m d^{-1}) and is sensi-

tive to grazing ($w_i = 0.8$). Calculate for each phytoplankton species (a) the grazing rates for optimal food algae (γ_{opt}) at which the population of that species would stop growing ($r = 0$) at a mixing depths of 0.5, 1, 2, and 10 m; (b) the mixing depth at which the population of that species would stop growing at γ_{opt} of 0.5, 1.0, and 2.0 d^{-1}.

5. In a *Daphnia* population consisting of only females, the average daphniid carries 6 eggs. The development time of the eggs is 4 days. Calculate the birth rate according to Paloheimo's equation. Calculate how the birth rate would change (a) if the egg number were doubled; (b) if the development time of the eggs were reduced by one-half.

6. You are investigating the population dynamics of the rotifer *Keratella cochlearis* in a lake. A net haul yields 375 females with one egg each, 180 females without eggs, and 25 *Keratella* eggs unattached. Three days later, using the same sampling system, you catch 880 females and 790 eggs. At the lake temperature of 20°C at the time of sampling, the time required for the development of a *Keratella* egg is 17 hours. Calculate the population growth rate (r), the birth rate (b) and the mortality rate (d). How many *Keratella* rotifers would you expect on the second sampling date if there had been no mortality?

7. The seasonal abundance of *Daphnia* in many lakes typically shows a bimodal pattern, as depicted in Fig. 5.8. How would you infer from field sampling which of the following factors caused this pattern: physical factors (e.g., high temperature), interference by inedible algae, fish predation, invertebrate predation? How would you design an experiment to test your hypothesis?

8. Fish are able to forage visually at very low light intensities, even at night. Why does it nevertheless pay the zooplankton to have diel vertical migration?

CHAPTER **6**

Interactions

It is well known that most species have a more limited distribution than would be predicted by their physiological tolerances. In many cases one can exclude historical events as the cause of an organism's distribution. Explanations then often involve interactions between populations. By consuming a resource, an organism may reduce the availability of the resource or it may itself become food for other organisms. The interaction between organisms that use a common resource is called *competition*. Such interactions may be indirect, involving the exploitation of common resources, or direct, where the competitors directly harm or disable one another. Interactions where one population serves as food for the another population are either *predator–prey* or *parasite–host* relationships. Competition and predator–prey relationships result in a negative impact on at least one of the partners. There are also *symbiotic* relationships, where both partners benefit. For example, nitrifying bacteria, *Nitrosomonas* and *Nitrobacter*, have metabolic end products that are resources for each other (Section 4.3.8).

6.1 COMPETITION FOR RESOURCES

6.1.1 Historic Concepts: Competitive Exclusion Principle—Niche

When organisms of the same or different populations use the same resource, the resource may become scarce, causing decreases in reproductive rate, physiological stress, and even death from starvation. This type of interaction is called *exploitative competition*, as it involves the exploitation of a common resource. Exploitative competition is an indirect interaction, since the reduction in the reproductive ability or vitality of the competitors is not caused by direct effects such as antibiosis and aggression. When two populations are considered to be competing, it is important to define which resource or resources are being competed for.

The first competition experiments under controlled laboratory conditions were carried out with the ciliates *Paramecium aurelia* and *Paramecium caudatum*, which competed for yeast as a common food resource (Gause 1934). Both species reproduced well and reached comparable population densities when grown separately. *P. caudatum* was suppressed by *P. aurelia*, however, in mixed cultures (Fig. 6.1). This and similar experiments led to the development of the *competitive exclusion principle*: If two or more species compete for the same resource in the same habitat, only one species will survive.

The concept of the ecological *niche* is derived directly from the competitive exclusion principle: To coexist for an extended period, different species must live in different habitats, use different resources, obtain the same resources from different sources (e.g., shallow and deep-rooted plants), or have different requirements in the physical environment. The sum of all of these requirements is defined as the species-specific niche (cf. Section 4.1.2). According to the competitive exclusion principle, only one species can occupy a niche. The evolution of distinct niches is thus a mechanism for avoiding competition.

There is much debate over the concept of the niche, with arguments on theoretical as well as empirical grounds. Critics concerned with its theoretical validity point out that because of the *n*-dimensionality of the niche, the competitive exclusion principle cannot be tested empirically (cf. Section 2.1). If a study comparing two species shows no differences in their habitats, their resource requirements, or their physical requirements, it can always be said that the decisive niche dimension that separates the two species has simply not yet been found.

It can also be argued that based on the niche theory there should be no competition today. Species that were once competitors should have developed differences during the course of their evolution, or, if they differed enough behaviorally, they would have avoided one another. Connell (1980) called this idea the "ghost of competition past." Actually, the concept that past competition would lead to an absence of competition in the present is a violation of Lyell's Principle, originally developed in geology and later

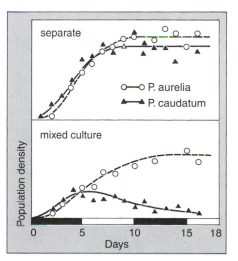

Figure 6.1 Competition experiments of Gause (1934). *(Above)* Growth of *Paramecium aurelia* and *P. caudatum* in separate cultures (food: yeast). *(Below)* Growth of both species in a mixed culture.

used by Darwin for the theory of evolution. According to Lyell's Principle, only those processes whose effects can be currently observed can be used to explain past events.

In reality, it is very difficult to conceive of applying the principle "one species—one niche" to plants. The several hundred thousand plant species with their different organs, seeds, and flower nectar could provide several million animal species with perfect resource specializations. Plants, on the other hand, have relatively few resources (light and nutrients), and these are almost always in excess. This problem is especially obvious for phytoplankton suspended in a turbulent medium, with no means of avoiding competition while using their resources. Hutchinson (1961) described this contradiction between the competitive exclusion principle and the large variety of species in the phytoplankton, even in tiny samples of water, as the *"paradox of the plankton."* His paper stimulated a series of theoretical and experimental investigations of interspecific competition that eventually led to the mechanistic theory of competition (Tilman 1982).

The controversies about the niche concept stimulated a great deal of empirical research. Much of the controversy, however, was caused by a misunderstanding, especially disregard for the time dimension. It can be easily seen using Gause's *Paramecium* experiment. Even when both species were in the same container and were using the same food the complete exclusion of one of the species required 16 days or almost 16 generations. Clearly, it takes time to replace the unsuccessful competitor. The slower the reproductive rate and the less the difference in the abilities in the competitors, the longer the exclusion will take. It is also easy to imagine that for two species with very similar competitive abilities, small changes in environmental conditions (e.g., temperature) could cause shifts in the dominance before the slightly inferior species is eliminated. One cannot conclude anything about the niche of species from distributional data, since the time dimension is lacking (Section 4.1.2). It is impossible to tell from a snapshot in time whether a species is stable or if it is about to be displaced. Even if one observes a declining trend, it is still possible that the particular species may eventually stabilize at a lower population density.

The possibility that under fluctuating environmental conditions there may be temporary or long-term coexistence suggests that competitors need not evolve toward divergence of features (niche specificity). Competitors may also evolve greater similarities, thereby minimizing the rate of exclusion. Species that differ greatly but share common resources need not have a higher probability of coexisting, since the rate of exclusion is higher. Very similar species that have identical requirements and similar capabilities can coexist due to a fluctuating advantage or the deceleration of the rate of replacement, whereas dissimilar species can coexist because of their differing requirements (niche specificity). Least likely to coexist are species with moderate differences, where they have essentially the same requirements but differ in their abilities to satisfy their requirements.

6.1.2 The Competition Model of Lotka and Volterra

The first mathematical description of competition was derived from an expansion of the logistic equation for growth (cf. Section 5.2.4). In its simple form, this equation describes the negative feedback effect of increasing density of a population on its own growth rate. The Lotka–Volterra model extends the effect of increasing density on growth rate to a competing species. The mathematical expression for two species is:

$$\frac{dN_1}{dt}\frac{1}{N_1} = r_1 \frac{(K_1 - N_1 - \alpha N_2)}{K_1} \tag{6.1}$$

$$\frac{dN_2}{dt}\frac{1}{N_2} = r_2 \frac{(K_2 - N_2 - \beta N_1)}{K_2} \tag{6.2}$$

N_1 and N_2 indicate species 1 and 2, and K_1 and K_2 are their respective carrying capacities. The coefficients α and β are a measure of the effect of one species on the other. Growing populations will follow the logistic equation and reach an equilibrium at the carrying capacity, where the growth rate is zero. Each species represents a certain proportion of the total population, since the carrying capacity depends on the available resources and the two competing species must share these resources. If one species does not reach its carrying capacity, that capacity can be filled by the other species. One can determine the relationship between the two species by solving Eqs. (6.1) and (6.2) for the condition of zero growth ($dN_1/dt = 0$ and dN_2/dt) as:

$$N_1 = K_1 - \alpha N_2 \tag{6.3}$$

$$N_2 = K_2 - \beta N_1 \tag{6.4}$$

One can create from these a graphic model that clearly demonstrates the result of competition between species with differing competition coefficients. Given a certain number of individuals of species 1, we can use Eq. (6.3) to calculate how many individu-

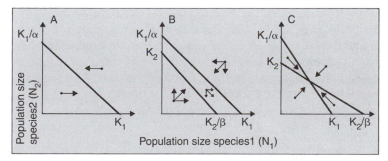

Figure 6.2 A graphic model of competition according to Lotka and Volterra. The lines connect all equilibrium combinations of species 1 and 2. The population size of each species is shown on the axes of the graphs. (A) Changes in the population density of species 1, which competes with species 2. When a combination appears that does not occur on the line (point), the population develops in the direction indicated by the arrow. (B) Development of both species. Here, the arrows indicate the direction of both species and the resulting direction for the total population. The population moves in the space between the lines for both species and then in the direction of the X axis. Species 1 wins. (C) If both lines cross, this intersection represents the point of stabile coexistence.

als of species 2 are necessary to bring the total number to the carrying capacity. A straight line called a *zero growth isocline* will be formed by placing all possible combinations where species 1 does not grow on a coordinate system with N_1 on the X axis and N_2 on the Y axis (Fig. 6.2A).

If no individuals of species 2 are present, from Eq. (6.3) $N_1 = K_1$; that is, species 1 will be at the carrying capacity. For the condition $N_1 = 0$, we can derive from Eq. (6.3) the point of intercept of the line with the Y axis:

$$N_2 = K_1/\alpha$$

If, by chance, there is a combination that does not fit on the straight line, the numbers of each species will increase or decrease until the carrying capacity has been reached; that is, the total population will move along the line.

A similar line can be constructed for species 2 using the β coefficient. The position of the line will determine the pattern of the competition. Figure 6.2 illustrates two such possibilities. In the first case (B) the equilibrium line of species 1 is higher than that of species 2. If the total population is below the carrying capacity for species 2, its population will increase until it has reached zero growth for species 2. Now species 2 will no longer grow, but species 1 can continue to increase. If the total population is above both lines, the populations of both species will decrease until they reach the topmost line. Species 1 will then remain constant, while species 2 continues to decline. Between the two lines species 1 will increase and species 2 will decrease. The result is always a movement of the total population in the direction of the starting point toward K_1 on the X axis, where only members of species 1 will remain. Species 2 will be excluded.

The crossing of the two lines (C) indicates a very different situation. In the example shown ($K_1 < K_2/\beta$ and $K_2 < K_1/\alpha$), the entire population tends to move toward the point of interception. Both species have a stable coexistence at that point. If the isoclines intersect, but $K_1 > K_2/\beta$ and $K_2 > K_1/\alpha$ (not shown in Fig. 6.2), the equilibrium is unstable. Arrows will move away from the intersection point. The initial conditions will determine which species drives the other to extinction. There are several additional possibilities, depending on the location of the lines. There can be coexistence, however, only if when one species is at its maximum density (K), the other species has a positive value. This can be expressed mathematically as:

$$\frac{K_1}{K_2} > \alpha \quad \text{and} \quad K_2 > \frac{\beta}{K_1}$$

This means that the density-dependent effect must be larger within a species than between species. The exclusion of species 2 by species 1 follows from:

$$\frac{K_1}{K_2} > \alpha \quad \text{and} \quad \frac{K_2}{K_1} < \beta$$

[164]

For the reverse, where species 2 excludes species 1:

$$\frac{K_1}{K_2} < \beta \qquad \text{and} \qquad \frac{K_2}{K_1} > \alpha$$

The α and β coefficients express some type of negative interaction between two species, but they do not take into account the actual mechanism of competition. They must be derived experimentally from a comparison of population growth rates in mixed-and monocultures. This makes it impossible to use the coefficients to make predictions in nature.

6.1.3 Tilman's Mechanistic Theory of Resource Competition

Early competition experiments (Gause) and the early mathematical models of competition (Lotka, Volterra) treated competitive abilities as a "black box." One could only identify the winner of a competition experiment at the end of the experiment. No additional definition of competitive ability, such as physiological ability to utilize a resource, could be derived from such experiments. Using circular logic, competitive ability was simply defined as the ability to exclude a competitor. Tilman's (1982) *mechanistic theory of competition*, developed from the principles of chemostat cultures (cf. Box 4.2), relieved this problem. The salient feature of this model is that the competitive ability of a species is derived from its physiological characteristics, such as the kinetics of its growth under resource limitation and its loss rates (mortality and emigration). A graphic model can be used to show the predicted outcome of a competitive interaction.

The simplest model is presented in Fig. 6.3: competition for *one common resource*. The two hypothetical competitors differ in their growth kinetics. Species A reaches higher reproductive rates at higher resource concentrations and species B at lower resource concentrations. The net growth rate is the difference between the reproductive rate and loss rate (cf. Section 5.2.3). It is assumed in this example that both species have the same loss rates. This can be accomplished experimentally in a chemostat culture by adjusting the flow-through rate to control population losses. Shortly after inoculation, when the organisms have not yet reduced the limiting resource, both species reproduce at nearly maximal rates. This allows the population of species A to grow most rapidly. As the resource is reduced with increasing population density, the reproductive rates decline.

When the loss rate is high (2), species B will first reach the concentration of its resource, where the rates of reproduction and loss are identical (R^*_{B2}). There is a zero net growth rate. If there were no competition, species B would reach a steady-state equilibrium in which there would be a balance between new production and elimination of organisms as well as between consumption and supply of the resource ("*steady state*"). The resource may be imported or supplied by remineralization (for inorganic resources) or new growth (for live resources). At the concentration of resources de-

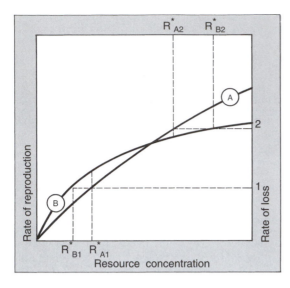

Figure 6.3 A graphic model of competition between hypothetical species A and B for a common limiting resource. In systems with low loss rates (1), species B wins; in systems with high loss rates (2), species A wins. The strength of the competition is defined by the equilibrium value (R^*, the point where the growth kinetic intersects the loss rate). The species with the lowest R^* wins, since the species that lost can only achieve a negative net growth rate at this resource concentration (loss rate > reproduction rate) (from Tilman 1982).

picted, species A still reproduces faster than its rate of loss, and its population therefore increases. As species A continues to increase, the resource decreases until it reaches the equilibrium concentration for species A (R^*_{A2}). At R^*_{A2} species B has a reproductive rate that is lower than its loss rate, meaning the growth rate is negative. B population is lost, and A maintains its steady state.

On the other hand, if the loss rate is lower (1), increasing consumption of resources would lower the resource concentration first to the equilibrium for species A (R^*_{A1}). At this level of resources, species B can continue to have a positive growth rate until the concentration of the limiting resource drops to the value R^*_{B1}. Species A is then excluded, and species B will stay in steady-state equilibrium. From this example we can see that the competitive strength of a species is defined by the equilibrium concentration of the limiting resource (R^*) at the point where the curves for growth and loss rates cross. If several species compete for the same limiting resource, the successful species will be the one that drives the concentration of the limiting resource to the lowest level (minimization of R^*). The R^* criterion is also valid if the competing species differ in their loss rates.

Sommer (1986) tested this theory experimentally. Monod uptake kinetics for phosphorus were determined for several green algae in Lake Constance. A natural inoculum from Lake Constance including several species was cultured in a chemostat at very low silica concentrations to exclude the diatoms. In every experiment, after several weeks only one particular species remained, irrespective of how abundant it had been in the inoculum. The winner of this competition was in fact the species predicted by the Monod kinetic model.

This clearly illustrates the major difference between the competition models of Lotka–Volterra and Tilman. In the Lotka–Volterra model the competitive abilities of the species involved are determined by the outcome of the experiment. In the Tilman model, the mechanisms of the competition (resource use) are used to predict the winner for a set of conditions. This is why we refer to it as a *"mechanistic"* model.

When the competition experiment with the Lake Constance phytoplankton was repeated with high concentrations of silica, the diatoms always became the dominants. The green algae were able successfully to compete for phosphorus as long as the diatoms were excluded by the shortage of silica. It is therefore necessary to expand the theory to include several resources. Would it be possible to balance the supply of Si and P so that green algae and diatoms can coexist? The predicted patterns of *competition for several resources* can also be presented graphically, as shown in Fig. 6.4 for two essential and two substitutable resources. The axes in the diagram (called the *resource plane*) represent the concentrations of both resources. Instead of an R^* value there is a *ZNGI line* (*zero net growth isocline*, line of zero net growth). This line (isopleth of growth) links all the concentration combinations at which the reproductive rate equals the loss rate. With only essential resources the ZNGI forms a right angle that does not cross either of the axes. The position of each of the lines is determined by the equilibrium concentration (R^*) of the respective resource. If one of the two resources is at a concentration to the left of or below the angle, the species cannot grow at all.

In a chemostat system resources are supplied at fixed rates (cf. Box 4.2). The concentrations of the two resources in the inflow can be plotted as a point on the resource plane. This point is called the supply point (Fig. 6.4). Note that the supply point is fixed, since the resource concentrations in the inflow are independent of what happens in the chemostat. The actual concentrations of the resources in the chemostat (*resource availability*) can change, depending on the biological activity, and will eventually reach a steady state. Without any consumption, the actual concentration must approach the supply point after the chemostat vessel has been flushed several times. In this case, the direction of movement of the actual concentration on the resource plane over time (trajectory) will be described by the *resource supply vector*, which always points toward the supply point.

If the chemostat contains organisms, these will consume resources as they grow. Essential resources are consumed in an *optimal ratio* (cf. Section 4.3.7), equivalent to the stoichiometric ratio of the resources in the growing tissue of the organism and consequently, to the ratio of the two Rs. Without a new supply of resources, the change of the actual concentrations would be described by a *consumption vector* with the slope of the optimum resource ratio. The real change in the concentrations (resource availability trajectory) is the result of the combined supply and consumption vectors. Contrary to the fixed slope of the consumption vector, the slope of the supply vector will change with the actual concentrations, so that the direction of the resultant vector will also change.

The location of the supply point determines the equilibrium conditions. If the supply point lies between the ZNGI and the origin of the graph, the species cannot grow and will disappear. However, if the supply point lies on the other side of the ZNGI, the concentrations of both resources are above R^*, and the population can grow. With continued population growth the resource trajectory eventually will hit the ZNGI. At this point, net population growth is zero, the population remains at the carrying ca-

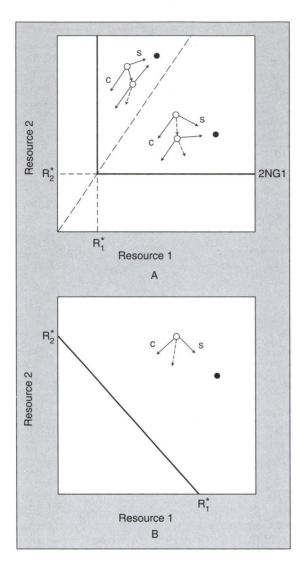

Figure 6.4 Limitation of one species by two resources. All ratios of the two resources form a resource plane. (A) Essential (nonsubstitutable resources) resources. ZNGI, zero-net-growth isocline. Solid dots are different supply points. Open circles represent resource availability points. The initial resource availability point will move in the direction resulting from the combined effects of the supply vector (s) and the consumption vector (c) when the population grows and consumes resources (dotted vector). The availability point (actual concentrations in the chemostat system) will follow a trajectory until it hits a ZNGI. The dashed line is the optimal resource ratio of the species that determines the consumption vector. (B) The same situation for perfectly substitutable resources.

pacity, and there will be no further reduction in the resource concentrations. The consumption vector will not cross the ZNGI. The location of the supply point determines where the ZNGI is hit, that is, which of the two resources will eventually be the limiting resource. Both resources can be limiting simultaneously if the consumption vector hits exactly at the intersection of the two lines forming the right angle. This is only possible when the supply point lies on the line of the optimum resource ratio. This line divides the resource plane into two fields that determine which resource is finally limiting. If the supply point is above the line, R_1 will be the limiting resource; if the supply point is below the line, R_2 will be limiting.

The ZNGI looks different for completely substitutable resources (Fig. 6.4B). Organisms may grow with either resource (cf. Section 4.3.4), but one of the resources may be more nutritious. Consequently, R^* for the more nutritious resource must be lower. The ZNGI is a line that intersects each of the axes in the positive region. Supply points and consumption vectors can also be constructed for this graph, but the slope of the consumption vector must be determined experimentally, as it cannot be predicted from the stoichiometry.

We can now use this model to predict the outcome of competition between two species for two resources by combining the ZNGIs of the two species into one graph. As the ZNGIs of all species differ, there are several possible combinations (Tilman 1982). For example, if the ZNGIs of competing species are parallel to one another, the successful species will be the one with the ZNGI closest to the axes (both R^* are lower). The most interesting case involves a tradeoff, so that the best competitor for R_1 is a poorer competitor for R_2 (Fig. 6.5A). In this case the ZNGIs cross and delimit four areas on the resource plane. None of the species can exist if the supply point lies between the axes and the combined ZNGIs (dark area). In the areas with incomplete overlap (lightly shaded), one of the two species is always below one R^*; hence it cannot grow, and the other species will monopolize the resources. Species 1 can grow in the vertically oriented area and species 2 in the horizontally oriented area.

Both species are above their respective R^*s for both resources (i.e., they can grow) in the area of complete overlap (white). As they grow they consume the resources according to their own optimum ratios. The resource availability trajectory will now depend on both consumption vectors and the supply vector. Contrary to the single species situation, the slope of the combined consumption vector will not remain constant. If the resource concentrations approach the ZNGI of one species, this species will no longer grow and it will stop consuming resources, but the second species that is not yet limited will continue to utilize the resources. Consequently, the combined consumption vector will approach the optimum resource ratio of the species still growing. The slope of the combined supply and consumption vector will then be deflected towards the intersection point of the two ZNGIs.

Again the location of the supply point determines the resource availability trajectory and the final equilibrium condition (Fig. 6.5B). The optimum ratios of the two species delimit three areas within the field of possible growth for both species. If the supply point lies above the optimum ratio for species 1, the trajectory will finally cross the ZNGI of species 2, and species 2 is replaced by species 1. The opposite will happen if the supply point lies below the optimum ratio for species 2. However, if the supply point lies between the two optimum ratios, the trajectory will always end at the intersection point of the two ZNGIs. The trajectory may be curved and may transiently cross an area where only one species can exist in the long run, but the combined effects of consumption and supply eventually draw it back to the intersection point. At the intersection point of the ZNGIs, both species are in equilibrium. This is a point of stable coexistence. The two species will finally coexist, if the supply point is located within the area delineated by the individual consumption vectors (optimum ratios). For essential resources, to which Liebig's law of the minimum applies, this is that region within which both species are limited by different resources.

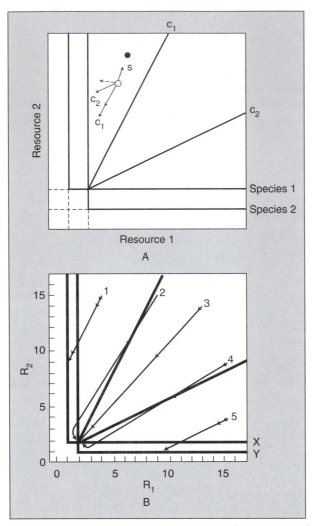

Figure 6.5 (A) Crossing ZNGIs for two species competing for two essential resources. Species 1 is the better competitor (lower R^*) for resource 1 and species 2 is the better competitor for resource 2. C_1 and C_2 denote the consumption vectors for the two species. The full circle is the supply point. The movement of the resource availability point (open circle) is determined by the combined effect of the two consumption vectors and the supply vector. (B) Resource availability trajectories originating at various supply points located in different areas of the resource plane.

There are now six areas defined on the resource plane (Fig. 6.6A). The outcome of the competition between the two species is predicted by the location of the supply point. None of the species can persist if the supply point lies in the area adjacent to the origin of the graph. Only one species can exist a priori in the rectangles between the ZNGIs. In each of the two areas delimited by the ZNGIs and the consumption vectors, one species will competitively displace the other. Finally, if the supply point lies between the consumption vectors, the two species can coexist. The closer the supply point is to the area of sole dominance of one species, the larger the proportion of this species when both are in equilibrium. A similar graphical model can be constructed for substitutable resources if the ZNGIs of the two species intersect (Fig. 6.6C).

Note that it is the *ratio of the limiting resources* and not the absolute amount of re-sources that defines the boundaries of coexistence and exclusion in the model. The maximum number of species that can coexist in equilibrium within the region of co-existence is equal to the number of limiting resources (extending the concept of the competitive exclusion principle). Even though no more than two species can coexist at a particular ratio of two resources, coexistence of more than two species is possible if there is a gradient in the resource ratio. Figure 6.6B and D illustrate the conditions under which this is possible: The intercept of two "neighboring" species must be in

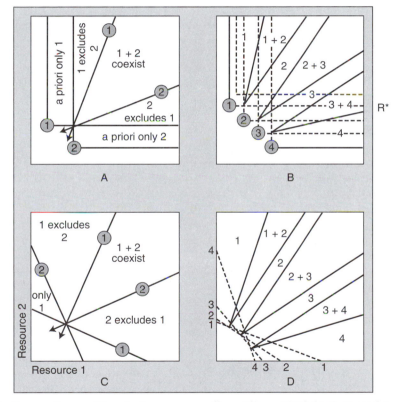

Figure 6.6 The theory of competition for two resources (from Tilman 1982) (A-C: Essential resources; D-F: Substitutable resources). (A) Competition between species 1 and 2 for two re-sources. Delineation of regions of exclusion and coexistence by ZNGIs and consumption vec-tors. (B) Several species (1,2,3,4) share the gradient of the R_1 to R_2 ratio. The numbers indicate the regions of coexistence and sole dominance. The dashed line represents the ZNGI of each species. Although the ZNGI crosses with species 2, 3, and 4, (C) Competition of two species for two substitutable resources. Regions of exclusion and coexistence are delineated by ZNGIs and consumption vectors (D) Several species (1,2,3,4) share the gradient of the R_1 to R_2 ratio; cf. (B).

the region where no other species can exist. This would occur if species 1 was the best competitor for resource one, but the poorest competitor for resource 2 and vice versa. Two interesting conclusions can be drawn from these graphs: (1) Species that coexist at a particular resource ratio are "neighbors," having relatively similar resource requirements; (2) the final outcome of the competition is not determined by the initial abundance of the competitors.

This competition model has also been recently tested many times using chemostat experiments with planktonic algae (essential resources; usually Si and P, but also N and P) and once with rotifers (substitutable resources; two species of algae as food). Figure 6.7 shows the first published example (Tilman 1977) using the diatom *Asterionella formosa* (higher Si and lower P requirements) and *Cyclotella meneghiniana* (op-

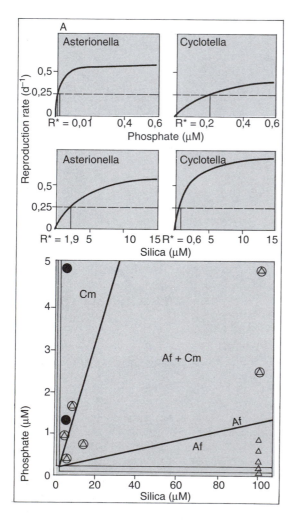

Figure 6.7 Summary of the competition experiments of Tilman (1977) with *Asterionella formosa* and *Cyclotella meneghiniana* at a dilution rate of 0.25 d^{-1}. *(Above)* Monod kinetics of P- and Si-limited growth, determination of the R^* values. *(Below)* A competition diagram with ZNGIs and consumption vectors. The symbols characterize the supply points (composition of the medium in the inflow) of each experiment and the taxonomic outcome of the competition. Circles: *Cyclotella* dominant; triangles: *Asterionella* dominant; combined symbols: coexistence.

posite) for silica and phosphorus as limiting resources. The ZNGIs were derived from the Monod curves for both species; the consumption vectors were drawn so that the resources were consumed in an optimal ratio for the algae. The supply points in the competition diagram correspond to the concentrations in the inflow to the chemostats (S_0). The predictions were supported in 11 of the 13 experiments. The diagram clearly demonstrates that the controlling factor is actually the resource ratio and not the absolute concentration.

Many other studies followed, some using two or more cultured species and some using natural phytoplankton as the inoculum. Taxonomic trends were consistent, irrespective of the geographical origin of the species used: Diatoms were dominant at high Si:P ratios. Within the diatoms there was a consistent order from centric to pennate species that followed the increasing ratio of Si:P. When N and P were manipulated as limiting resources, cyanobacteria became the dominants at low N:P ratios and high temperatures ($>15°C$) and green algae and diatoms dominated at high N:P ratios, depending on the availability of silica.

Figure 6.8 illustrates an experimental test for substitutable resources. The two rotifer species, *Brachionus calyciflorus* and *Brachionus rubens*, can both feed on the flagellated green alga *Chlamydomonas* as well as on the coccal green alga *Monoraphidium*, but they differ in their ability to utilize these resources (Rothhaupt 1988). *B. calyciflorus* grows better with *Chlamydomonas*, and *B. rubens* grows better with *Monoraphidium* with the same carbon content of each food. This results in lower R^* values for the more efficiently used resource. ZNGI's and consumption vectors were used to predict that *B. rubens* would dominate at low ratios, *B. calyciflorus* would dominate at high ratios, as well as the regions where the two species would coexist (Fig. 6.8). Long-term experiments then tested the competitive abilities at two rates of loss. The results of the experiments agreed with the predictions in 11 of the 12 cases. There was only one experiment where coexistence was predicted, but *B. rubens* dominated.

As algae absorb light, they "consume" this resource. Models based on Tilman's concept can therefore be used to describe competition for nutrients and light. In a mixed water column, on average all algae will be exposed to the same light conditions, ranging from high light intensity at the surface to low light intensity in the deep water. Analogous to the R^* concept, each species has a minimum light intensity, I^*, needed to maintain its population size. As algal populations grow, they reduce the light intensity in the deep water. Under these light conditions, the species with the lowest I^* would win the competition. Graphical models can be constructed for simultaneous competition for light and nutrients (Huisman and Weissing 1994), although they are more complicated than the graphs for two nutrients. The basic difference between the two types of resources is that nutrient consumption depends on resource concentration (amount per unit volume), whereas light absorption is related to a resource flux (irradiation per unit area). Therefore, the consumption vectors depend on the mixing depth (z) and on the turbidity caused by nonalgal particles. Despite these differences, the predictions of the models are quite similar to those of the standard Tilman models; that is, coexistence is possible when one species is light limited and the other is nutrient limited.

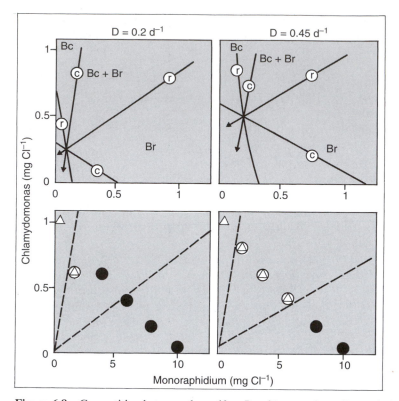

Figure 6.8 Competition between the rotifers *Brachionus rubens* (Br, r, circles) and *B. calyci-florus* (Bc, c, triangles) for the substitutable resources *Chlamydomonas* and *Monoraphidium* (from Rothhaupt 1988). *(Above)* Region near the origin; position of the ZNGIs and the consumption vectors, as well as predictions of the coexistence and exclusion regions for the dilution rates of 0.2 and 0.45 d^{-1}. *(Below)* Actual results of the experiment; the symbols show the location of the supply points as well as the outcome of the competition (circles: Br; triangles: Bc; combined symbols: coexistence). The dashed lines delineate the regions of coexistence and exclusion defined in the upper graph.

6.1.4 Competition Under Variable Conditions

Equilibrium conditions do not appear in nature, as they do in chemostat cultures. In addition to the regular circadian rhythm of light, there are a number of less regular or irregular fluctuations in the supply of resources as well as in boundary conditions that influence phytoplankton competitions. Some fluctuations cause indirect effects. Temperature changes in the epilimnion, for example, influence the mixing depth and therefore the availability of light, the addition of nutrients to the epilimnion, and the sedimentation losses. Zooplankton feeding, an important cause of phytoplankton mortality, varies with the time of day (vertical migration) and with the changes in the abundance of the zooplankton. The supply of nutrients from zooplankton excretion also fluctu-

[174]

ates. The excretion of nutrients by zooplankton creates microzones of elevated concentrations of nutrients that can be exploited by nutrient-starved cells with higher uptake rates (cf. Section 6.4.4). Such temporal variability in resource supply can lead to shifts in the competitive advantage of different species.

Some authors believe that competition theory cannot be applied to natural systems because they are too variable. Accordingly, many species can coexist because the competitive advantages of different species are constantly changing. Harris (1986) argued against this concept asserting that there is no competition under variable conditions. One might also propose that the complete exclusion of a species occurs only as the final result of the process (competition) and that competitions do not always reach this final condition. This raises the question whether the patterns observed in chemostats would stay the same if the conditions for the competition were varied temporarily.

The speed of the changes in the environmental conditions is of critical importance. "Fast" and "slow" are relative concepts that must be viewed in relationship to the rate of exclusion of the losing competitor. If changes in conditions come very infrequently, they may not occur in time to provide the inferior species with some possible advantage that would slow its exclusion. If the conditions fluctuate too rapidly, there may be short-term physiological changes, but these may have little influence on the competitive interaction, since all the species may simply respond to the average conditions. This leads to the hypothesis that both very slow and very rapid fluctuations offer no changing competitive advantages that will lead to few coexisting species, but intermediate frequencies of disturbance should promote the coexistence of the most species and exclusion of the fewest. This is the major point of the *intermediate disturbance hypothesis* of Connell (1978).

This hypothesis was tested with phytoplankton in chemostats where the addition of nutrients and removal of cells was done at specific intervals, rather than continually (Gaedeke and Sommer 1986). When the dilutions were done at intervals less than one generation time, the experiment did not differ from steady-state experiments in the number of coexisting species and the diversity index (see Section 7.4.1). The maximum diversity occurred at intervals of approximately three mean generation times. This result lies exactly on the line of the intermediate disturbance hypothesis, but must be tested with further experiments.

Support for this hypothesis also comes from experiments in microcosms (Fig. 2.1A), which tested the coexistence of different clones of *Daphnia pulex*. The greatest clonal diversity occurred also at intermediate disturbance intervals, independent of the intensity of the disturbance (Weider 1992).

Fluctuating additions of limiting resources allow for additional strategies for minimizing R^*. Species with high maximum reproductive rates can use a brief, high supply of a resource to increase reproductive rates in order to compensate for losses at times when resources are in short supply. Another method of taking advantage of favorable conditions is to form reserves that enables species to reproduce adequately as resources decline. This does not work for all types of resources: Algae can store phosphorus in great quantities and to some degree also nitrogen and the products of photosynthesis (polysaccharides, lipids), but not silica, which is essential for diatoms. The storage of reserves (e.g., lipids) is very important for animals. By living longer, animals are more likely to encounter a new supply of the limiting resource. For many animals that are unable to increase their reproduction rapidly when food conditions improve, the formation of reserve energy pools allows them to maximize food utilization during brief periods of food abundance.

One can recognize different strategies from the response of populations to fluctuating resources. Species with high maximum rates of reproduction follow resource maxima with rapid pulses of growth, whereas species that are capable competitors for low resources and specialize in forming reserves are more likely to have stable population densities. This can be demonstrated with a competition experiment using continuous cultures of phytoplankton (Sommer 1985). Phosphorus was withheld from the continuously added nutrients and added separately in pulses once a week. In this experiment both P and Si were potentially limiting resources. If these nutrients had been added continuously, two species should have coexisted. The pulsed addition of nutrients, however, led to a significant increase in the number of species (Fig. 6.9). Four species had nearly stable population densities (phosphorus storage), but the densities of two species fluctuated regularly following each pulse of phosphorus. The pulsing of nutrients not only resulted in an increased number of species; it also caused a clear shift in the species composition. Despite such shifts, the general pattern for equilibrium conditions held qualitatively: the tendency of the diatoms to increase with increasing ratios of Si:P, especially the Fragilariaceae.

The robustness of this pattern can be seen in the seasonal shifts in the composition of the phytoplankton community in two lakes (Fig. 6.10). In both lakes, the Fragilariaceae followed the Si:P ratio with a lag time of 1 to 2 weeks. This correlation is surprisingly close, considering the potentially masking effects of numerous other factors that could alter the competitive outcome in nature, such as the normal seasonal variability in environmental conditions as well as species-specific differences in losses due to grazing and sinking and vertical gradients in the availability of resources. Cyanobacteria abundance is correlated with the P:N ratio in Lake Schöhsee only after June, when the water has warmed up adequately.

Resources can also vary spatially as well as temporally. The resources of animals are frequently distributed heterogeneously. The algae of the aufwuchs community often form mosaics, and plankton can occur in clouds or patches. Consumers moving through resources heterogeneously distributed in space are much like those moving through time. The effects of spatial and temporal variability are identical. If the consumer lives in a *fine-grained environment* where the spatial variability is small relative to its body size, it is able to integrate the variability over short time intervals. In a *coarse-grained environment*, the consumer may never leave its large resource patch and, in fact, never perceive that there is heterogeneity. At intermediate "grains" the consumer is faced with changing resource availability. In this case, the outcome of the competition may depend on the ability of the consumer to make use of resources as they suddenly become available as well as its ability to survive a period of starvation after it has left the resource patch.

Spatial and temporal heterogeneity are especially pronounced in streams. Resources may differ profoundly from rock to rock in a small stream, depending on the type of

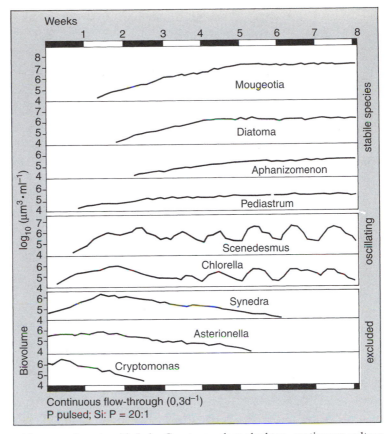

Figure 6.9 Competition experiments with Lake Constance phytoplankton, continuous culture ($D = 0.3$ d^{-1}), addition of P in weekly pulses, Si:P ratio 20:1. Biomass changes (logarithmic scale) of the species that persisted and of three of the species that were excluded (from Sommer 1985).

local disturbance. Even the stream bed is constantly changing due to floods and deposition of sediments. These conditions led many researchers to believe that competition was a relatively unimportant process in streams. There is growing experimental evidence, however, that competition can strongly influence the distribution of stream organisms (Hart 1983). For example, the species richness of the fauna dwelling on stones in an upland stream was found to be highly correlated with stone surface area and structure. Larger stones and grooved stones harbored more species of invertebrates, indicating that more microhabitats allowed more species to coexist (Douglas and Lake 1994). Often there is direct competition (interference) for space (cf. Section 6.2), but exploitative interactions also appear to be important. These interactions are difficult to demonstrate in situ.

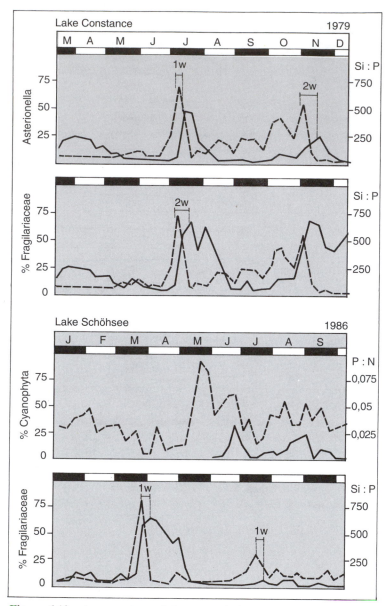

Figure 6.10 A comparison of the seasonal changes in the Si:P ratio and the percentage biomass of diatoms (*Asterionella formosa* or Fragilariaceae) in Lake Constance and the Fragilariaceae and in Schöhsee. Also, the seasonal changes in the percentage of cyanobacteria in Schöhsee are compared to the P:N ratio. Algal taxa: ———; nutrient ratio: - - -.

6.1.5 Keddy's Theory of a Unified Competitive Ability

Tilman's model requires that organisms have specialized competitive abilities for different resources. For example, of the two diatom species, *Asterionella* is the better competitor for silica, and *Cyclotella* is the better competitor for phosphorus. These two species can coexist because neither is the better competitor for both resources

(Tilman et al. 1982). This trade-off has also been demonstrated for other phytoplankton (Sommer 1989). Its physiological basis lies in the problem of allocation: Material and energy that is invested in increasing the efficiency of uptake of one resource is not available for improving an organisms efficiency in acquiring another resource. After successfully applying his theory to plankton, Tilman expanded it to include higher plants. Allocation of energy into leaves promotes rapid growth under noncompetitive conditions, allocation into stems increases competitive abilities for light, and allocation into roots increases competitive abilities for mineral nutrients (Tilman 1988).

Keddy (1989) considered allocation a minor problem, relative to the striking differences in productivity between organisms. According to Keddy, an organism that is more productive because it is competitively superior for one resource will have more material and energy available for investment into competition for other resources. This concept assumes a *unified competitive hierarchy* for all resources; that is, a species is either a strong or poor competitor. It cannot be a strong competitor for one resource and a poor competitor for another resource.

Keddy tested his theory with macrophytes from North American wetlands. He measured the competitive strength of plants by using *Lythrum salicaria* as a phytometer. A phytometer is a test plant that is planted into stands of other plants whose competitive strength is being estimated. Control plots have the same environmental conditions, but only the phytometer plants. The competitive strength of a plant is measured by the growth depression of the phytometer, relative to the control. There was no change in the competitive hierarchy in experiments with different plots and different limiting factors. Good competitors were good competitors everywhere and poor competitors were poor competitors under all conditions tested. Keddy then determined that some easily measured traits such as biomass and height of the plant (Fig. 6.11) could be used to predict its competitive strength. In natural vegetation, the strongest growth depres-

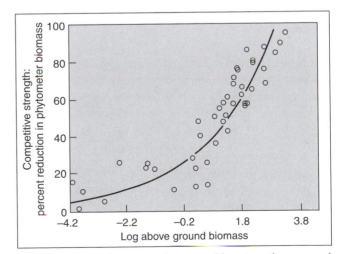

Figure 6.11 The effect of above-ground plant biomass on the competitive strength, measured as the percent reduction in the biomass of a test plant species (phytometer). (Data from Gaudet and Keddy 1988; figure after Keddy 1990).

sion of the phytometer (greatest competitive strength) occurred where the total biomass was the highest.

The relationship between biomass and competition led to the theory of the *centrifugal organization in plant communities* (Keddy 1990). In each large-scale unit of vegetation (e.g., wetlands of a certain region) there are optimal sites where plant production and biomass are highest. Here, distances between individual plants are minimal and mutual shading is maximal. Competition is most intense, and the best competitors become dominant. In the case of the wetlands studied by Keddy, the dominant competitors were cattails (*Typha* spp.).

Sites of maximal productivity are considered to be centers of the large-scale units. Gradients of decreasing biomass originate from this center and move toward sites with lower biomass. Such low biomass may be caused by numerous factors, such as scarcity of soil resources and physical disturbances (e.g., ice scouring along shorelines and shore erosion). The dominant species along each gradient are characterized by decreasing importance of competitive ability and increasing ability to resist particular stresses or disturbances. Seven such gradients for wetlands in southern Canada and northern United States are shown in Fig. 6.12.

6.1.6 A Comparison of the Theories of Keddy and Tilman

There is experimental evidence to support the theories of both Keddy and Tilman. Can this apparent contradiction be resolved? It is possible that competition between plankton follows different laws than competition between macrophytes, but this explanation has become less plausible since Tilman's theory has been supported from experiments with terrestrial plants.

One difference in the two theories is the *time scales* of the experiments used for their validation. Time scales used by the Tilman school require several generations of population growth until competitive exclusion has been achieved, or nearly so. With terrestrial plants, this may require long-term studies. Keddy's experiments, on the other hand, are completed after one season's growth or less for annual and perennial plants. Because of the different experimental approaches used, the theories actually measure different aspects of a species competitive ability, that is, the ability to exert competition and the ability to withstand competition.

The *ability to exert competition* is the potential to cause an adverse impact on other competitors. If competition is purely exploitative, it results from making a resource unavailable to competitors by consuming it. This is directly related to the functional response (cf. Section 4.3.2) and can be measured in studies of individuals. Growth depression of a phytometer is a simple and direct measure of the competition exerted by neighboring individuals that does not require the sometimes complicated measurement of limiting resources and resource depletion. However, a phytometer experiment does not answer the question of long-term persistence of either the phytometer or its competitors.

The *ability to withstand competition* is defined as not being excluded from a competitive interaction. An organism must reproduce sufficiently under competitive pres-

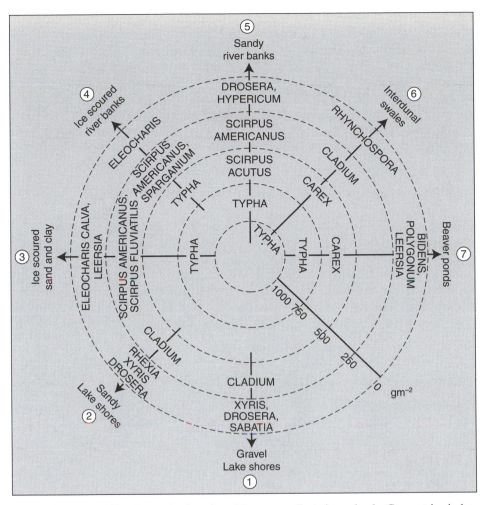

Figure 6.12 Centrifugal organization along biomass gradients in wetlands. Concentric circles: biomass levels; (1) Wilson Lake; (2) Axe Lake, (3) Luskville; Ottawa River; (4) Ottawa River at four locations; (5) Westmeath, Ottawa River; (6) Presquile Park, Lake Ontario; (7) beaver ponds, Lamark County, Ontario. (Original graph in Moore et al. 1989, cited from Keddy 1990).

sure to avoid exclusion (birth rate > death rate). Ultimately, this ability depends on the numerical response to resource scarcity caused by competition. For steady-state conditions it is defined by R^*, sensu Tilman. This response can only be measured by observing many generations of the competing populations. Such experiments are on a much larger time and space scale than the Keddy neighbor–neighbor interactions.

Intuitively, one might think that the abilities to exert and withstand competition should be closely linked; for example, species that capture more of a resource, thereby exerting a strong competitive pressure, should also grow more rapidly. This does not necessarily follow, however, for the resource requirements for growth may differ for different species. For example, a weakly silicified diatom might require less silica per capita to reproduce than a heavily silicified diatom. Thus, under Si-limited conditions, the species with the lower silica requirements might develop a larger population than the species with the high silica demand. The consumption of silica by the larger population with the smaller requirement will eventually impose a stronger exploitative pressure on the smaller population with the higher per capita consumption. However, a phytometer exposed to the competitive pressure of equal-sized populations of both species will be depressed more by the population with the high requirements.

6.1.7 Other Concepts of Competition for Substitutable Resources

Unlike plants, animals usually compete for substitutable resources, since they ingest their food in "packages" that contain many substances. If animals cannot eat their optimal food, they sometimes can compensate by eating a larger amount of a poor food. Although one should be able to apply the Tilman concept to animals, until now this has only been done with rotifers, which can be easily cultured in chemostats because of their small size and simple life cycle (Rothhaupt 1988). For example, it is not so easy to maintain a constant death rate with animals that have a more complex life history.

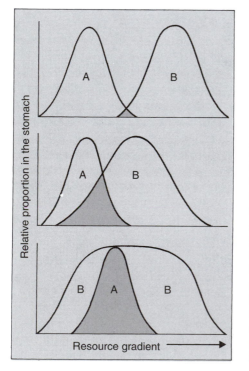

Figure 6.13 Overlap of the food spectra of two species (A and B) as a measure of potential competition. Various food resources are arranged along the X axis (e.g., according to size). The curves indicate the utilization of the resources by both species. (*Above*) Both species are very specialized. There is no significant overlap. (*Middle*) The food spectra overlap; i.e. both species use a common part of the resources and may compete. Species A is more affected than species B. (*Below*) A case of "inclusive overlap." Species B is a generalist with a very broad food spectrum, whereas B is specialized. The overlap is 100% for species A and it is under very strong competitive pressure.

When the loss rate is small, R^* approaches the threshold concentration for zero growth (cf. Section 4.4.3). Under these conditions one can compare competitive abilities of different species that use the same resources (e.g., filtering zooplankton) by examining the minimal food concentration for growth. The species with the lowest threshold value will always be the better competitor. An example of such a comparison is shown in Fig. 6.34.

Animals normally have a broad spectrum of food available to them in nature. This makes it impossible to determine average ZNGIs for possible competitors. Instead, attempts are sometimes made to estimate the intensity of competition from the so-called *niche overlap* (Giller 1984). Every species utilizes a specific part of the entire spectrum of resources (Fig. 6.13). The resource gradient in this schematic representation could consist, for example, of invertebrates of increasing size. Two benthos-feeding fish species may prefer different sized prey, but also can eat both larger and smaller forms. There is no competition for food between these two species if their spectra of food do not cross. However, the more similar the spectra, the greater the overlap. The extent of the overlap can be used as a quantitative index of competition.

The food spectrum can be measured easily through an investigation of the stomach contents of coexisting species, as Zaret and Rand (1971) demonstrated with 11 tropical fish in a small stream in Panama. Most species overlapped very little in their food spectra, although they ranged from 0 to 90%. There was much seasonal variation. The overlap was small during the dry season when the stream had little water and there was little food available, for at that time the fish were more specialized.

The overlap of food spectra need not be symmetrical. Many filter-feeding zooplankton have very similar food spectra (algae) that may differ greatly, however, in their breadth. Kerfoot et al. (1985) describe a case of "inclusive overlap" (cf. Fig. 6.13) for coexisting zooplankton. They revealed through gut analysis that *Daphnia pulicaria* used a very broad spectrum of algae, whereas other zooplankton were more specialized. The food spectrum of *Daphnia* completely encompassed the spectra of the other species. In this case, the specialized species had complete overlap with *D. pulicaria*, but *Daphnia* only had a partial overlap with the other species. One can conclude, therefore, that the daphniids were important competitors for the other species, but not the reverse.

One must be careful not to equate the overlap index with the α and β coefficients of the Lotka–Volterra equation. The overlap index defines only a single niche dimension, whereas the competition coefficient describes the result of all of the interactions between two species. It is impossible to be certain that the overlap index one has measured represents the entire interaction. A large overlap may mean nothing, if the resource measured is not limiting. In fact, it is easy to imagine that species would be more likely to use similar resources when there is a surplus of those resources. Furthermore, the overlap index is dependent on the availability of the resources and therefore changes along environmental conditions.

Individuals of the same species might be expected to have very similar food spectra (large overlap). Such competition is called *intraspecific*, in contrast to *interspecific* competition between different species. We have already learned about intraspecific competition as the basis of the logistic growth curve. These definitions are oversimplified, for competition actually occurs between phenotypes rather than between species. The effect of competition is usually seen, however, only at the population level. There are many examples illustrating the importance of intraspecific competition. During the "clear water phase" (cf. Section 6.4.1) intraspecific competition between daphniids becomes so strong that they completely exhaust their resources and the *Daphnia* population collapses. An important question for fisheries management is whether fish populations are controlled by intraspecific competition or by external factors. If intraspecific competition is important, the harvesting of more fish would reduce the competition and result in improved production.

Intraspecific competition differs from interspecific competition, for, although it may cause large population fluctuations, it cannot result in a complete exclusion of the species. Hamrin and Persson (1986) give an example of this with *Coregonus albula*, a small whitefish that occurs throughout much of Northern Europe. Regular fluctuations of a population of this obligate plankton feeder were followed over 8 years in a Swedish lake. Only the first two year classes are numerically important. The fish spawn in their second year. The abundance of fish alternated from year to year by a factor of 10. Every other year the first year class was abundant and the second year class was scarce, and in the alternate years it was reversed. The consistent pattern of fluctuations was the result of intraspecific competition, since these fish overlapped greatly in their food spectra within a year class as well as between year classes with an index of overlap of 0.8–0.9. A strong year class of juveniles resulted in strong competition for the second year class. As a result the older fish grew slowly and produced few offspring that were still relatively small at the end of the growing season. Thus in the following year the adult year class was correspondingly weak; the survivors from the previous year therefore experience little intraspecific competition. They grow rapidly and produce many offspring, repeating the cycle once again. The cycle is maintained by this intraspecific competition.

6.2 DIRECT INTERACTIONS OF COMPETITORS

6.2.1 Chemical Factors

The direct effect of competitors through the release of chemicals that inhibit other species is called *allelopathy*. One of the best known examples of this type of interaction is the suppression of bacteria by the fungus *Penicillium*, which secretes penicillin into its surroundings. Terrestrial organisms have some advantages over aquatic organisms in the use of allelopathic substances. If, for example, a plant excretes an allelopathic chemical over its own roots, the substance will stay concentrated in this area and provide a protective zone. Substances released into water, on the other hand, are quickly diluted and dispersed. This restricts the effectiveness of this mechanism in water, so that one might not expect that this process would be so important in this environment.

There are many anecdotal observations, but few are documented. It is difficult to extrapolate the results of laboratory tests showing biological effects of a chemical substance into nature. Cyanobacteria, for example, produce algal-inhibiting substances. If a certain strain of the cyanobacterium *Anabaena* is grown as small colonies on an agar plate and then covered with a layer of another strain of *Anabaena*, clear regions of inhibition will form around the outer edge of the colonies of the first strain (Flores and Wolk 1986). The substance involved can be isolated and chemically defined, but it is difficult to determine whether the substance actually provides *Anabaena* with a competitive advantage in nature.

The benthic cyanobacterium *Fischerella* produces a similar substance that, when isolated, can inhibit or at high concentrations kill other cyanobacteria. In nature, *Fischerella* covers rocks as a thick layer. Quite possibly, allelopathy may be responsible for the development of such *monocultures* (Gross et al. 1991).

There is also much speculation concerning the importance of chemical inhibition in aquatic macrophytes. These higher plants usually provide an excellent substrate for periphytic algae. Such periphyton inhibit the exchange and nutrient uptake by the macrophyte leaves and may also causes reductions in light. Sand-Jensen (1987) estimated that between 7 and 70% of the light that reached the surface of the submerged macrophyte *Littorella uniflora* was absorbed by periphyton, depending on the periphyton biomass. A chemical defense to keep the surface clean would therefore be an advantage to the plant. Quite surprisingly, young macrophytes are often free of periphyton, even though the algae can grow more rapidly than the macrophyte. It is possible that the young plants "protect" themselves chemically. Inhibitory chemicals have also been isolated from aquatic plants. For example, water milfoil (*Myriophyllum*) contains numerous phenolic substances that inhibit the growth of cyanobacteria (Aliotta et al. 1992). Also Cannell et al. (1988) purified the antibiotic pentagolloylglucose, an inhibitor of the enzyme α-glucosidase, from the filamentous green alga *Spirogyra*.

There are few, if any, known allelopathic interactions between aquatic animals. Seitz (1984), for example, suggests that *Daphnia* produce fewer offspring when grown in water that had contained another *Daphnia* species; nothing is known, however, about possible mechanisms involved.

Chemical ecology is very poorly developed in aquatic ecology compared to what is known for terrestrial systems. It would be extremely interesting to know whether allelopathic processes are involved in the development of communities and whether they are involved in the regular changes in algal species that occur during seasonal succession.

6.2.2 Mechanical Interactions

Competition for space is an important process in marine littoral communities. In freshwater, sessile species, especially in fast-flowing streams, form similar-type communities where space can become a limiting resource. The overgrowth of encrusting algae on the surface of rocks can lead to the slow exclusion of certain species. Sponges and bryozoans may also overgrow and thereby exclude other sessile species. Such interactions are so slow that they can only be seen from maps or photographs documenting the changes over long periods.

Some native bivalves in Lake Erie have become endangered by being overgrown by the introduced zebra mussel *Dreissena polymorpha* (cf. Section 5.7). The planktonic larvae of the zebra mussels require a hard substrate on which to settle before they metamorphose. With a limited amount of hard surface available, many larvae settle on the shells of larger bivalves (e.g., Anadontidae). The burden of growing zebra mussels on the shells reduces the fitness of the larger bivalves by mechanical interference as well as exploitative competition for suspended food particles (Haag et al. 1993).

Insects also may *directly defend their territories*. The larvae of the caddisfly *Leucotrichia pictipes* live in silk cases that are attached to the rocks. They defend a region around each case against intrusion by larvae of their own species as well as other species (Hart 1985a). This leads to a uniform spatial distribution (cf. Section 5.5). The larvae graze on the aufwuchs algae. This defense provides each larva with its own protected "pasture." More algae tend to grow in such "private territories" than in areas where there are many grazing larvae. The size of the territory is related to the body size of the larva. The area can be viewed as a resource since resource availability and area are closely coupled. This can be demonstrated nicely with the following experiment: If one removes algae from the territory of a caddisfly larva, it responds by enlarging its territory.

Predatory stonefly larvae catch fewer mayflies when they live on rocks where other larvae of their own or other species are present. This mutual disturbance occurs only at intermediate prey densities, and not when prey are rare (extremely long search time) or when prey are in excess (Peckarsky 1991).

Another example is the blackfly larva *Simulium piperis*, which sits attached to rocks and places its filtering fans in the current to catch drifting particles (cf. Section 4.3.11). These larvae also defend the space from which they can filter out particles. It is interesting that the larvae are only aggressive toward larvae that sit upstream, as these are the only ones that can "filter away" their food. It can also be demonstrated experimentally that the larvae actually eat more, once they succeed in driving away the larvae sitting upstream (Hart 1986).

The competition between large cladocerans and rotifers for common resources (algae) has been investigated extensively. It is often observed in the field that large daphniid species and rotifers are mutually exclusive. Rotifers often appear early in the season, followed by *Daphnia*. When *Daphnia* populations fluctuate, the rotifers become abundant at times when the *Daphnia* are scarce. When planktivorous fish decimate the large *Daphnia*, the rotifers dominate. Even when the daphniids are killed by an insecticide, the rotifers quickly develop enormous populations (Hurlbert et al. 1972). At first, it was suspected that this was a case of exploitative competition. It was assumed that the *Daphnia* were the more efficient competitors for the algae, and, in fact, rotifers do need higher food concentrations than the *Daphnia*.

Recently, however, it has been demonstrated that mechanical *interference* is also involved (Gilbert 1988). While filter-feeding, *Daphnia* sucks small rotifers into its filtering chamber, where they are damaged or even ingested along with the algae. *Daphnia* densities of as few as five animals per liter can result in heavy mortalities for the rotifers. This requires, however, that the *Daphnia* are large (at least 1.2 mm) and the rotifers are small. Large rotifers, such as the armored *Keratella quadrata*, are damaged very little, and such resistant rotifers tend to co-occur with *Daphnia*. Large rotifers, on the other hand, have a higher food threshold for reproduction than the small species, and are often excluded by exploitative competition under food-limited conditions. It is difficult to determine whether exploitative competition or mechanical interaction is more important in any specific situation, since both processes lead to the same result. In the laboratory, it has been shown that the mechanical effect is more important for large *Daphnia*, but small cladocerans such as *Ceriodaphnia* suppress rotifers by exploitative competition (MacIsaac and Gilbert 1991). The reverse situation, where rotifers suppress cladocerans, has never been observed.

6.3 PREDATOR–PREY RELATIONSHIPS

6.3.1 Causes of Mortality

The fitness of an individual is determined by its mortality, in addition to its reproduction. Except for unicellular organisms, which are potentially immortal, all other organisms have a division of labor between somatic and germ cells and must therefore die. *Natural death* from aging is of relatively minor importance in an ecological context, since under natural conditions the vast majority of individuals never reach an age where they would die without outside influences. Most individuals die as a result of some clash with either the abiotic or biotic environment. Such external mortality also occurs in potentially immortal single-celled organisms.

Abiotic factors are frequently a source of mortality when they exceed the physiological limits of an organism. Massive die-offs may occur when unusual climatic events cause the temperature to go above the tolerance range or the oxygen concentration to fall below the minimum level. Erosion and scouring caused by flooding can cause mechanical mortality for stream organisms.

Lack of resources can be another cause of mortality. Life requires a supply of energy. Microorganisms will die if they do not have sufficient substrate and algae will die if the are placed in the dark; animals starve without food. Mortality can also result from negative interactions between organisms, which we referred to earlier as lowered reproductive rates resulting from competition for common resources (Section 6.1) or direct effects (Section 6.2).

Almost every organism is potential food for some other organism. Thus "being eaten" is perhaps the most important biotic cause of mortality. In the classical sense, a *predator* is a carnivore, an animal that eats other animals. This concept is too narrow however, to be functional. We will therefore consider under the general heading *predator–prey relationships* all interactions that result in energy transfer from one organism to another, and with few exceptions, represent a mortality factor for the prey. This definition can be applied to *flagellates* that consume bacteria, *herbivores* that eat algae or macrophytes, *carnivores* that feed on other animals, and *parasites*. This broad definition is justified, since each of these interactions has the same end result for the prey, namely, mortality. There are basic similarities between a population of algae that are eaten by filter-feeding zooplankton and chironomid larvae that are hunted by fish: The populations become decimated, creating a selection pressure that favors those individuals that are best protected against the mortality factor and that have a higher relative fitness. Even though the mechanism of feeding is very different for the various nutritional types, the effect of the interaction on the population and the evolutionary consequences are the same.

6.3.2 Predator–Prey Models

Lotka and Voltera also constructed a simple model for predator–prey relationships that was similar to their competition model (cf. Section 6.1.2). They wrote two equations describing the population growth rates of the prey and the predator. Each equation contains both growth and loss terms. Changes in the population of the prey follow

$$\frac{dN_1}{dt} = bN_1 - pN_1N_2$$

Growth is the product of the unlimited birth rate in the absence of predation losses (b) and the number of prey present (N_1). The losses are proportional to the number of prey (N_1), the number of predators (N_2), and a coefficient of predation (p).

Changes in the predator population are described by

$$\frac{dN_2}{dt} = apN_1N_2 - dN_2$$

Predator growth is the product of the losses of the prey (pN_1N_2) and a coefficient a, that defines how many prey a predator must eat in order to produce offspring. The losses are proportional to the number of predators (N_2) and the death rate (d).

The ability of this model to predict *cycles* of alternating prey and predator abundance has aroused much interest. If both populations are in equilibrium so that $dN_1/dt = 0$ and $dN_2/dt = 0$

$$bN_1 = pN_1N_2$$

and

$$dN_2 = apN_1N_2$$

The constant equilibrium densities will be

$$N^*_1 = \frac{d}{ap} \quad \text{and} \quad N^*_2 = \frac{b}{p}$$

N^*_1 is the number of prey at which the death rate of the predator can be compensated by growth. N^*_2 is the number of predators at which the growth rate and death rate of prey balance. When $N_2 > N_2^*$, the prey population will decrease; when $N_1 > N_1^*$, the predator population will decline. If by some chance the populations move away from the equilibrium point, they will oscillate around these densities without dampening.

The primary difference between the predator–prey model and the competition model are the coefficients (Section 6.1.2). The coefficients a, b, p, and d of the predator–prey model are physiological parameters of the organisms that can be measured experimentally, whereas the competition coefficients α and β describe the combined effects of one species on the other. One can easily measure the effects of temperature and food conditions on the birth rate (b) or the effect of prey density on the predation coefficient (p). The latter parameter describes the capture success of a predator under specific conditions. Coefficient a corresponds to the gross production efficiency (cf. Section 4.4.3). The death rate of the predator can also be measured experimentally and independent of the predator–prey experiment. The following sections will go into some detail concerning the mechanisms involved in predator–prey interactions.

Regular fluctuations of population densities are easy to explain, if a time lag exists between the response of the prey and the predator. As the prey can be considered a resource for the predator, the model of logistic growth with a time lag (Section 5.3.4) can be applied to a predator–prey system. The result is an oscillation of the predator population around the carrying capacity, as depicted in Fig. 5.3.

The Lotka–Volterra model is quite unrealistic because it is oversimplified. There is, for example, no *feedback* between the prey and their food resources. Such feedbacks would affect b. Also, the death rate of the predator (d) is independent of predator density. It would also be more realistic to assume that the death rate of the predator is in turn controlled by a higher-level predator in a density-dependent way. The model does not include a time lag between predator and prey. Finally, the model assumes a linear correlation between the per capita feeding rate of the predator and the prey. Since many of these assumptions are not valid (cf. Sections 6.4 and 6.5), it has been replaced by better models.

More general predator–prey models are based on the concept proposed by Rosenzweig and MacArthur (1963). Included in their model is self-regulation of prey and predator, as well as the functional response of the predator. The growth rate of the predator population is dependent on its density. With increasing prey density the per capita predation rate and thus prey mortality reaches a plateau when the predator is satiated (cf. Box 4.1), unlike the Lotka–Volterra model, where predation rate is independent of the prey density. This model is very flexible, for it allows one to build into it the specific types of functional responses as well as threshold values that represent a refuge for the prey. The model can also include changes in the behavior of the predator due to changing prey densities, such as the shifting to alternate prey species.

One can predict the changes in the predator and prey populations by examining their isoclines of zero growth. These two lines are analogous to the ZNGIs (Section 6.1.3) and connect all possible numerical combinations of predators and prey at which there is no net growth of either the predator or prey population. Where the two zero-growth isoclines cross there is an equilibrium point where neither population will change. One can see in Fig. 6.14 what happens when the equilibrium is disturbed, as, for example, by a disease that decimates one of the populations. If the population returns to the equilibrium point, it will become stable once again; both predator and prey will live in stable coexistence.

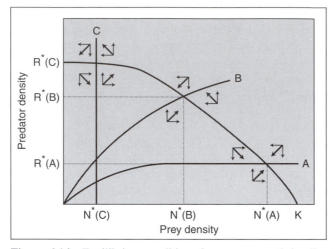

Figure 6.14 Equilibrium conditions for a zero-growth isocline for prey with self- regulation (solid line) and various zero-growth isoclines of predators (dotted line). The intersection points indicate the equilibrium densities for prey (N^*) and predators (R^*) at which both populations are constant. The trajectories show the directional movements of predator–prey densities following a disturbance that pushes the population out of equilibrium. The zero-growth isoclines of the predators are shown for the following cases: (A) predators regulated by external factors; (B) predators with self-regulation; (C) predators without self-regulation that switches from negative to positive population growth at a fixed prey density.

Box 6.1 illustrates how one can construct zero-growth isoclines. The characteristic features of predator–prey systems produced by the Rosenzweig–MacArthur model are shown in Fig. 6.14. The prey zero isocline is combined with various zero isoclines of the predator. A steeper predator zero isocline indicates a more efficient predator, requiring fewer prey to grow. This could occur if the predator is a more efficient hunter or utilizes its prey more efficiently. A leveling off of the isoclines (bending to the right) indicates there is self-regulation by the predator, as, for example, would occur if the predators begin to interfere with one another.

Predator and prey populations will either increase or decrease, that is, move toward their zero isoclines, depending on their respective densities. If, for example, the population of predators is controlled by other factors and remains constant at low prey densities (A), every combination of predator and prey will move directly to the equilibrium point. Thus, whenever anything disturbs the predator–prey ratio, it will try to move directly back to its original equilibrium condition without fluctuations. In case (B), the trajectory of the predator–prey ratio does not move directly toward the equilibrium point. At first it misses the equilibrium, but then gradually approaches it. This means: Following a disturbance the predator and prey would at first oscillate around the equilibrium density, but the oscillations dampen with time, eventually reaching once again the equilibrium condition. The duration of the oscillations is directly related to the steepness of the slope. If the predator zero isocline is very steep, the trajectory may form a closed cycle around the equilibrium point. This is referred to as a

| BOX 6.1 | CONSTRUCTION OF ZERO-GROWTH ISOCLINES |

For the Prey

If a prey population grows exponentially without any losses (Section 5.2.4), its reproductive rate follows an optimum curve related to the population density (heavy line).
The intersection on the right with the X axis gives the capacity (K). The net population growth rate is the difference between the rate of reproduction and mortality rate.
The mortality rate is determined by the consumption rate (dashed line). The predator
shows a functional response to increasing prey density (Fig. 6.1.1). For simplicity, only
the increasing portion of the curve is shown in the figure. The population growth rate
is zero where the functional response curve intersects the reproductive rate. As the
predator becomes more abundant, the consumption rate increases, increasing the
steepness of the curve. There is a point of intersection for every predator density, indicating zero growth of the prey population. It is therefore possible to define for all
prey densities the consumption rate (predator density) at which the prey achieves only
zero growth. The line connecting these points is the zero-growth isocline for the prey.
At any given predator density the prey population will move toward the isocline. If
the prey population is above the isocline, it will decrease, and if it is below, it will
increase (arrows).

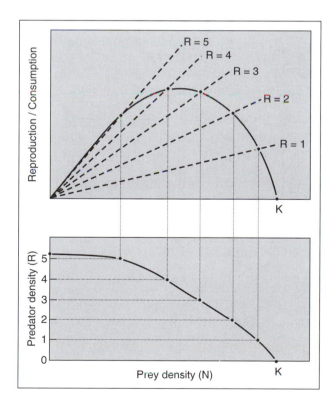

BOX 6.1	(*Continued*)

For the Predator

It is easy to see how in an equilibrium the changes in predator population depend on the number of prey, for each predator requires a certain amount of prey to grow (Fig. 6.1.2). The zero-growth isocline for this situation is therefore an increasing straight line (A). At higher prey densities the predators can have density-dependent effects on one another that result in some form of self-regulation. This causes the zero-growth isocline to begin to level off, bending downward (B). Finally, the predator density may be regulated by factors that have nothing to do with the prey, such as the availability of space. In this case, above a certain prey density the isocline becomes horizontal and the maximum number of predators (population growth rate is zero) is independent of the prey density.

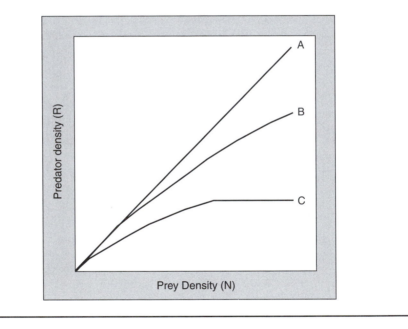

stable limit cycle within which the predator and prey oscillate continually around the equilibrium, without ever reaching it. The amplitude of the oscillations depends on the size of the disturbance. A final case can be considered, where at low population densities of the prey its zero isocline becomes lower. This could occur, for example, if the prey can only find mates at some threshold density. If the predator isocline crosses to the left of the maximum of such a prey isocline, the result is a spiral that becomes increasingly larger, moving further and further away from the equilibrium point. This type of system is unstable.

Two conclusions can be drawn from this model:

1. The more efficient a predator is, the more persistent the population oscillations and the greater the likelihood that one of the populations will go extinct.

2. The stronger the self-regulation by the predator, the more the oscillations are dampened, until eventually they disappear. Self-regulation therefore leads to a stabilization of the predator–prey system.

It is uncommon to find regular cycles of predator and prey in nature, and in freshwater systems they are very rarely found. McCauley and Murdoch (1987) analyzed numerous time series of *Daphnia* and their "prey," algae, where *Daphnia* was not regulated by predators. In such instances, there were cycles caused by the predator–prey interactions and not by outside factors. In the normal seasonal succession in a lake only the beginning of an oscillation is seen (cf. Fig. 8.15). In the spring, the population of algae (prey) then grows to a maximum, followed by a pulse of filter-feeding zooplankton (predators). Eventually, the zooplankton become so abundant that their feeding exceeds the algal growth rate and the algae decline (cf. clear-water phase). The growth rate of the zooplankton then decreases and their population also declines. The algae once again can increase, but the cycle becomes interrupted at this point, for now, in early summer, the zooplankton, especially the *Daphnia*, can no longer increase, as they are under other controls, such as fish predation and food limitation caused by the inedibility of or interference by large algae (Carpenter et al. 1993).

The Rosenzweig–MacArthur model is also mechanistic, since all its parameters can be experimentally defined. With such a model one can attempt to determine why predator–prey cycles are so rare in nature. A simple model of this sort, with functions that describe the effects of environmental conditions on the coefficients, can serve as a basis for complicated simulation models (cf. Section 2.5).

6.3.3 Evolution of Defense Mechanisms

All prey are identical and have the same probability of being eaten in the predator–prey models. In reality, this is, of course, not correct. Individual prey differ genotypically and phenotypically, and some of this variation will influence the success of the predator (coefficient p). Prey with characteristics that lower the predator's success (e.g., protective coloration, rapid escape) have a smaller probability of being killed. The gene pool of the prey will thus shift in the course of evolution toward predator-resistant genotypes. This causes the predator to become more and more inefficient. To prevent its own eventual extinction, the predator must either undergo a similar evolution or find an alternative prey.

The first strategy involves an "*arms race*" between predator and prey. Individuals in the predator population also vary in their efficiency and thus their selective advantage. For example, if a prey becomes more transparent, the predator must evolve a better sense of sight. These processes can take place almost simultaneously, since a more efficient predator has a selective advantage, even if the prey has not yet evolved changes. Such *coevolution* is, in fact, most often the result of predator–prey interactions. Many features of the morphology, life history, and behavior of aquatic organism can be explained as defense mechanisms (O'Brien 1987, Walls et al. 1990).

The process of adaptation for each of the partners in coevolution generally stops before it is complete, since fitness is influenced by many factors and every adaptation has its costs as well as its physiological and genetic constraints. In such cases the prey may actually be removed from the immediate effects of the predator so that predator and prey no longer interact. It then becomes impossible to say whether a particular feature that prevents the predator from attacking the prey was the result of coevolution or whether it was an adaptation to some other environmental factor. Copepods, for example, that have a summer diapause are not affected by fish predation. One should not assume, however, that this is an avoidance of predation, for the diapause may well have evolved as a response to the drying out of small ponds in the summer. The fact that the diapause disappears in the absence of fish suggests strongly that fish are at least necessary to maintain this behavior.

There is a rapidly growing body of evidence that predator–prey interactions play a key role in aquatic ecosystems. Our better understanding of the mechanisms involved has allowed us to interpret the possible defense function of many features of aquatic organisms, from protists to vertebrates. This also led to the development of a priori hypotheses that can be tested experimentally.

6.4 GRAZING

6.4.1 Grazing in the Plankton—Quantitative Aspects

Grazing refers to the predator–prey interactions in water where algae and bacteria are the prey organisms (cf. Section 5.2.6). The importance of grazing by herbivorous zooplankton on the development of phytoplankton populations became recognized by the 1970s. Demonstration that grazing was directly involved in the *clear water phase* provided valuable field support for this concept (Lampert 1978). The clear water phase describes the very regular occurrence of a minimum density of phytoplankton in the middle of the growth period (in north temperate latitudes usually in May or June), most frequently in meso- and eutrophic lakes. In lakes with a moderate amount of turbidity due to abiotic particles this period can be recognized by a summer maximum water transparency, similar to the transparency normally seen in the middle of winter. There is usually an abrupt decrease in the phytoplankton density, in eutrophic lakes commonly dropping within a week to about one-tenth the densities found during the spring maximum (cf. Fig. 4.23, first week of June).

The clear water phase occurs during the time of maximum solar radiation and longest day length. During the spring phytoplankton maximum there is usually only weak evidence of nutrient limitation, but during the clear water phase the nutrient concentrations actually increase. Also, during the clear water phase phytoplankton have their highest rate of specific photosynthesis (photosynthesis/biomass), indicating that at least the dominant algal species have high rates of reproduction. Thus the decline in the phytoplankton densities can only be attributed to higher mortality. The clear water phase also corresponds to the time of the yearly maximum of herbivorous zooplankton, implicating zooplankton grazing as the major cause of phytoplankton mortality. Recently, there is direct evidence that the feeding rates of the zooplankton during the clear water phase exceed the production rates of the phytoplankton (Fig. 6.15).

Herbivorous zooplankton feed on algae by phagocytosis (protozoans), filtration (cladocerans and rotifers), and raptorial capturing of specific cells (copepods). It is common to refer to *filtration rates*, since cladocerans are often the functionally dominant zooplankton in many lakes. Filtration rates are determined by measuring the uptake of radioactive food in short-term (minutes) experiments or by measuring the decrease in food suspensions in long-term (hours) experiments (Peters 1984). It is necessary to correct for the simultaneous growth of food when live particles are used. If some of the food particles survive passage through the gut and are resuspended, the two methods can produce very different results.

The ingestion rate (I) tells how many food particles (N) or the mass of food (M) that is eaten by an individual per unit time (t), with the dimensions: N ind^{-1} t^{-1} or M ind^{-1} t^{-1}. To estimate the total feeding rate, which can be compared to the phyto-

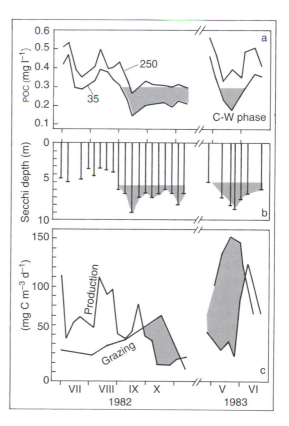

Figure 6.15 The effect of zooplankton biomass on the phytoplankton in Schöhsee. (a) Particulate organic carbon (POC, includes both phytoplankton and detritus), particle sizes <250 and <35 μm. Particles under 35 μm are considered edible. (b) Water transparency. (C) Photosynthetic rates of phytoplankton (dashed line) and the feeding rates of the zooplankton (solid line). Periods where the grazing rates of the zooplankton exceed the production rates of phytoplankton are shaded. The clear-water phase occurs in mid-May (from Lampert 1989).

plankton production rate, the ingestion rates are summed for all individuals of all species in a specific water volume (V) or under a specific unit surface area of the lake (A), with the dimensions: $M\ V^{-1}\ t^{-1}$ or $M\ A^{-1}\ t^{-1}$.

The filtration rate (F) describes the volume of water that contained the amount of food eaten. This measure does not actually refer to the quantity of water passing through the filtering appendages, since it is calculated from the particles taken out of the water or the decrease in particle number and the particles are usually retained with <100% efficiency. It is also often applied to species that do not actually filter. *Clearance rate* is another expression for filtration rate that better describes that it is the volume of water cleared of particles per unit time. The individual filtration rate can be calculated by dividing the ingestion rate by the food concentration, with the dimension $V\ ind^{-1}\ t^{-1}$:

$$F_{ind} = I/C$$

or determined directly from the decrease in particle concentration (after correction for the control without animals):

$$F_{ind} = V\ \frac{\ln C_2 - \ln C_1}{NT}$$

where C_1 and C_2 are the particle concentrations at the beginning and end of the measurement, N is the number of animals in the experiment, and V is the volume of suspension used in the experiment (Peters 1984).

One can determine the filtration rate of a population (F_{pop}) by multiplying the individual filtration rates of each age category (size classes) by the abundance of each age category and adding these products. The sum of the filtration rates of the entire herbivore population is the *community grazing rate (G)*. This *grazing rate* (shortened form) provides a measure of the grazing-induced mortality rate for algae that have no defenses (γ_{opt}). It has the dimension of inverse time and expresses the proportion of the water volume that is "filtered" per unit time (cf. Section 5.2.6).

The community grazing rate can also be measured in situ with specialized incubation chambers (Haney 1973) that allow one to measure spatial (depth distribution) and temporal (diel rhythms) variations of grazing under natural temperature, light, and food conditions. These measurements can then be integrated over the entire day. During the clear water phase grazing rates reach values from 1.0 to 2.5 d^{-1}, which are greater than the maximum growth rates of many algal species. Such high rates, however, appear only for the few weeks during the clear water phase; more often in the summer they are around 0.2 d^{-1} (Lampert 1988a).

6.4.2 Feeding Selectivity of the Herbivorous Zooplankton

Even if all phytoplankton were eaten with the same efficiency, grazing would still exert a selective pressure on the composition of the phytoplankton, similar to increasing the flow-through rate in a chemostat (cf. Fig. 6.3). Uniformly high grazing rates would favor those phytoplankton with high maximum growth rates (μ_{max}), whereas uniformly low grazing rates would select for species that have a steep initial rise in their growth kinetic curve. Zooplankton do not actually eat all phytoplankton with the same efficiency, since many algae can avoid either being eaten or digested. Both methods of protection result in lower mortality, compared to species that are optimally eaten. This can be expressed as the selection coefficient $w_i = \gamma_i/\gamma_{opt}$.

The selection coefficient is not a fixed characteristic for a particular phytoplankton species. It varies with the zooplankton species and age class and therefore depends on the composition of the zooplankton. Figure 6.16 shows a comparison of the selection coefficients for several algal species with differing morphology when eaten by *Daphnia magna*.

Grazing selectivity also implies a selection factor in evolutionary terms. One can expect the evolution of mechanisms that reduce the vulnerability to grazing, since resistant genotypes of algae have lower mortality losses. Many features of phytoplankton do appear to be defense mechanisms against grazing. Particle size (cell or colony size including the gelatinous sheath) is an important attribute of grazing resistance. The

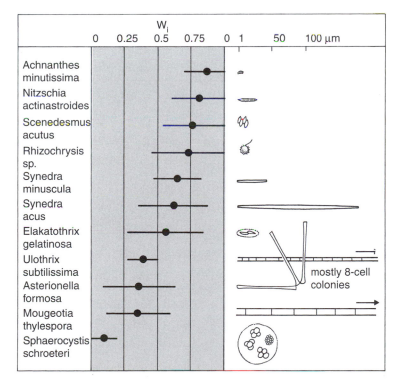

Figure 6.16 Selectivity coefficients (W_1, mean value and standard deviation) for various phytoplankton species in microcosm experiments with *Daphnia magna* as the grazer. The sketches of the algal species are scaled according to actual size; large and species with gelatinous sheaths are grazed poorly. (Data from Sommer 1988c.)

[197]

lower boundary of edible particle sizes for filter-feeding zooplankton is determined by the mesh width of the filtering apparatus (Fig. 4.25, Gliwicz 1980). For filter-feeding cladocerans this is the distance between the secondary spines (setules) of the filtering comb of the third and fourth pair of thoracic legs (cf. Fig. 4.25b). The mesh widths of various species of cladocerans range from 0.16 to 4.2 μm (Geller and Müller 1981). The cladocerans with the finest mesh widths (*Diaphanosoma brachyurum, Chydorus sphaericus, Ceriodaphnia quadrangula, Daphnia magna*) can even filter free-living bacteria, whereas the species with coarse meshes (*Holopedium gibberum, Sida cristallina*) have lower size limits in the range of small phytoplankton. The majority of *Daphnia* species have an average mesh width of approximately 1 μm, which allows some protection for the smallest phytoplankton ("picoplankton," approx. 0.5–2 μm) and the bacterioplankton (Brendelberger 1991). Such small particles are the preferred food of many protozoans such as zooflagellates and ciliates that feed by phagocytosis. Copepods generally eat larger particles than cladocerans, and rotifers have pronounced species-specific differences in food.

The other end of the particle size spectrum that can be eaten is determined by the opening width of the mandibles and/or the opening width of the carapace gape in cladocerans (cf. Fig. 4.26). The upper boundary for small cladocerans and rotifers is about 20 μm and for larger cladocerans and copepods about 50 μm (Burns 1968). This upper size can be achieved by large single-celled organisms (*Peridinium*), body extensions (*Staurastrum*), colony formation (*Pediastrum*), or the production of mucilage (*Planktosphaeria*). Size alone, however, cannot account for the differences in edibility. Particles that exceed the upper size limit in only one dimension (needlelike cells, thin, straight threads) can be eaten when they are oriented lengthwise. Fragile colonies (e.g., *Asterionella, Dinobryon*) may be broken during the process of feeding. Cells that are mechanically stable and that exceed the critical size in at least two dimensions are almost perfectly protected. They may, however, be attacked by specialists. For example, the highly specialized rotifer *Ascomorpha* can eat the large and generally inedible dinoflagellate *Ceratium hirundinella*.

Unlike filter-feeding zooplankton that select their food primarily on the basis of size, raptorial herbivores are capable of selecting on the basis of chemical qualities ("taste"). Cladocerans (with the exception of *Bosmina*) that are offered a mixture of synthetic particles and algae of the same size will ingest the algae and the artificial particles at the same rate. Herbivorous copepods, however, can test each particle and decide whether they will eat it. Figure 6.17 illustrates how the copepod *Eudiaptomus* spp. eats algae, but rejects the artificial particles. The copepods ingest the same plastic microspheres if the particles have been incubated for some hours together with natural algae to "flavor" them. This supports the hypothesis that the particles are discriminated by taste (DeMott 1986). *Eudiaptomus* can even distinguish between living and dead algal cells.

Another form of grazing protection are toxins, produced in fresh water primarily by cyanobacteria (*Microcystis, Anabaena, Aphanizomenon*). The chemical structure of some of these toxins have been described (microcystin, anatoxin, saxitoxin), since blooms of toxic cyanobacteria have created serious problems for drinking water for humans and cattle (Carmichael 1992). Only certain strains of cyanobacteria produce

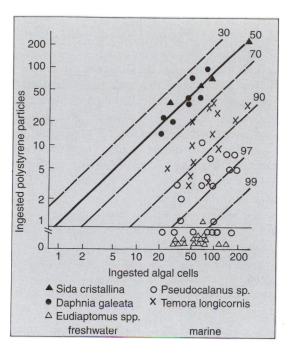

Figure 6.17 Selectivity in the food selection for cladocerans (*Sida, Daphnia*) and copepods (*Eudiaptomus, Temora, Pseudocalanus*). An equal mixture of approximately the same sized artificial particles and algal cells were offered as food. Diagonal lines: percentage of the algae in the food ingested; 50% means nonselective uptake (from DeMott 1988).

toxins, and both toxic and nontoxic strains can occur in the same lake. Zooplankton that eat toxic cyanobacteria either die or have reduced growth. Often they avoid eating such cyanobacteria, apparently by recognizing their taste. By selecting individual particles copepods can reject toxic cells in a mixture and eat the other cells at a normal rate (DeMott and Moxter 1991). Cladocerans that "sort out" the algae and must simply reduce their filtering rates and increase mechanical rejection of all food when they detect toxic cells (Lampert 1987a, Haney et al. 1995).

Size, bulkiness, bad taste, or toxins can protect cells from ingestion. Cells that are ingested still can defend themselves by having a thick cell wall or a stable mucilaginous sheath that prevents them from being digested and allows them to be egested while alive. This protective mechanism is especially common in the sheathed green algae (*Planktosphaeria, Sphaerocystis*) and the cyanobacteria (*Chroococcus, Microcystis*). These algae can even take up nutrients during the passage through the gut. Such forms of phytoplankton often dominate in field experiments with high zooplankton densities (Porter 1977).

6.4.3 Adverse Effects of Inedible Algae on the Zooplankton

Phytoplankton that are resistant to grazing benefit by being protected, but also interfere with the ingestion of edible phytoplankton by filter-feeding zooplankton (Gliwicz 1980). If highly inedible algal species, especially filamentous cyanobacteria and large single-cells (*Ceratium*), are drawn into the filtering chamber and then into the food groove of a cladoceran, they must be removed by cleaning movements of the postab-dominal claw. This also removes edible algal species. The more frequent such inter-fering algae, the more they must be removed and the more frequently the filtering process is interrupted. Cladocerans can reduce the gape between their two carapace halves to avoid collecting large algae. This also reduces the water current and prevents the ingestion of some of the larger edible algae. Large, inedible algae reduce the fil-tration rate, in any case. The otherwise highly competitive *Daphnia* are especially sus-ceptible to such disturbances by filamentous algae and other large species, whereas small cladoceran species (e.g., *Ceriodaphnia*) that have a narrow carapace gape already are much less disadvantaged (Gliwicz and Siedlar 1980).

The grazing pressure by zooplankton leads to an increase in the abundance of grazing-resistant algae toward the late summer growth period (Sterner 1989). The in-crease in grazing-resistant, interfering algae, in turn, provides a competitive advantage for the smaller, less sensitive zooplankton. This leads to a reverse size trend between phyto- and zooplankton. Larger phytoplankton and smaller zooplankton tend to dom-inate in mid- to late summer. It is difficult, however, to clearly demonstrate that the trend for decreasing zooplankton size is caused by algal interference, since increasing predation pressure by size selective fish produces the same effect (cf. Section 6.5.3).

6.4.4 Nutrient Regeneration by Herbivorous Zooplankton

When zooplankton feed on algae they also release a portion of the nutrients contained in the algae. This release can occur in several ways: The nutrients may come from al-gae that were damaged during the feeding process (*sloppy feeding*), from feces, or di-rectly from animal excretion (Lampert 1978). Release of nutrients by zooplankton is often the most important source of *regeneration of dissolved nutrients* during the sum-mer stagnation. Because of this the zooplankton influence the gross growth rates of the phytoplankton as well as their mortality. Assuming that both the grazing rates and nu-trient regeneration increase linearly with zooplankton density, three hypothetical cases can be postulated (Fig. 6.18):

1. If a phytoplankton species is not nutrient limited, the gross growth rate (μ) will not respond to the zooplankton biomass. Since the grazing rate (γ) is increasing, how-ever, the net growth rate (r) must decrease.

2. The gross growth rate of nutrient-limited phytoplankton will increase with the zoo-plankton biomass as a nonlinear saturation curve because zooplankton excretion pro-vides nutrients. If the feeding selectivity coefficient for this phytoplankton is high, the grazing rate will increase faster with the zooplankton biomass than the gross growth rate. There will thus be a nonlinear decrease in the net growth rate with the zooplankton biomass; this decrease is at first slow and then becomes more rapid.

3. If the phytoplankton species is nutrient limited, but has a low feeding selectivity coefficient, at low zooplankton densities, its gross growth rate may increase more rapidly than the grazing rate. This would produce a unimodal curve of net growth rate with a maximum at a specific zooplankton biomass. In this regard, algae that are poorly eaten can even benefit by being grazed (Sterner 1986).

Figure 6.18 shows an example of net growth rates actually measured in a field experiment where the zooplankton densities were manipulated. Promotion of certain phytoplankton species, as in case C, would only be possible if there were also a sufficient quantity of edible food organisms present, to provide a source of nutrients that the zoo-

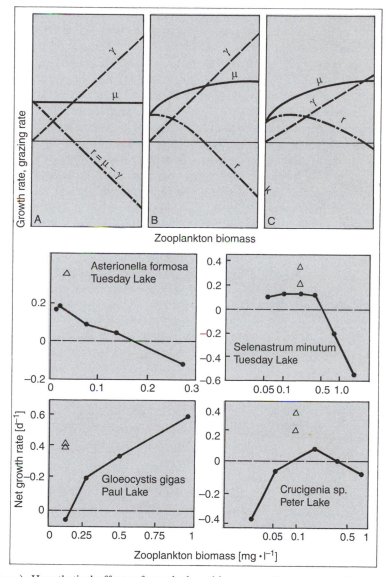

Figure 6.18 (*Above*): Hypothetical effects of zooplankton biomass on the gross growth rate (μ), grazing rate (γ), and net growth rate (r) of phytoplankton. (A) non-nutrient-limited; (B) nutrient-limited, highly edible; (C) nutrient limited, poor edibility. (*Middle and below*) An example of net growth rates based on field experiments with manipulated densities of zooplankton. The net growth rates with unmanipulated zooplankton densities and addition of nutrients (triangles) indicate nutrient limitation for all of the species shown (after Elser 1988).

plankton could excrete. In this way, zooplankton transfer nutrients from the more edible to the less edible organisms.

The redistribution of phosphorus from bacteria to algae by grazing protozoans is a particularly efficient mechanism. Usually bacteria have a higher P content than algae as well as a low k_s for phosphorus, allowing bacteria to compete effectively with the algae when phosphorus is limiting. Figure 6.19 shows the results of a chemostat experiment that gives insights into the interactions between bacteria, grazers, and different algae (Rothhaupt 1992). A chemostat was inoculated with bacteria isolated from a lake, the elongate diatom *Synedra*, and the small *Cryptomonas*. Grazers were not added. The diatom, which has a very low k_s (cf. Fig. 4.20) and the bacteria grew immediately, while *Cryptomonas*, with a high k_s, remained at low numbers. After about 1 week, the diatom remained at a constant density, as it now became silica limited. Bacteria increased further, but *Cryptomonas* was nearly driven to extinction. Before this happened (week three), a heterotrophic flagellate (*Spumella*) that feeds on bacteria was added to the culture as a grazer. It quickly reduced bacterial numbers and reached an equilibrium when the bacteria were at low numbers. As soon as *Spumella* began grazing, *Cryptomonas* also began to grow. The silica-limited *Synedra* did not respond. By the end of the experiment, *Cryptomonas* became the dominant organism in the chemostat. There was no increase in dissolved phosphorus following the addition of *Spumella*, since the released P was immediately incorporated into *Cryptomonas* biomass. This experiment clearly demonstrates how a grazer can alter the outcome of a competition by keeping the superior competitor in check and redistributing the phosphorus.

Zooplankton regenerate different nutrients with different efficiencies. Phosphorus is excreted primarily as dissolved phosphate and nitrogen mainly as ammonium. Both forms are readily available for plants. Silica, on the other hand, is egested as particles of siliceous algal detritus, which sink rapidly out of the epilimnion before the silica can redissolve. Zooplankton feeding activity therefore tends to lower the essential re-

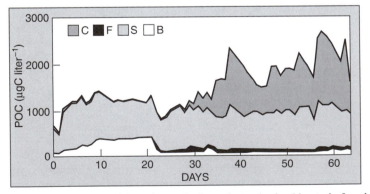

Figure 6.19 Competition of two algal species and mixed bacteria for phosphorus in a chemostat experiment. Cumulative biomass of *Synedra* (S), *Cryptomonas* (C), and bacteria (B). The addition of the heterotrophic nanoflagellate *Spumella* on day 21 results in a reduction of bacterial biomass and facilitates growth of *Cryptomonas* (from Rothhaupt 1992).

source ratios Si:P and Si:N as well as the substitutable resources $NO_3:NH_4$. This changes the competitive conditions for the phytoplankton (cf. Section 6.1.3).

This zooplankton-induced shift in the competitive conditions to the detriment of the diatoms can be demonstrated in culture experiments (Fig. 6.20). Light and dark chemostat chambers were connected in a circulation system. Phytoplankton grew and competed in the light chamber and were continuously transported into the dark chamber containing *Daphnia* that ate the phytoplankton and excreted nutrients. The inflow back into the light chamber from the dark chamber contained excreted nutrients and the algae that were not eaten. As predicted, the Si:P ratio in the recycled water was lowered, and the diatoms were replaced by the green algae. The possibility that this succession

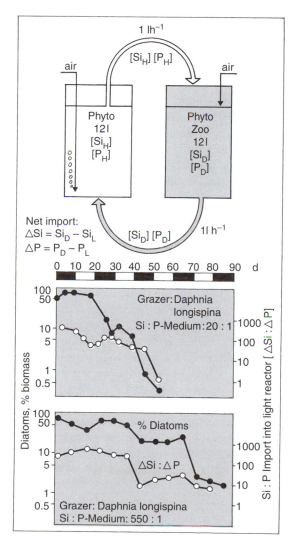

Figure 6.20 Experiments on the interaction between grazing and competition for nutrients. (*Above*) Diagram of the experimental apparatus. (*Below*) Decrease in the Si:P ratio in the outflow in the light chamber and the decrease in the percentage of the diatoms in the phytoplankton biomass (from Sommer 1988).

was a direct effect of selective grazing can be excluded, for the dominant diatom (*Asterionella formosa*) and the dominant green alga (*Mougeotia thylespora*) have almost the same selectivity coefficient (Fig. 6.16). This leaves the shift in the competitive conditions caused by zooplankton excretion as the most plausible explanation.

Tiny diffusion zones with elevated nutrient concentrations form around the zooplankton and their feces. Zooplankton excretion produces small-scale variability or "micropatchiness" in available nutrients. Goldman et al. (1979) speculated that phytoplankton could have brief periods of high rates of nutrient uptake when they encounter such microzones. This could lead to higher gross growth rates than would be predicted from the Monod equation, using average nutrient concentrations. This speculation has been challenged, and there is still no strong experimental evidence to support the growth-promoting effect of nutrient micropatches.

6.4.5 Periphyton

One might assume that the same principles that apply to grazing in the plankton would also apply to grazing on microalgae that grow on submersed surfaces (periphyton, aufwuchs). Unfortunately, we have much less knowledge about the quantitative importance of grazing and its selectivity in this community. There are much greater differences in the body size and taxonomic origin of the grazers on periphyton (from protozoans to snails and insects) than in the plankton. Thus one should not expect as clear a picture as for the selectivity of the herbivorous zooplankton. In addition to protective sheaths and particle size, the edibility of periphytic algae also depends on their ability to adhere to the substrate. The stream snail *Theodoxus fluviatilis*, for example, can only scrape off and break up periphyton with its radula if the algae is growing on a hard, rough surface (Neumann 1961).

In Lake Mephremagong, Canada, Cattaneo (1983) examined the dynamics of the aufwuchs algae on natural and artificial aquatic plants. She found the same pattern for both substrates: In June there was a maximum biomass of periphyton that quickly collapsed with the invasion of oligochaetes and chironomids. Thereafter, the periphyton biomass remained low, possibly due to snails that then appeared. Exclusion experiments with fine-mesh cages suggested that the periphyton biomass was actually kept at low levels by grazers. The species composition of the periphyton also changed in these exclusion experiments. Cyanobacteria dominated in the controls, but green algae and diatoms became abundant in the cages.

The importance of grazing for epiphytic algal communities on macrophytes has been clearly demonstrated by Underwood and Graham (1990). These authors examined epiphytes on the shoots of water milfoil (*Myriophyllum*) from locations with and without grazing snails. Epiphytes on grazed plants were dominated by the oval-shaped diatom *Cocconeis*, which typically grows flat on plant surfaces. Plants from nongrazed locations hosted different diatoms and cyanobacteria. In the laboratory, both epiphyte communities were kept with and without grazing snails. After 35 days, the epiphyte communities changed as predicted. Grazed treatments had a low epiphytic biomass and were dominated by *Cocconeis*, while cyanobacteria were abundant in ungrazed treatments. The differences were due to selective grazing and resistance of some algal species to digestion by the snails. Over 60% of the algal taxa present in the diet of the snails were able to survive gut passage, resettle, and grow. Thus grazing not only changes the epiphyte composition; it also benefits the macrophytes by removing competitors.

Only recently has the importance of grazing by benthic invertebrates (insect larvae and snails) in streams been recognized. It is difficult to quantify these effects as the water current often interferes with grazer-exclusion experiments. Experiments that have succeeded have shown that grazing invertebrates and fish (cf. Section 7.3.3) can keep the periphyton on the rocks in check. Feminella et al. (1989) placed artificial substrates in a stream in California. Some of the tiles were laid on the stream bottom so that they were completely assessable to grazers. Other tiles were slightly raised off the bottom so that crawling grazers were excluded, except for a few drifting mayfly larvae that colonized these tiles. This experiment clearly demonstrated that light controlled the periphyton in the sections of stream that were under tree canopy, but in the sunny, open sections the grazers were the dominant controlling factor.

Artificial streams have been used to study grazing effects on periphyton under more controlled conditions (McCormick et al. 1994). Periphyton biomass was lower in the presence of grazers (snails), but grazers indirectly also affected the periphyton downstream. Bare tiles exposed downstream of the grazed periphyton were colonized by algae at a faster rate than in control plots below ungrazed algae. This was not simply due to higher numbers of drifting algae released by the grazers. Instead, under grazing pressure, the algal cells become physiologically more active and grow faster. Cells exported from grazed reaches may have higher division rates once they attach to a new substrate than cells exported from ungrazed reaches.

Benthic grazers can also be selective. Hart (1985b) reports an example in a small stream in Michigan where he observed sections that varied between slowly growing aufwuchs algae (mainly diatoms) and dense covers of a filamentous cyanobacteria (*Microcoleus vaginatus*). The sections that were free of the cyanobacteria were the feeding territories of the caddisfly larvae (*Leucotrichia pictipes*). When the larvae were removed, the areas quickly became covered with *Microcoleus*. Interestingly enough, the caddisfly larvae did not actually eat the cyanobacteria, for they simply removed them, much like a gardener, to allow for better growth of the diatoms that provided their food.

Dodds (1991) observed in a Montana stream some complicated interactions between filamentous green algae (*Cladophora*), their epiphytes (diatoms), and invertebrate grazers (insect larvae). There was no apparent nutrient competition between filamentous algae and epiphytes. However, when diatoms grew on the algal filaments they increased the frictional resistance for the water flowing past and reduced the current velocity amongst the algal strands. The poor exchange with the water caused a reduction in the photosynthesis of the green algae. Grazers eliminated 75% of the epiphytes and excreted nutrients that were available to the green algae. In this case the green algae benefitted from the activities of the grazers in a fashion similar to the grazing-resistant phytoplankton and the zooplankton.

[205]

6.5 PREDATION

6.5.1 Components of Foraging

For a predator successful hunting represents an increase in fitness through higher reproduction rates and improved chances for survival of its offspring, whereas for the
prey it represents decreased fitness through mortality. Predators and prey have coe-
volved complex characteristics and tactics with which they can effectively capture prey
or be protected against predation. The process of foraging can be characterized by four
successive phases (Gerritsen and Strickler 1977):

1. Predator and prey encounter one another.
2. The predator recognizes the prey and attacks it.
3. The predator captures the prey.
4. The predator ingests the prey.

This sequence can be presented in the following way:

Movement of prey and predator relative to one another

$\downarrow P_E$

Encounter

$\downarrow P_A$

Attack

$\downarrow P_C$

Capture

$\downarrow P_I$

Ingestion

At each level there is a particular *probability* (P) that the process will continue. It then
follows that the probability for a successful predatory interaction (P_{SI}) is:

$$P_{SI} = P_E P_A P_C P_I$$

The same relationship also applies to the prey except that the success of the prey increases with the decrease in probability. The likelihood that the next step will follow
depends on both the adaptive abilities of the predator and prey that occur at each step.
By dividing the entire process of prey capture into subcomponents, one can systemat-
ically examine the mechanisms of adaptation.

In order to have an interaction, predator and prey must first encounter one another.
The *probability of encounter* depends on the mean velocity of movement of both preda-

tor and prey. It would be most beneficial for the prey if it remained totally still, but since animals must also search for food and mates, this is not possible. The predator has two choices. It can either wait quietly and ambush the prey as it passes by, or it can actively search for the prey. Examples of ambush predators in freshwater are the pike, *Chaoborus* larvae, and dragonfly larvae. Active searching-type predators include perch, predatory copepods, and certain stoneflies.

A predator must encounter a prey and also detect its presence. Once again, the characteristics of the prey and the predator are important. We can define a *detection radius* around the predator within which it can recognize the prey. The larger the detection radius, the greater the likelihood that the prey will be pursued. The detection radius depends primarily on the sensory capabilities of the predator. For visually orienting predators such as carnivorous fish, not only is their visual acuity important, but also characteristics of the prey such as visibility, size, coloration, and swimming movements. Environmental conditions may also affect the detection radius. For example, an visually orienting predator would have a small detection radius in turbid water.

Once the prey has been detected, the predator must decide whether it will attack. Here learning can play a role. For example, fish have a wide detection radius for conspicuously colored water mites, but they quickly learn not to eat them because of their bad taste (Kerfoot 1982).

Even after detection, the prey still must be captured, subdued, and eaten so that there is still a chance that it may escape. It can, for example, flee at the last moment or be protected by armored structures that interfere with the predator's ability to handle it. In freshwater, various species of rotifers (Fig. 6.21) that come in the vicinity of a predatory copepod have different behaviors. The soft-bodied, slow *Synchaeta* has little chance of escaping, while *Polyarthra* has a number of paddlelike processes that can suddenly beat synchronously, propelling the rotifer forward in a rapid jump, allowing it to escape. *Filinia* spreads out three long spines that make it difficult for the predator to handle, and *Keratella* depends on a powerful spiny armor for protection. The most decisive factor in determining whether the prey will escape is the size relationship of predator and prey.

6.5.2 Selectivity

All components of the foraging model restrict the types of prey organisms that can be eaten by the predator. If the stomach contents of a predator are compared with the composition of the possible prey animals in its environment, there are great differences; thus predators are selective. There are a number of *selectivity indices* (see Box 6.2) that allow one to characterize the "preference" of a predator for a particular prey species. These indices are all based on the relative frequency of a prey species in the environment and in the diet of the predator.

Based on the method of prey detection employed, aquatic predators can be divided into two groups with completely different patterns of prey selection. Vertebrates, in freshwater essentially fish and occasionally salamander larvae, use visual orientation in most cases. A critical step in the foraging process is the detection of the prey. In

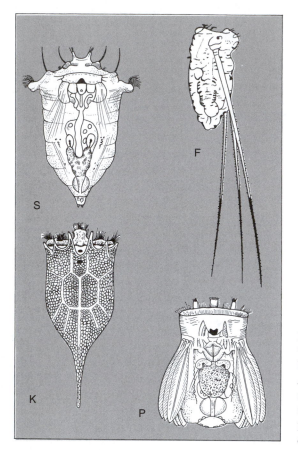

Figure 6.21 Rotifers with various "defense mechanisms." The soft-bodied *Synchaeta* 9S) cannot escape. *Filinia* (F) can extend three long spines. *Polyarthra* (P) can make rapid jumps with its paddle-like appendages to escape predators. *Keratella* (K) is protected by an armor.

piscivorous fish, the size of the mouth limits the maximum size of prey fish that they ingest. Planktivorous and benthic feeding fish are so large (with the exception of young stages) that the handling of prey is not a problem. For a fish, larger prey are visible at a greater distance. Thus the probability of being eaten is greater for larger prey. As a result, vertebrate predators eat the larger items from a mixed offering of food.

Invertebrate predators such as carnivorous zooplankton and insect larvae orient with mechano- or chemosensors rather than visually. Here the limiting step for success is not the visual detection, but rather the handling, of the prey. Invertebrates are relatively small and therefore require prey that are smaller than themselves. They normally select smaller objects from a spectrum of prey.

In this way vertebrate and invertebrate predators have very different effects on the community of the prey. Vertebrates prefer the relatively large invertebrate predators, further compounding the effects of predation on plankton community structure (cf. Section 6.5.6).

| BOX 6.2 | QUANTIFICATION OF SELECTIVITY |

This and the following chapter demonstrate the importance of selective feeding by predators and grazers on the community structure by imposing differential mortality on the prey. An appropriate measure is needed to express quantitatively such differential mortality. A number of indices have been developed, all of which are based on a comparison of proportion of a specific prey category or particle class relative to all available prey found in the food consumed and in the environment. To determine, for example, whether a particular size class of food is preferred by a predator, one must first estimate what proportion or percentage this class is of the entire prey population and then the proportion of the same size class in the gut of the predator.

Using the relative proportion of a particular class in the environment (p) and the proportion in the gut (r), the electivity index (E) can be calculated (Ivlev 1961):

$$E = (r - p)/(r + p)$$

E can vary from -1 to $+1$, where -1 to 0 means negative selection of the prey item considered and proportionally fewer of the prey type were eaten. 0 to $+1$ means positive selection and that this prey is eaten more frequently than other classes. When the proportion of a prey type is the same in the stomach and in the environment, there is no selection, and $E = 0$.

The classical Ivlev index is not independent of the relative proportion of the prey species and therefore cannot be used to test whether the relative frequency of prey has an effect on the prey selection by a predator (frequency-dependent selection). Jacobs (1974) recommended a modified index (D) that does not have this disadvantage and has thus been widely used. D also ranges from -1 to $+1$, with 0 indicating nonselective feeding:

$$D = (r - p)/((r + p) - 2rp)$$

Another often-used index (α) developed by Chesson (1978) uses i as the food class considered and j for all other classes. Then the proportion of the item of interest in the food is r_i or r_j and the proportion in the environment is p_i or p_j. The electivity index for class i is then:

$$\alpha = \frac{r_i/p_i}{r_i/p_i + r_j/p_j}$$

α ranges from 0 (negative selection) to 1 (positive selection) and nonselectivity is 0.5.

Example

There are three size classes of food organisms available to a fish. A sample of the environment shows:

Class	Number	Proportion
A	600	0.60
B	250	0.25

BOX 6.2	(*Continued*)

C	$\frac{150}{}$	0.15
Total	1000	

In the stomach of the fish we find:

Class	Number	Proportion
A	15	0.47
B	8	0.25
C	$\frac{9}{}$	0.28
Total	32	

The stomach contents give the impression that the fish favors eating class A. When we calculate the indices according to Jacobs and Chesson, we find:

$$D_A = \frac{0.47 - -0.60}{} = -0.257$$

$$\alpha_A = \frac{0.47/0.60}{0.47/0.60 + 0.53/0.40} = 0.372$$

Correspondingly, we get for B: $D_B = 0$; $\alpha_B = 0.5$ and for C: $D_C = 0.376$; $\alpha_C = 0.688$. Food type C rather than A was preferred, and B was eaten without selection and simply chosen in proportion to its relative abundance.

6.5.3 Vertebrate predators

Planktivorous fish

Almost all fish go through a phase when they eat zooplankton, even those that feed on benthos or become *piscivores* as they grow up. Some fish remain *planktivores* throughout their life. Pelagic coregonids (white fish), alewife (*Alosa*), and smelt (*Osmerus*), for example, feed almost exclusively on zooplankton, even as adults.

Most planktivorous fish orient visually, catching individual prey. There are, however, a few fish such as the adult alewife (*Alosa pseudoharengus*) and the gizzard shad (*Dorosoma cepedianum*) that feed by filtering (Drenner et al. 1982). These fish feed by pumping water through their mouths and retaining the zooplankton with their gill rakers. The size of the zooplankton collected depends on the distance between the gill projections. Larger particles are more easily retained than small particles that may pass through the gill filters. This results in a selection for larger zooplankton.

Visually orienting fish can also retain only those particles that are larger than the gill raker width. Here the selection process simply occurs sooner. A prey must be first detected. The *reaction distance* can be used as a measure of the detectability of the prey. This is the distance at which a fish reacts by attacking the prey. A larger reactive distance indicates a more visible prey. The reactive distance depends on both the size and type of prey. Large prey can be located at a greater distance than small prey. Prey recognition actually depends on the image formed on the retina of the fish eye. A small zooplankter that is close to the fish can appear to be the same size as a large zooplankter that is farther away; the zooplankter is thus said to have an "apparent" size. Contrast with the surroundings and type of movement are additional factors involved in prey detection. Prey that have striking colors and movements are more easily detected (O'Brien et al. 1985).

The reaction distance is shorter in turbid or highly colored water. Light intensity is extremely important. The reactive distance quickly decreases when the light falls below a threshold intensity of several lux. Fish must then be very close to their prey in order to locate them (Fig. 6.22). For the zooplankter, this means it has a lower probability of ebing eaten. Different fish species have different threshold light intensities for feeding. Some fish species have low light thresholds and can feed efficiently under moonlight conditions. For this type of fish, all these factors lead to easier detection of larger prey.

The capture success of filtering and visually feeding fish depends also on the ability of the prey to escape. Both feeding modes involve the sucking of water containing the prey into the mouth. Some zooplankton, such as copepods and the cladoceran *Diaphanosoma*, can make powerful jumps against the current to escape capture. The sucking by the fish can be simulated by siphoning water from an aquarium through a tube and observing how the zooplankton behave when they are caught in the suction. Figure 6.23 gives an example of such an experiment. For all the zooplankton tested, they were caught more efficiently when they were near the tube. At a distance of 8 mm from the tube, all of the *Daphnia* were sucked out of the aquarium, whereas only 20% of the copepod *Diaptomus* were captured at a much shorter distance (3 mm). The end result of such a manipulation is that the *Daphnia* are sucked out of the aquarium first and the copepods last. One can predict from these results that it should be easier for fish to capture *Daphnia* than copepods. This, in fact, is the case (Brooks 1968).

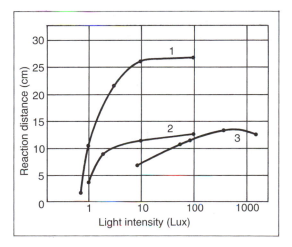

Figure 6.22 The reaction distance of fish for 2 mm large *Daphnia pulex* at various light intensities. 1, sunfish (*Lepomis macrochirus*); 2, white crappie (*Pomoxis annularis*); 3, brook trout (*Salvelinus fontinalis*) (from O'Brien 1987).

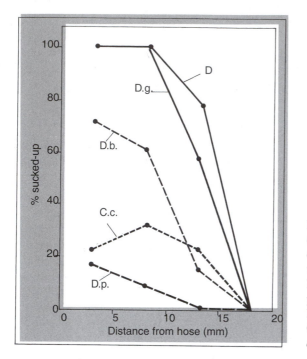

Figure 6.23 Experiments on the escape behavior of zooplankton. Percentage of the zooplankton sucked up by a hose at a fixed distance (from Drenner et al. 1978). D, dead zooplankton; D.g., *Daphnia galeata*; D.b., *Diaphanosoma brachyurum*; C.c., cyclopoid copepods; D.p, *Diaptomus pallidus*.

When prey animals are so plentiful that searching time becomes unimportant, energetically, fish should concentrate on obtaining the best food parcels. This effect is seen with young perch, which reach a feeding plateau with increasing concentration of prey (Fig. 6.24a and b, cf. Section 4.3.2). The fish can eat only a certain number of *Daphnia* per unit time, irrespective of their size. Far fewer *Cyclops* are captured during this time, simply because they are better at escaping. The rate of energy consumed is much greater if the fish concentrated on the large *Daphnia*, since a large *Daphnia* contains much more energy than a small *Daphnia*. It is beneficial for the fish to eat the small *Daphnia* only when the density of the large prey becomes so low that the search time is too long. It can be assumed that there has been an evolution in the direction of such optimal feeding behavior, since greater energy consumed generally results in greater fitness.

Fish sometimes switch between filtering and particulate feeding strategies, especially under poor light conditions (Jansen 1980). All morphological, sensory, and behavioral factors lead to the same conclusion: Planktivorous fish should selectively eat the largest and most easily captured zooplankton. This hypothesis has been confirmed many times by field studies. One can find further support for this idea by comparing the size distribution of *Daphnia* in the lake with those found in the stomachs of young perch, rainbow trout or bleak (cyprinids). The larger *Daphnia* are over-represented in the stomachs (Fig. 6.25).

Planktivorous fish can also become specialists. As long as *Daphnia* are present, coregonids have hardly any copepods in their stomachs, even when they are abundant.

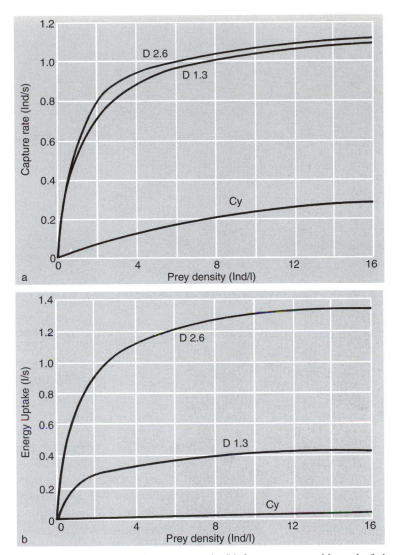

Figure 6.24 Capture rate (a) and the resulting energy gain (b) for a one-year-old perch, fed small (1.3 mm) and large (2.6 mm) *Daphnia* (D) and the copepod *Cyclops* (Cy) (after Persson 1987).

In an experiment with mixed zooplankton (Brooks 1968), the *Daphnia* were quickly eaten, followed by the large copepods and then finally the small copepods. Such pronounced specialization cannot be completely explained by the visibility and escape behavior of the prey and is probably related to the learning ability of fish. The fish knows after a few trials that when it sees both a *Daphnia* and a copepod it will be more likely to catch the *Daphnia*. This type of learning by further strengthens the selective process.

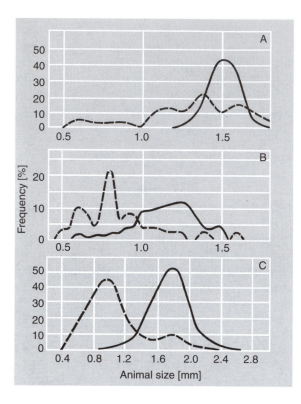

Frequency [%]

Animal size [mm]

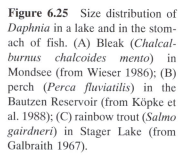

Figure 6.25 Size distribution of *Daphnia* in a lake and in the stomach of fish. (A) Bleak (*Chalcalburnus chalcoides mento*) in Mondsee (from Wieser 1986); (B) perch (*Perca fluviatilis*) in the Bautzen Reservoir (from Köpke et al. 1988); (C) rainbow trout (*Salmo gairdneri*) in Stager Lake (from Galbraith 1967).

These general rules apply to fish that are large relative to their prey size, which for young fish would be after they are several centimeters long. There are also ontogenetic changes in food preference. Larval fish and extremely small young fish are restricted to small-sized food by the size of their mouth. Small fish larvae may at first depend entirely on the smallest zooplankton, such as protozoans and rotifers. As they get larger, their food preference shifts next to stages of copepods and finally to cladocerans. Fish, during their early development period follow the rule of eating the largest prey that can be captured, rather than the largest prey available. This is a relatively brief phase in the life of fish and therefore does not affect the zooplankton as much as the feeding by larger fish that persists throughout the summer.

The size-dependent response of fish can be nicely demonstrated with the predatory cladoceran *Bythotrephes cederstroemi*, which in recent times invaded the Laurentian Great Lakes. *Bythotrephes* has a body size of about 2 mm, but carries a very long tail spine of about 10 mm. The tail spine is an effective defense device against young fish. Small rainbow trout (4–8 cm) require 8 times as long to handle a spined prey, compared to a despined one. Often the small fish attack the spined *Bythotrephes*, but then reject them (Barnhisel 1991). Larger planktivorous fish, however, have no difficulty handling the spiny prey and even prefer them over other zooplankton because of their large size and high visibility. Alewife more than 16.5 cm long in Lake Ontario were more efficient in finding *Bythotrephes* than limnologists. The fish had *Bythotrephes* in their stomachs, although the density in the lake was too low to find specimens regularly in the plankton tows (Mills et al. 1992).

Benthic and piscivorous fish

Some fish species change to feeding on benthos as they get older. They search for prey on plants and stones or in the sediments. Stream fish that do not eat insects that fall into the water must depend on macroinvertebrates. *Benthivorous* fish are selective and prefer the largest prey, although the concept of reactive distance is not so easy to apply here, as prey in the benthos can hide. Even macroinvertebrates are small enough that they are potentially edible for all fish. If they are small enough, however, they can avoid detection. Brook trout, for example, were able to change the size structure of invertebrate communities in streams. The fish reduced the abundance of large epibenthic and drifting invertebrates, but did not affect animals smaller than 2 mm and those living in rock intersticies (Bechara et al. 1993).

The larvae of the salamander *Dicamptodon* in streams behave similar to fishes. Larger prey are overrepresented in their guts compared to abundance in the stream. Direct observations showed that the reactive distance was greater for large prey (i.e., they are easier to detect). Although handling time increased with increasing prey size, it was still profitable for the larvae to select large items as the ratio of prey biomass to handling time was higher for large prey (Parker 1993).

This example indicates that selection of the largest food depends not only on whether the prey can be detected, but also on whether it is beneficial to spend a limited amount of time searching for larger prey. This can explain, for example, the size selection of roach (*Rutilus rutilus*) that eat the zebra mussel (*Dreissena polymorpha*). These mussels are an excellent food, but fish must break the shells with their pharyngeal teeth. The benefit derived depends on the energy content of the mussels, which increases with increasing size, and the cost of breaking the shell, which also increases with the size of the mussel. The result is that small mussels (<10 mm) are not eaten, even though they are the most numerous. Roach only begin to eat mussels when they themselves are over 16 cm (Preijs et al. 1990). It is only after this size that it pays to break the mussels. The costs of handling the mussels is relatively small for larger fish, so that they select the largest mussels they can eat, which depends on the size of their mouth (approx. 60% of the mouth size). It simply does not pay the large fish to search for small mussels. For the small fish that could eat the small mussels, based on their mouth size, breaking the shells is too difficult. The end result is that the small mussels are not eaten at all.

There is a similar optimal prey size for fish-eating (piscivorous) fish. It would not be profitable, for example, for a large bass to chase after a small minnow. On the other hand, this might be a worthwhile prey for a small bass. The largest prey are determined by the mouth size of the predator. The large mouth bass (*Micropterus salmoides*), for example, prefers forage fish that are about one-third its own size (Werner and Hall 1988).

Limitations and tradeoffs

A fish cannot always search for the optimal prey, for it may also be at risk while search-ing for food. A planktivorous fish would benefit most if it searched for food in bright light where its reaction distance to prey would increase. At the same time, however, it would be in greater danger of being eaten by predatory fish and fish-eating birds that also forage visually. This situation requires a risk assessment. However, each fish can-not estimate the actual risk of being eaten. Instead, during the course of evolution the species has developed a behavioral strategy that is on the average optimal.

Bohl (1980) measured the distribution of young cyprinid fish in lakes in Bavaria, Germany, with an echosounder. Surprisingly, the young fish entered the pelagic zone only at night when they ate zooplankton. During the day they stayed in swarms near the shore, but as soon as it became dark the schools dissipated and the young fish be-gan to feed in the open water.

Young fish are better protected from predatory fish when they are in schools. There would be a large risk for them to stay in the open water during the day. Better food availability during the day could possibly compensate for this greater risk. The zoo-plankton prey are indeed more visible during the day, but at that time they are deep in the lake and not in the surface waters (cf. Section 6.8.4). Thus the prey concentration is low during the day. At night, the zooplankton migrate into the surface water. Ap-parently it is more beneficial for the fish to hunt at night when the prey densities are high, even though the visibility is poor. This, of course, requires that the eye of the fish be able to adapt to very low light intensities (cf. Fig. 6.22).

A striking example of the effects of size-selective predation by piscivores has been reported by Bråbrand and Fåafeng (1993). A Norwegian lake was dominated by roach (*Rutilus rutilus*) that are planktivorous as juveniles, but benthivorous as adults. Juve-nile roach lived in the pelagic region as well as in littoral areas of the lake. Larger size classes were littoral. After the lake was stocked with pikeperch (*Stizostedion luciop-erca*), a piscivorous relative of the walleye, young roach, which are easy prey for the pikeperch, disappeared from the pelagic region, but numbers did not change in the lit-toral. Protected by their size, large roach, in contrast, moved to the open water. In the littoral, however, the young roach became available to the predacious perch (*Perca flu-viatilis*). As a consequence, the dominance in the littoral fish community shifted from >95% roach to >50% perch.

Just as the risk of being eaten changes during the life of a fish, its behavior can also change. Werner and Hall (1988) were able to explain the behavior of sunfish (*Lepomis macrochirus*) in Michigan lakes by estimating the profit and risk components. Sunfish build their nests in the littoral. After the fry hatch they move into the pelagic zone where they feed on zooplankton. As soon as they grow to a length of 12–14 mm, they move back into the littoral for several years to live among the plants and feed on lit-toral animals. When they reach a length of about 8 cm, they return to the open water and feed on zooplankton once again, at first in the water column above the plants and then later in the real pelagic region. The authors demonstrated that plankton was the most profitable food in every stage of their life. Intermediate-sized sunfish grew bet-ter in cages suspended in the open water than in the littoral, where they are normally found. Why would they return to the littoral for several years? The answer is that they are prey of the large-mouth bass that feeds in the open water. The bass is a predator that is limited by the size of its mouth and eats prey that are from 15 to 40% its own length. It becomes a piscivore and a threat to sunfish when it is about 35 cm long. However, there are relatively few large bass longer than 20 cm in the lake. Apparently, when a sunfish in the littoral reaches a size of 8 cm, the profitability of eating plankton outweighs the risk of being eaten by the few predatory fish; so they return to the open water.

6.5.4 Invertebrate Predators

Types of predators

In addition to plant eaters, there are many species of invertebrates that are also omnivores and predators. These often have dramatic effects on the populations of their prey. Their effect on communities, however, is completely different from fish. Visual orientation is generally unimportant (except for a few insects), for aquatic invertebrates do not have as highly developed eyes as the vertebrates. Invertebrate predators usually detect their prey by mechano- or chemoreception. Also, invertebrate predators are relatively small compared to fish. Their food selection depends, therefore, on their ability to capture the prey. The only invertebrate predators that can capture prey larger than themselves are those that inject poisons (e.g., predacious diving beetle *Dytiscus*). Fish and amphibians are the only important vertebrate predators in freshwater, whereas the invertebrate predators consist of many different taxonomic groups with a correspondingly diverse methods of prey selection and prey capture.

The predatory cladocerans *Leptodora* and *Bythotrephes* are typical invertebrate predators in the pelagic region. Their thoracic appendages are shaped into a capturing basket. They move through the water collecting mainly small cladoceran prey into their thoracic basket. Another predatory cladoceran, *Polyphemus,* is restricted to littoral regions, where it forms dense swarms. *Polyphemus* catches small zooplankton, using its enormous compound eye for visual orientation and prey detection. The larvae of the phantom midge *Chaoborus* are ambush predators. They can remain stationary in the water column by means of modified tracheal "swim bladders." As a prey swims by, *Chaoborus* detects it by mechanoreceptors and suddenly attacks, grasping the prey with its mouthparts (Swift and Fedorenko 1975). Cyclopoid copepods transform from herbivores to carnivores during their larval development. In addition to eating algae, the fourth and fifth instars of many cyclopoid copepods eat rotifers and small crustaceans (nauplii, copepodid stages, cladocerans). Occasionally they also attack larger prey, such as fish fry. One can see from high-speed cinematography that copepods can track their prey in the water by detecting trails of microturbulence produced by the prey (Strickler 1977). The calanoid copepods are generally considered to be herbivorous, but there are also predatory (e.g., *Epischura*) and omnivorous species within the calanoid copepods. In fact, more and more there are reports of "herbivorous" copepods (*Diaptomus*) also eating small zooplankters (rotifers, protozoans) (see also Section 6.5.5). One of the most studied predatory rotifers is *Asplanchna* (cf. Section 6.5.5). It catches other rotifers by means of extendible, pincherlike trophi. Water mites (*Piona*) are often important predators in tropical waters.

Important invertebrate predators in streams include numerous insect larvae, such as caddisflies without cases (Trichoptera), stoneflies (Plecoptera), and dragonflies. The larvae of the predacious diving beetle (dytiscids) and the dobsonfly (*Sialis*) are easily recognized as predators by their powerful, pincherlike mandibles. The backswimmer *Notonecta*, which hunts from just below the water surface, can have an important effect on the prey populations in certain lakes and ponds. Leeches and flatworms (turbellarians) are often important invertebrate predators in the benthic community.

Food selection

The most important step for visually hunting predators is the detection of prey. In contrast, the success of invertebrate predators depends more on the *probability of encounter* and *handling* of the prey. The predator finds larger prey more easily, but the capturing becomes more difficult and the handling time longer for larger prey, thereby reducing the capture efficiency (the fraction of prey actually eaten per capture attempt). This relationship has been especially well studied for the phantom midge larva *Chaoborus*, which feeds on *Daphnia* (Pastorok 1981). Figure 6.26 summarizes the com-

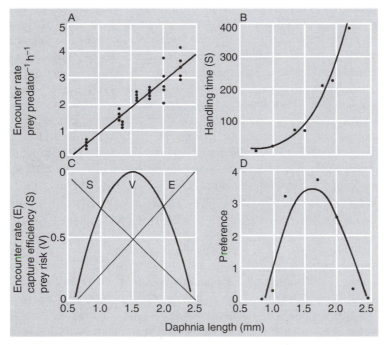

Figure 6.26 The basis of preference of fourth instar *Chaoborus* larvae for medium-sized *Daphnia*. (A) Encounter rate increases with the size of the prey. (B) Handling time increases with prey size. (C) Model predictions of the risk to the prey (*V*), estimated from the encounter frequency and the capture efficiency. The model predicts a preference for approx. 1.5 mm *Daphnia*. (D) Experimental measurement of preference (from Pastorok 1981).

ponents of the predation by *Chaoborus* larvae. The probability of encounter increases linearly with size of the *Daphnia* (A), but the handling time increases exponentially (B). There is an almost linear decline in capture efficiency with the size of *Daphnia*, since the capture efficiency decreases with the handling time. The product of encounter rate times the capture efficiency is a measure of the prey's risk of being eaten. Prey animals of intermediate size have the highest predation risk (C). Experiments with prey selection using *Chaoborus* and *Daphnia* have actually found a strong preference for the intermediate size classes (D) (cf. Section 6.5.5).

This predation model can also be applied to benthic predators. The predatory stone-fly larvae (*Hesperoperla*) select prey in accordance with their own body size: Small stonefly larvae prefer small prey and large larvae prefer larger prey. Allan et al. (1987) were able to demonstrate experimentally that both the probability of encounter and handling time increase with prey size, relative to the size of the stonefly larvae. The relative importance of these two processes can vary. The size of the prey alters the handling time, but has little effect on the probability of encounter, if the prey are sessile, such as blackfly larvae (*Simulium*). The encounter frequency increases with size for prey that are highly mobile, such as mayfly larvae. Predators that are offered a mayfly and a *Simulium* larva simultaneously find the *Simulium* larvae less frequently than the mayfly larvae, but are more successful in capturing the sessile larvae that do not try to escape. The result is a preference for the *Simulium* larvae. The point of this example is that the "preference" of a predator for a certain prey need not be a conscious recognition, but rather may be simply a result of the characteristics of prey and predator and the difficulties encountered in handling the prey.

6.5.5 Defenses of the Prey

Risk reduction
Predation is a very strong selective factor. One might expect that prey have thus evolved many defense mechanisms that lower their mortality and thereby increase their fitness. The type of adaptation depends heavily on whether the predator is vertebrate or invertebrate.

Defense mechanisms can be aimed at any of the steps of prey capture (Section 6.5.1). At the first step, the detectability of the prey can be reduced. For visually orienting predators, the prey benefit by becoming as "invisible" as possible. Benthic animals can hide or take on adaptive coloration, mechanisms not available to plankton. The plankton can, however, become transparent or move into the deep water where they cannot be seen. The predatory cladoceran *Leptodora* is an extreme example of transparency; all visible parts (eye, gut) are strongly reduced. It would be impossible for an animal of this size (10 mm) to survive with planktivorous fish if it were not so transparent.

Lakes with visually orienting predators usually have zooplankton that are transparent and inconspicuous, whereas in temporary ponds that lack fish the zooplankton are highly pigmented. There are several color variations of the calanoid copepods (e.g., *Diaptomus leptopus*). They can be pale, bright red, or blue. The red pigment is a carotinoid, and the blue pigment is a carotinoid–protein complex. The pigment protects the copepods against damaging light. Unpigmented copepods die more quickly than pigmented copepods when they are placed in direct sunlight. In nature, unpigmented copepods can avoid such effects by migrating into deep water. The most brightly colored copepods are found in alpine lakes, where solar radiation is especially intense. Given an opportunity, fish feed preferentially on the colored copepods. Colored copepods are thus relatively rare and found only where there are no fish or where the protection against light outweighs the disadvantages of increased predation mortality (Byron 1982).

Hairston (1981) presents an interesting example. Two nearby lakes in Washington State differed in their salinity. Soap Lake had a salinity (17%) much higher than Lake Lenore. The copepod *Diaptomus nevadensis* occurred in both lakes. In Soap Lake it was bright red, but in Lake Lenore it was almost colorless. At first glance, one might attribute the color difference to differences in an abiotic factor (salinity). The actual reason for the difference is the presence of salamander larvae that hunt copepods visually in Lake Lenore. The high salinity in Soap Lake excluded the salamander larvae and allows, indirectly, the copepods to form protective pigments.

Invertebrate predators detect their prey by mechanoreceptors. Prey that produce little turbulence are more difficult to detect. Small cladocerans (*Bosmina* and *Chydorus*) take advantage of this fact when they are pursued by predatory copepods (*Cyclops*) by tucking their antennae into the carapace opening and "playing dead man." This allows them to sink slowly, producing little turbulence. Time-lapse photographs show that a *Cyclops* has great difficulty finding a *Bosmina* that is "playing dead" (Kerfoot 1978).

Filter-feeding animals produce considerable turbulence, especially when feeding. This creates an optimization problem. Greater filtering activity means greater food intake, but also increases the likelihood of being detected by an invertebrate predator. This appears to be the reason why herbivorous copepods reduce their filter-feeding activity when predatory copepods are present. Filter-feeding copepods probably likewise recognize their predators by turbulence and possibly chemical communication (Folt and Goldman 1981).

Once detected, the prey may still survive by escaping. There are various ways of solving the problem of escape, such as fleeing before contact has been made or preventing successful handling by the predator (cf. Fig. 6.21). Interference with handling is very effective against invertebrate predators. This can be achieved by simply increasing body size or by interfering structures, such as spines or armor. The soft-bodied rotifer *Conochilus unicornis* forms colonies made up of many animals that are attached at the base in a gelatinous ball and thus are much larger than a single animal. The cladoceran *Holopedium gibberum* employs a similar mechanism by encasing itself in a thick gelatinous mantle that effectively increases its volume many times. There is only a narrow opening for the carapace and antennae. *Chaoborus* larvae that attack a *Holopedium* will be unsuccessful, even if the animal is an ideal sized prey without its gelatinous covering (Stenson 1987).

The gelatinous mantle of *Holopedium* is an excellent example of the new type of thinking now established in ecology that emphasizes evolutionary concepts and biological interactions. The classical interpretation of the "meaning" of the gelatinous covering was that it reduced excess weight to prevent sinking. Actually, the gelatin does have a specific density of almost 1.0, which reduces the excess weight of the entire animal (cf. Section 4.2.6). To test the hypothesis, the thickness and sinking speed of animals with and without the gelatinous covering were measured. Narcotized animals without gelatin sank at a speed of 0.31 cm/s at a density of 1.014, whereas animals with gelatin sank at 0.22 cm/s, since their density was only 1.0015. By only considering the abiotic effects, one could only conclude that the advantage of the gelatin was that the animal would require less energy to stay suspended in the water. No attention was paid, however, to the energetic costs of producing the gelatin and its disadvantages to swimming and filter feeding, which raises the question of whether there really would be a net energetic advantage. In recent years there has been an emphasis on the interactions of the individual with its biotic environment. It was only after the mechanisms of prey selection and the importance of invertebrate predators on zooplankton became known that a more convincing interpretation of the gelatinous covering as a defense against predators was developed.

Unlike zooplankton, benthic organisms can hide or be cryptic. Their color often matches the bottom, or they may use material from the bottom to build cases. Damselfly larvae actively hide behind leaves or stems of aquatic plants on the opposite side of their enemies. Many stream organisms stay under rocks during the day and only come to the upper side at night to graze on the algae that grow there. This behavior protects them from visual predators such as fish. Avoidance of visually hunting predators thus causes a diel cycle of activity that is, for example, also reflected in the organismic drift (cf. Section 4.2.5). Flecker (1992) showed that the timing of the drift has been selected for by fish. Insect larvae in the Andes showed diel drift cycles in streams inhabited by fish, but the cycle was lacking in fishless streams. The diel cycle (response to light) was genetically fixed as it persisted when fish were experimentally excluded. A similar genetically fixed antipredator response was found in mayfly nymphs in New Zealand. Introduced brown trout (*Salmo trutta*) have eliminated endemic fish species (*Galaxias*) from many streams in New Zealand. Trout and *Galaxias* differ in their feeding behavior; trout are visual hunters that prominently feed on drifting insects, whereas *Galaxias* are active at night. Although the trout have only recently been introduced, mayflies in trout streams show their greatest activity and drifting at night, which is an advantage in the presence of visual hunters. However, they maintain this pattern in the presence of *Galaxias*, or in the absence of any predator, which is a disadvantage. In contrast, mayflies from fishless streams or streams with *Galaxias* show no nocturnal drift maximum in the absence of fish. These mayflies are also more likely to leave food patches at night when *Galaxias* are present and in the day when there are trout. Hence, their behavior is more flexible (McIntosh and Townsend 1994). There is evidence that mayflies from streams with native salmonids in other parts of the world evolved more flexible predator avoidance strategies. They can change their diel activity cycle in response to the presence or absence of the predator. This is the most profitable strategy, since nocturnal activity is no advantage in the absence of visual predators and causes reduced growth due to a restricted time budget for feeding (Peckarsky et al. 1993). The insects may detect the presence of fish predators by a chemical cue (Cowan and Peckarsky 1994) (cf. Section 6.5.5).

Benthic invertebrates that live in soft substrates can burrow into the sediments. For example, when they are small, chironomid larvae (*Stictochironomus*) forage near the sediment surface in lakes, but as they get larger they burrow deeper. During the fourth larval instar, they live below 2 cm deep in the sediments, where they are out of reach of bottom- feeding bream (Van de Bund and Groenendijk 1994).

Some aquatic beetles, water bugs, and mites also use chemical defenses; they either have poison glands or are distasteful. Fish quickly learn to avoid these organisms (Kerfoot 1982).

Other defense mechanisms must be used against invertebrate predators in the benthos, since they orient either by mechanoreception or chemoreception. Small organisms may retreat into the interstitial spaces in the sand where the larger predators cannot follow. Also, cases such as those used by caddisflies protect against biting predators. Mainly, however, there are adaptations that assist the prey in a timely escape from the predator.

Mayfly larvae recognize predatory stonefly larvae by mechanical detection, just as they are also recognized by the stoneflies. They escape by running away or by releasing themselves into the current to drift downstream. The fact that stonefly larvae selectively eat certain mayflies is not because they prefer them, but rather due to the differing abilities of mayflies to escape (Peckarsky and Penton 1989). The effects of invertebrate predators in streams may be caused by different mechanisms. For example, Lancaster (1990) showed that the presence of predatory stonefly nymphs (*Doroneuria*) reduced the abundance of both chironomid larvae and mayfly larvae (*Baetis*). Reductions in the density of chironomids were attributed to being eaten by the stoneflies, but *Baetis* numbers decreased due to predator-induced increases in their drift rate.

Snails may even leave the water when attacked by a leech (Brönmark and Malmqvist 1986). This, of course, is only a short-term solution, for eventually the snail would dry out or fall prey to other predators, such as birds.

Such specialized behavior is especially well suited to avoid predators, since it need only be employed when there are actually predators present. One must assume that there are costs associated with the avoidance of predators (e.g., the prey has less time to spend searching for food). The costs of predator avoidance will be mentioned in several of the subsequent chapters. It has been frequently demonstrated that organisms foraging under predation risk suffer sublethal consequences of restricted time budgets, suboptimal habitats, and excess energy allocated to defenses. The avoidance of predatory stonefly nymphs, for example, results in lower growth and fecundity in mayflies (*Baetis*), similar to those caused by food shortage (Peckarsky et al. 1993). There are a number of other phenomena described elsewhere that can be considered predator avoidance strategies, for example, swarming behavior (Section 5.5), vertical migration (Section 6.8.4), and habitat shifts (Section 6.5.3).

Chemical induction

It has been recently established that numerous defense mechanisms are induced chemically (Larsson and Dodson 1993). Prey organisms are capable of detecting the presence of a predator by means of chemical substances, either released directly by the predator or by injured prey. The prey may respond by changing its behavior or even by developing morphological defense structures. The diffusion of chemicals in the water allows the prey to detect the predator before they encounter one another. This is an especially effective means of allowing for the employment of defenses only when they are needed.

It has long been known that the skin of injured minnows release an *"alarm substance"* that causes other members of the same species to flee in panic (von Frisch 1941). In this example the prey release a chemical signal that is detected by other prey. Other chemical signals called *kairomones* are released by predators and detected by the prey. The discrimination between the two types of signals is easy. If an alarm substance is involved the prey will react to the injuries of a conspecific in the absence of the predator. If the signal is a kairomone the prey will not react to the conspecific's injuries, but rather to the smell of the predator. In the past few years many examples have been found where the prey respond to a chemical signal from a predator ("smell") with a morphological change or a specific behavior. For example, crucian carp (*Carassius carassius*) change their body shape in the presence of northern pike (*Esox lucius*). They develop a larger ratio of body height to body length. Pike are limited by their mouth gape and are thus less successful feeding on prey that have a high profile compared to slender fish of the same length. This induced body shape change therefore protects the crucian carp from this specific predator. The change in the pattern of growth is not induced by an alarm substance from injured crucian carp, but by a chemical released from the pike (Brönmark and Pettersson 1994).

Within the zooplankton, the rotifers and cladocerans exhibit the most phenotypic variation in the presence of specific predators. The best-known example is the induction of spines by the herbivorous rotifer *Brachionus calyciflorus* and *Keratella testudo* by the predatory rotifer *Asplanchna*. *Brachionus calyciflorus* appears in two different forms, with and without long, extendible caudal spines (Fig. 6.27a). *Keratella* also can either have long spines or a rounded base (Fig. 6.27b). There are also intermediate forms.

When *Brachionus calyciflorus* are cultured in water that has no predators, they have no spines. However, if the large, saccate rotifer *Asplanchna* is introduced into the culture, the first offspring will have long spines. This also works when the *Brachionus* come in contact with only the water in which the *Asplanchna* were living; the spine formation must therefore be caused by substance dissolved in the water (Gilbert 1966).

The defensive value of the spines becomes apparent when *Asplanchna* are fed with spined and unspined *Brachionus*. *Asplanchna* has no difficulty eating the *Brachionus* without spines. When it grasps a spined animal, it extends the spines so wide that it can no longer be swallowed by the predator and it must be released. Halbach (1969) designed experiments to examine the importance of spines in the interaction between two rotifers, *Brachionus rubens*—which cannot form spines—and *Brachionus calyciflorus* (Fig. 6.28). Both species show typical population fluctuations when cultured individually on a limited supply of the alga *Monoraphidium* (cf. Section 5.2.4). *Brachionus rubens* won when mixed populations of both species competed for the algae. The situation changed when *Asplanchna* were added. At first the predator could eat both rotifer species. Quickly, *B. calyciflorus* began to form spines that protected it against the predator. The unprotected *B. rubens* then died out entirely. The spine de-

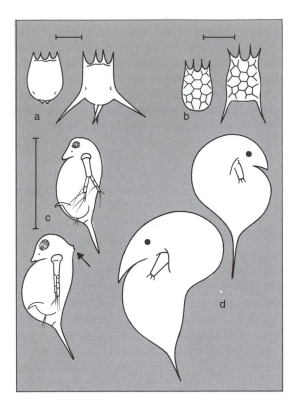

Figure 6.27 Chemical induction of defense mechanisms in zooplankton. The morphs produced in the presence and absence of predators are placed next to one another. (a) Induction of moveable spines in the rotifer *Brachionus calyciflorus* by the predatory rotifer *Asplanchna*; scale, 100 μm (after Halbach 1969). (b) Induction of stiff spines in the rotifer *Keratella testudo* by *Asplanchna*; scale, 100 μm (from Stemberger 1988). (c) Induction of "neck teeth" in *Daphnia pulex* by *Chaoborus* larvae; scale = 1 mm (from Havel and Dodson 1984). (d) Induction of a helmet in *Daphnia carinata* by the backswimmer *Anisops*; size of *Daphnia* approx. 5 mm (after Grant and Bayly 1981).

fense was so effective that finally after the *B. rubens* had disappeared, the predator starved. Then *B. calyciflorus*, the final winner, once again lost its spines.

Daphnia respond to the presence of the phantom midge larvae *Chaoborus*, with morphological changes. Some species (*Daphnia pulex, D. hyalina, D. ambigua*) form small jagged protrusions or "neckteeth" on the dorsal side of the head (Fig. 6.27c). Many studies have demonstrated that these neckteeth are chemically induced. Various clones of *D. pulex* have different sensitivity. Some clones respond to a very dilute *Chaoborus* factor, whereas others show no reaction (Spitze 1992). *Daphnia* with neckteeth have a better chance of escaping attacks by *Chaoborus* larvae (Havel and Dodson 1984). Thus they suffer less mortality, although it is not clear yet whether the neckteeth themselves provide protection or whether they are simply the visible part of some other protective mechanism. Figure 6.29 illustrates that the expression of neckteeth is correlated with the risk of predation. The second instar produces the most pronounced neckteeth in this clone of *Daphnia pulex*. *Daphnia* lacking the neckteeth also suffer the highest predation risk at this intermediate size (cf. Fig. 6.26). The induction of neckteeth leads to the greatest reduction in predation risk in this instar and thus equalizes the risk for all size classes.

As neckteeth are only produced when the predator is present, we can expect costs associated with the protection. These seem to be demographic costs. They have a

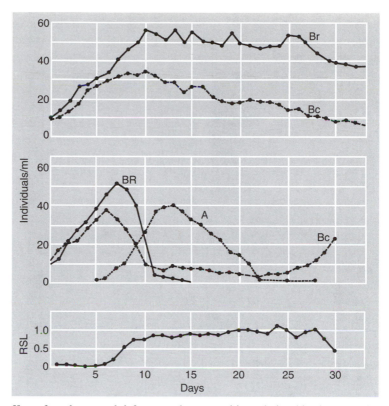

Figure 6.28 The effect of predators and defense on the competitive relationships between the rotifers *Brachionus rubens* (Br) and *Brachionus calyciflorus* (Bc). *(Above)* Competition experiment without predators. *(Middle)* Addition of the predatory rotifer *Asplanchna* (A) on day 5. *(Below)* Induction of spines in *B. calyciflorus*. The relative spine length (RSL) is the ratio of the length of the caudal spine (cf. Figure 6.27a) to the length of the lorica (shell) (redrawn from Halbach 1969).

smaller population growth rate because of a slight shift in the time of first reproduction (cf. Section 5.4). But other factors may be involved, as no clear association between the potential of clones to produce neckteeth and life-history changes has been found (Spitze 1992).

The *Chaoborus* factor evokes a variety of responses in different zooplankton. It induces the high helmets of *Daphnia cucullata* (Fig. 6.30) (Tollrian 1990). These helmets are also modified by other factors (cf. cyclomorphosis). *Holopedium gibberum* increases the size of its protective gelatinous sheath in response to *Chaoborus* without physical contact (Stenson 1987). *Daphnia* can also respond to *Chaoborus* factor by shifting its life history parameters (Stibor and Lüning 1994). Backswimmers (notonectids) induce a spectacular crest formation in the Australian *Daphnia carinata* (Grant and Bayly 1981). These relatively large (>5 mm) *Daphnia* live in fish-free

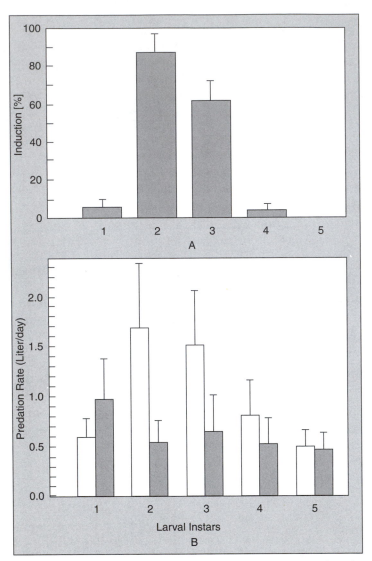

Figure 6.29   Induction of neckteeth by a *Chaoborus* kairomone in a clone of *Daphnia pulex* and protective capacity. Upper panel: Index of induction in different instars. Most pronounced neckteeth are found in instar two (about 1 mm body length). Lower panel: Predation rate of *Chaoborus* on different instars. The predation rate is the volume of water cleared by one *Chaoborus* larva per day; thus it measures the probability of a *Daphnia* being killed. Open bars are animals bearing neckteeth. The second instar is most endangered when unprotected, but also exhibits the highest degree of protection after induction of defensive structures (after Tollrian 1995).

ponds. They develop gigantic crests on their heads in the presence of the backswimmer *Anisops* (Fig. 6.27d). The predators attack *Daphnia* with and without helmets with the same frequency. However, only half as many animals with helmets that are attacked die, compared to the normal animals. The costs of defense have two components. First,

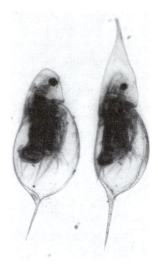

Figure 6.30 Cyclomorphosis in *Daphnia cuccullata*; round-headed (April) and helmeted (July) animals (photo by R. Tollrian).

the crest requires material for its initial construction and for every molt thereafter. Over the daphniids lifetime this is equivalent to forming 60 eggs. Second, the *Daphnia* become larger, resulting in a changed pattern of energy allocation to somatic growth and reproduction and in lower reproductive output (Barry 1994).

Zooplankton also respond to fish kairomones. For example, *Daphnia lumholtzi* grows an enormously long helmet and tail spine (Tollrian 1994). Various *Daphnia* species change their life histories (Stibor 1992). A chemical stimulus also provides the motivation for diel vertical migration (cf. Section 6.8.4).

New discoveries are constantly being made in the field of chemical communication between predator and prey, not only in metazoans, but also in protozoans. A ciliate (*Euplotes*) recognizes the presence of its predator, another ciliate (*Lembadion*) by a chemical cue. In response, it changes to the "winged" morph with dorsal and lateral extensions that make it less accessible to the predator (Kuhlmann and Heckmann 1985). Green algae (*Scenedesmus*) change their growth form from single cells to colonies of 4 or 8 cells in the presence of a chemical released by *Daphnia* (Hessen and Van Donk 1993). Colonies are better protected against grazing by *Daphnia*, but also sink faster due to their larger size. It is to be expected that chemical cues play an important role in many more predator–prey interactions.

Kairomones are chemical signals that are of advantage to the prey, but not to the predator. Little is known about their chemical composition, although different classes of chemicals appear to be involved. The *Asplanchna* factor may be a peptide (Gilbert 1967). The factor released by *Lembadion* is a protein with a mass of about 40 kDa (Kusch and Heckmann 1992). The fish factor that induces vertical migration in *Daphnia* (Loose et al. 1993) and the *Chaoborus* kairomone that induces neckteeth (Tollrian and von Elert 1994) are both small (<500 Da), moderately lipophilic, nonvolatile substances.

6.5.6 Effects on the Spectrum of Prey

Predator–prey relationships are generally very strong interactions. They thus are important in regulating the composition of the communities. The differential effect of invertebrate and vertebrate predators on populations can be demonstrated especially well with plankton communities (Fig. 6.31). It applies similarly to benthic communities.

Relatively large invertebrate predators can develop large populations in a lake without vertebrate predators (A), as, for example, a lake that regularly becomes anoxic under the ice, causing winter fish kills. They exert a strong mortality on the small zooplankton, whereas the large zooplankton can develop in the absence of fish-induced mortality. This causes the size spectrum to shift to large plankton. If the lake has a large population of planktivorous fish (C), as most eutrophic lakes do, these fish eat the large invertebrate predators as well as the large filter-feeding zooplankton. This reduces the mortality rates on the small zooplankton, and the size spectrum of the zooplankton community shifts toward the small species. If there are enough predatory fish to control the planktivorous fish (B), populations of invertebrate predators can also develop. This causes predatory mortality on both ends of the size spectrum, giving the

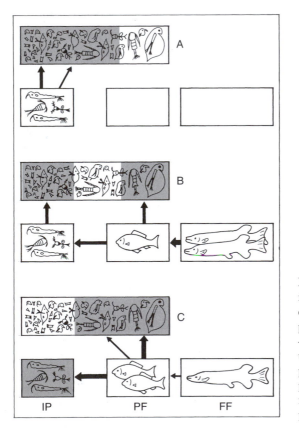

Figure 6.31 The different effects of invertebrate and vertebrate predators on the size composition of the zooplankton. (A) No planktivorous fish → large zooplankton; (B) large population of predatory fish → few planktivorous fish → medium-sized plankton; (C) few predatory fish → many planktivores → small zooplankton (from Lampert 1987).

intermediate-sized zooplankton the best chance of survival. This condition has been verified so frequently in the field that it can be regarded as a general rule. One can make predictions about the ratio of piscivorous to planktivorous fish in a lake by measuring the body size of the summer zooplankton (Mills et al. 1987). These interactions affect the entire ecosystem and will therefore also be dealt with in the chapters that follow.

6.6 PARASITISM

6.6.1 General Features

Parasitism is a predator–prey relationship in which the "predator" (parasite) is considerably smaller than its prey ("host") and lives in close association with its host. Parasites that live within the host are referred to as *endoparasites*, and those living on the outside of the host are called *ectoparasites*. Most parasites are more highly specialized for certain hosts than predators that eat their entire prey. Parasitic microorganisms are referred to as disease vectors.

Populations of parasites can grow inside or on the outside of the body of the host (e.g., viruses, bacteria) or they may be involved in an *alternation of hosts* (e.g., tapeworms). Most parasites can only live as long as they receive nutrition from their living host. If an especially virulent parasite rapidly kills its host, its population can only survive if it is able to infect another host population. This leads to a selection for decreased virulence in parasites that are not highly mobile. The simultaneous selection within the host population for decreased sensitivity to the parasite leads to a coevolution between host and parasite that results in a minimization of damage to the host. The many "gentle" parasites (e.g., roundworms, tapeworms, lice) are a result of this coevolution. Parasitism always must cause a reduction in the growth of the body or population, since even the most gentle parasite uses energy and matter from its host. Every type of parasitism requires a direct or indirect (through intermediate hosts) infection of new hosts. Many textbooks assume that the frequency of parasitism increases with the host density, since the probability of parasite transmission increases with the population density of the host. However, there are few documented cases to support this assumption.

There are many well-known cases of parasitism in aquatic habitats, from bacteriophages, viruses that infect cyanobacteria, and algae to worms that parasitize fish. Some of the best-studied examples involve the regional extinction of economically important species by parasites, as, for example, the widespread eradication of the European crayfish (*Astacus astacus*) by the fungus *Aphanomyces astaci* ("crayfish plague"). Many of the parasites of land animals have a stage that passes through water, such as the cercaria larvae of trematodes, which require an aquatic snail as an intermediate host.

The parasites of fish as especially well known because of their commercial value. Some parasitic copepods live on the gills of fish. The "carp louse" (*Argulus foliaceus*), a crustacean, and the fish leech (*Piscicola geometra*) are often found on the skin of fishes. The glochidia larvae of the pond mussel *Anodonta* live on the margin of the fins of fish. Many endoparasites, including roundworms, tapeworms, and acantho-cephalans, live inside the body of fish. Fish also serve as intermediate hosts for many parasites with complicated life cycles. One example is the fish tapeworm (*Diphyl-lobothrium latum*), which lives in the intestines of mammals, including humans. Eggs contained in feces must come in contact with water, where they hatch into ciliated lar-vae (coracidia). These larvae swim in the plankton and are eaten by their first inter-mediate host, a copepod, in which they develop into a procercoid stage. Infected cope-pods often have a brownish-red color that makes them highly visible. After being eaten by a small fish, the procercoid bores through the intestinal wall and develops into a larger plerocercoid in the liver or muscle tissue. Finally, the fish is eaten by the final host, a mammal, where the parasite forms a mature tapeworm in the intestines.

It is not always clear if a "parasite" really harms the host. Cladocerans and cope-pods are often infested by epibiont diatoms or peritrich ciliates (Threlkeld et al. 1993), but their effect on the zooplankton is difficult to measure. Even an endoparasitic trema-tode living inside *Daphnia* has no effect on the daphniid's fecundity as long as there are no more than two worms per female daphniid (Schwartz and Cameron 1993). Of-ten the relationship may be better described as commensalism when the "parasite" lives on the host, but does not feed on it. For example, the oligochaete *Chaetogaster lim-naei* lives in the shell of the stream snail *Ancylus fluviatilis* (Fig. 4.9). There may be 10–15 worms on a 5-mm-long snail. Just as the snail, these worms eat algae and do not appear to harm their host (Streit 1977). Some such "guests" may be highly spe-cialized. Chironomid larvae that live on dragonfly larvae prefer to live in a very spe-cial place under the wing pads. Conceivably, chironomid species may even compete for the best locations on the dragonfly (Dudgeon 1989).

In the best-known cases of parasite–host relationships, the investigations have been limited to identifying the parasite and its life cycle and the pathology of the host. There are few examples where the effects of the parasite on the population dynamics of the host have been studied. The following section presents as an example, the parasitism of chytridiomycete fungi on planktonic algae.

6.6.2 Example: Fungal Parasitism on Phytoplankton

The parasitism of planktonic algae by lower fungi is an example of a lethal parasitism that can lead to the rapid decimation of the host population; almost every infected cell is killed. The parasites produce large numbers of swarming spores that apparently are so mobile that there has been no selection for reduced virulence.

One of the best-studied examples is the parasitism of chytridiomycetes on phyto-plankton. These parasitic chytridiomycetes have a generation switch between asexual and sexual reproduction (Fig. 6.32) that does not appear to be tied to a change in hosts. Most species attack only one or a few closely related species of algae. The infectious agent is the zoospore, which attaches with its flagellum to the cell wall of the host. Af-ter the attachment, rhizoids develop and penetrate into the host cell, sucking out its contents. The zoospore grows into a sporangium or a gametangium. Resting spores are then produced through gametangiogamy (van Donk 1989).

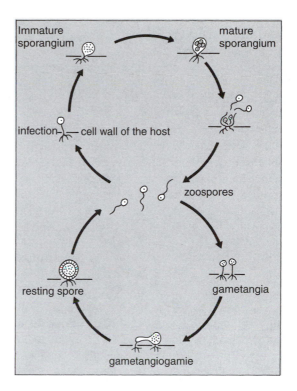

Figure 6.32 Life cycle of a parasitic chytridiomycete (after van Donk 1989).

If the zoospore finds a new host cell in a negligible amount of time, the gross growth rate of the parasite (μ_p) can be calculated from the number of zoospores per sproangium (N_z) and the development time of the sporangium (D):

$$\mu_p = \ln N_z/D.$$

The increase in the proportion of infected cells in the host population (I) can be calculated from the growth rates of the parasite and the host (μ_w):

$$I_2 = I_1 e^{(\mu_p - \mu_w)(t_2 - t_1)}$$

The advancement of the epidemic depends on the difference between μ_p and μ_w. One might conclude that a nutrient-limited host population (lower μ_w) would succumb sooner to a fungus epidemic than a population that is nutrient saturated. It has been shown, however, that in the host–parasite relationship between *Asterionella formosa* and *Rhizophydium planktonicum* that the fungus produces fewer zoospores when it infects P-limited algae (smaller μ_p) (Bruning and Ringelberg 1987). However, the difference between μ_p and μ_w is always positive at all phosphorus levels. That means that the parasite is capable of eradicating any *Asterionella* population, no matter how rapidly it grows. It is only at a temperature of less than 3°C that *Asterionella* can grow faster than the parasite (van Donk 1989).

Chytridiomycetes can exert an enormous influence on populations of algae. Within a 2-year period of observation, every population peak of the diatom *Synedra acus* was infected by the chytridiomycete *Zygorhizidium planktonicum* in Schöhsee (H. Holfeld, unpubl.). The infection rate of *Synedra acus* could not be controlled by adding nutrients to plastic bags suspended in the lake (Fig. 2.1b), but only by adding a fungicide. This demonstrates that the fungus would completely regulate the algal population if it were not controlled. This is not to say, however, that all population maxima of potentially infected algae are actually parasitized. Parasites do not always decimate host populations that have adequate resources. Parasites may be controlled by hyperparasites, or grazing by zooplankton may cause higher mortality to the zoospores of parasites than to their host cells. Also, it is theoretically possible that host populations may evolve immune clones, although this has not yet been demonstrated.

6.7 SYMBIOSIS

Interactions may be positive for both parties; that is, both participants benefit. This form of interaction is referred to as *symbiosis* or *mutualism*. Only a slight modification of the competition equations of Lotka and Volterra (cf. Section 6.1.2) is needed to model this type of interaction: The negative interaction coefficients α and β need only be replaced by positive values.

Symbiosis is usually considered to be relatively unimportant in structuring the communities in freshwater, although there are numerous cases of *endosymbiosis*, such as endosymbiotic algae (zoochlorella) in sponges, *Hydra*, and ciliates.

Bacteria consortia are important symbiotic relationships of independent bacteria populations in which the metabolic product of one bacterium serves as a resource for a second bacterium. Examples were discussed earlier, such as the nitrifying consortium *Nitrosomonas* and *Nitrobacter* (cf. Section 4.3.8) or the methane-forming consortium *"Methanobacterium omelianskii"* (cf. Section 4.3.10).

6.8 INTERACTIONS OF COMPETITION AND PREDATION

6.8.1 The Size-Efficiency Hypothesis

In a paper published in 1962, Hrbáček demonstrated that the size of zooplankton in a lake was closely related to the abundance of planktivorous fish. In lakes with many planktivorous fish the zooplankton were small. In lakes with few or no planktivorous fish large zooplankton were abundant. This phenomenon first became generally known after Brooks and Dodson (1965) published a similar example in the journal *Science*. They described the effect of a herring-type fish (*Alosa pseudoharengus*) on populations of zooplankton by following a "natural experiment," the immigration of *Alosa* in

[232]

the lakes of Connecticut. Plankton samples had been taken in 1942, before the *Alosa* had entered the lakes. Samples were taken again in 1964, after the fish were established in the lakes. Brooks and Dodson observed that the composition of the zooplankton had shifted dramatically. The large species had disappeared and the small species had become the dominants (Fig. 6.33). The dominant zooplankton were (including the size at sexual maturity):

1942	*Mesocyclops*	1.5 mm	
	Daphnia	1.3 mm	
	Epischura	1.7 mm	
	Cyclops	0.9 mm	
	Diaphanosoma	0.8 mm	Average body size 0.8 mm

1964	*Cyclops*	0.9 mm	
	Ceriodaphnia	0.6 mm	
	Tropocyclops	0.45 mm	
	Asplanchna	0.4 mm	
	Bosmina	0.3 mm	Average body size 0.35 mm

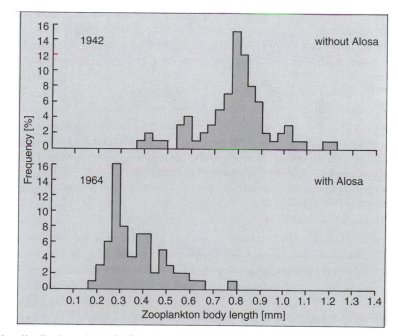

Figure 6.33   Size distribution of zooplankton in a lake in New England before (above) and after (below) the immigration of a planktivorous fish (*Alosa*) (redrawn after Brooks and Dodson 1965).

Tests of the idea in other lakes showed the same effect that was described in Section 6.5.6: When planktivorous fish were abundant, the zooplankton was small; if there were no fish, the zooplankton was large. Brooks and Dodson formulated the size-efficiency hypothesis (SEH) to account for this phenomenon. Due to its simplicity and predictive power, the SEH became a paradigm in aquatic ecology. It was later expressed in considerable detail by Hall et al. (1976):

The composition of the zooplankton is the result of competition and predation.

1. Filter-feeding zooplankton compete for small particles.

2. Large zooplankton filter more efficiently and also can eat larger particles.

3. Therefore, the large zooplankton dominate when there is little predation pressure by fish.

4. Under intense predation pressure by fish the small zooplankton dominate because the large zooplankton are eliminated.

5. Moderate predation pressure has a greater effect on the large zooplankton, permitting the coexistence of both large and small species.

There is no doubt about the importance of fish in this hypothesis; they are responsible for the elimination of the large zooplankton species. The competition side of the hypothesis is less clear. Why do the large plankton suppress the small species in the absence of the fish? If large cladocerans compete with one another in the laboratory, the largest do not always win. Expanding the SEH further, Dodson (1976) provided the simple explanation that the shift toward the large zooplankton in nature is caused by invertebrate predators (cf. Fig. 6.31). The large invertebrate predators (*Epischura*, *Leptodora*) did, in fact, disappear from the Connecticut lakes after the immigration of *Alosa*.

It still has not been satisfactorily explained why the large zooplankton are "more efficient." One can argue that the large zooplankton use less energy per unit body weight than the small species (cf. Section 4.5), but relationships with body size of respiration and assimilation both have exponents of about 0.75, meaning that energy uptake and consumption increase at the same rate with body size. One can demonstrate experimentally that when there is little food and a negative energy balance, the large zooplankton can survive longer. This would give the large species an advantage when the food fluctuates strongly. A comparison of the threshold food concentrations needed for growth provides another argument for the competitive superiority of the larger species (Gliwicz 1990a). Clearly, the larger cladoceran species have a lower threshold concentration than the smaller species (Fig. 6.34) and, in the absence of predators, should be more successful competitors for food (cf. Section 6.1.5).

While the size-reducing effect of fish is relatively simple and clear, there probably are many other factors involved in this general trend. Large cladocerans, for example, can outcompete the much smaller rotifers and the cladocerans do have a much lower food threshold for growth than the rotifers. The trend is reversed, however, within the rotifers. The small species are more efficient than the large species (Stemberger and Gilbert 1987). On the other hand, the small rotifers are also mechanically inhibited by

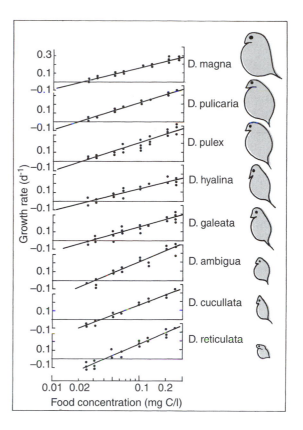

Figure 6.34 Shift in the minimal food concentration for growth (growth rate, zero) as a function of size for 8 species of cladocerans (*Daphnia* and *Ceriodaphnia*). The threshold concentration decreases with increasing size of the animals (from Gliwicz 1990).

large *Daphnia* (cf. Section 6.2.2), so that it becomes difficult to distinguish between exploitative competition and interference competition (Gilbert 1988). The predictions of the SEH are continually verified, and they have become the basis for biomanipulation in lakes (cf. Section 7.3.3), even though we do not understand all the mechanisms that are involved.

6.8.2 Evolution of Life History Strategies

Reproduction

The fitness of an individual is the result of its entire life history and cannot be determined at a specific stage in its life (Stearns 1992). This includes the probability of survivorship to sexual maturity as well as the number of offspring produced (cf. Section 5.4). An individual is exposed to a wide range of selection pressures during the course of its ontogenetic development. The result of size-selective predation is that the young zooplankton are under predation pressure from invertebrate predators, whereas the larger adults are eaten primarily by fish. Many organisms also change their nutrition as they develop. Many cyclopoid copepods are herbivorous as nauplii and copepodids,

[235]

but are carnivorous as adults; stream insects may graze on algae as larvae and consume nectar when they leave the water as adults; as they mature, fish depend on entirely different food, mainly as a function of mouth size. Abiotic factors, food limitation, and predation take on differing importance during an individual's life. The relative fitness of an individual can depend on whether it managed to optimize its food intake as a juvenile as well as how well it protected itself against predation as an adult. The manner in which an individual divides its life cycle to achieve maximum fitness depends on the relative importance of the various selective factors. The solution to the problem of how, when, and how often an individual should reproduce must therefore be a compromise. There may be many possible solutions for the same problem. Thus, depending on the environmental conditions and phylogenetic limits, we find for freshwater organisms a host of adaptations related to the time, duration, and type of reproduction.

The shape of the survivorship curve may depend on whether it is more profitable for an animal to reproduce once in its life and then die or to produce offspring several times. Closely related species may show both strategies. The Pacific salmon (*Oncorhynchus*) spawns only once, then dies. The young develop first in freshwater and then migrate to the ocean, where they grow and after 3 or 4 years return to the home stream to spawn. The Atlantic salmon (*Salmo salar*) has a similar life cycle, except the adults do not die after spawning and migrate back to the ocean and may return many times to spawn again. Multiple reproduction can be viewed as a type of insurance against the loss of an entire brood through unfavorable conditions. This also involves compromises. An animal has only so much energy to invest in reproduction. The alternatives are either to put everything out at once or distribute the energy into several broods. If there is a considerable danger that the animal will not survive to breed a second time, the second strategy would not be profitable (e.g., insects).

Opportunistic species that have the highest possible reproductive rate to colonize free space rapidly (cf. Section 5.6) are favored by rapidly changing conditions. Parthenogenesis is a successful strategy for short-term maximization of the population growth rate, since all offspring are females that can immediately contribute to further reproduction. This is the normal type of reproduction for rotifers and cladocerans. The occasional change to bisexual generations (cyclic parthenogenesis) guarantees maintenance of genetic variability and it is the basis for the production of resting eggs (cf. Section 5.7).

The other extreme life history can be found in bivalves, which are among the very few long-lived organisms in freshwater. The freshwater pearl mussel *Margaritifera margaritifera* has a life span of over 100 years. It grows extremely slowly, living in a nutrient-poor environment (clean streams), but produces about 200 million larvae during a reproductive period of 75 years. Larval mortality is extremely high, as the larvae (glochidia) live parasitic on the fins of fish. The probability of random encounter with a fish is very low (Bauer 1987).

Rotifers and cladocerans, as well as some mollusks, are also examples of organisms that have shortened development time and reduced mortality by minimizing the larval stages, which are especially vulnerable to predation. Females carry the eggs until the young hatch. The juvenile stages are very similar to the adults, except they are smaller.

This strategy is very common in freshwater, since the function of larvae as a means of dispersal in oceans is not needed in inland waters.

Growth and reproduction

Competition for common resources and predation can compel species to "design" highly specific life histories. Just as the morphology and behavior of a species represents the "tailoring" of evolution, life histories are also the result of selection factors such as competition and predation. The resulting life cycle often acts to minimize the effects of these selective forces, giving the appearance of a "smart" species, such as those that form a resting stage at times of the year when there is a high probability of excessive mortality due to predators. It must be emphasized that this does not mean that individuals decide to avoid periods when predators are especially prevalent.

We have already discussed the differences in life histories of various species in connection with r and K selection (Section 5.6). The onset of sexual maturity is a significant event in the life of every animal, for it is an important demographic parameter (cf. Section 5.4) and thus an important focal point for selection. Whether it is better for an individual to reach sexual maturity sooner or later is a question that may be determined by both competition and predation.

The cladocerans provide a useful example. *Daphnia* grows throughout its entire life. It grows more slowly after it is sexually mature, for it invests 80% or more of its production into offspring. A large-sized *Daphnia* produces many more eggs than a small one. Thus there are two alternatives for the onset of sexual maturity (Taylor and Gabriel 1992), if one assumes that there should be a maximization of the population growth rate (r):

1. The animals may become sexually mature when relatively small (early); then they can begin producing eggs earlier and thus produce more broods; if they invest most of the available energy into reproduction, the adult grows slowly and thus remains small, so that each of its broods contains relatively few eggs.

2. The animals become sexually mature after they are larger. They begin to have broods later and thus have fewer broods, but more eggs per brood.

Calculations from models show that the question of which strategy should be employed depends on whether there is size-dependent mortality (Fig. 6.35). When there is greater mortality on the larger size classes (fish predation), the largest r comes from animals that begin reproduction as soon as possible. In the absence of size-selective mortality it is more profitable to wait longer before reproducing. Field observations support this model: There are many reports that the same *Daphnia* species in lakes with many planktivorous fish become sexually mature at a smaller size than in lakes with few fish. In a single lake there can be seasonal changes, where in the summer, when fish are especially active, *Daphnia* become sexually mature at a smaller size.

Strong support for the validity of the model comes from the discovery that the life-history shift in a population is not necessarily driven by selection for different genotypes. Even a single clone of *Daphnia hyalina* can adopt different strategies phenotypically in response to a chemical signal (kairomone) from a particular predator

[237]

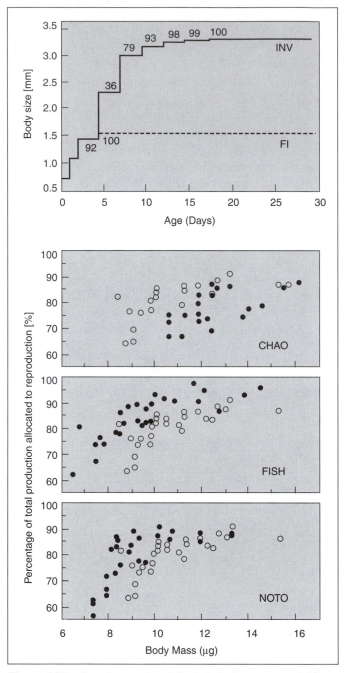

Figure 6.35 Growth strategies of *Daphnia* in the presence of different predators. (A) Results of an optimization model for the best life-history strategy for *Daphnia pulicaria*. The numbers denote the percentage of invertebrate predators. Daphniids maximize their population growth rate when they first invest in somatic growth, mature late, and become large. The best strategy in the presence of fish is to invest early in reproduction, mature early, and remain small. (B) Response of *Daphnia hyalina* to chemical cues of various predators (CHAO, *Chaoborus* larvae; NOTO, *Notonecta*). Percentage of net production allocated to reproduction at different sizes. Open circles are controls without a chemical cue.

(Stibor and Lüning 1994). The proportion of net production invested in reproduction can be measured in *Daphnia* by weighing its body and eggs separately. This investment increases with increasing body size from about 60% to over 80% (Fig. 6.35B). Juvenile *Daphnia* exposed to the "smell" of *Chaoborus* larvae or fish react as predicted by the optimization model. In response to the *Chaoborus* kairomone, the daphniids invest less into reproduction and start reproducing at a larger size than controls without the kairomone. In contrast, the daphniids start reproducing earlier and put more energy into reproduction when the chemical signal comes from fish. Interestingly, the response to chemicals released by backswimmers (*Notonecta*) is very similar to the reaction to fish, although *Notonecta* is an insect. Unlike *Chaoborus*, however, *Notonecta* is not gape-limited, as it holds its prey and then pierces it with its beak and sucks out its contents. Hence, like other true bugs, *Notonecta* does not select for small prey and, in fact, prefers larger daphniids. In this respect it behaves like a fish. This example illustrates how the chemical signals must be sufficiently specific to inform the prey about the capture mechanism used by the predator.

Benthic organisms appear to have similar reproductive strategies (Crowl and Covich 1990). The snail *Physella virgata* lives together with the crayfish *Orconectis virlis* in small streams in Oklahoma. The crayfish eats the snails, but it is very selective, since it can only break the shells of the small individuals. Snails in streams with no crayfish become sexually mature after 3-1/2 months and at a size of 4 mm and grow very little more. Snails in streams with crayfish must develop for 8 months and grow to 10 mm before they begin reproducing. This effect is also elicited by a chemical substance. The snails will also change their life cycle without predators, if they are placed in water that had contained crayfish feeding on snails. These are two examples of organisms that have a broad reaction norm and have the ability to invoke different life history strategies.

Timing strategies

Many organisms occur only at certain times of the year. This is often a direct effect of limiting resources for very small organisms that have high reproductive rates. The development of the spring algal maximum, for example, follows the availability of light. Maxima of protozoans and rotifers then follow the algal maximum. These plankters can grow so rapidly that they can utilize the resources optimally (cf. Section 8.7.3).

Such briefly appearing organisms are often referred to as spring, summer, or fall forms. These terms can lead to the intuitive conclusion that the seasonal appearance of a species is tied closely to abiotic factors, such as temperature. In fact, we often refer to warm water and cold water forms. This can be very misleading. The so-called "summer depression" of *Daphnia* is a good example. *Daphnia galeata* usually has a bimodal population pattern in eutrophic lakes at temperate latitudes, with its highest density in May and a second, smaller peak in September/October. Between these peaks is a summer minimum ("depression") (cf. Fig. 5.7). It would be incorrect to conclude from this that these *Daphnia* prefers cool temperatures. If you collect *Daphnia* from the lake at 10°C and test it in the laboratory, it has optimal growth at 20°C. Its restricted occurrence in the spring and fall in the field is the result of a low food availability and high predation that does not permit population growth in the summer.

Some species of aquatic organisms actually achieve their physiological optimum at low temperatures. The question remains, however, whether there are physiological constraints or whether the organisms optimized their metabolism in a temperature range that they were forced into by other factors (competition, predation).

Many species have "programmed" life cycles that can be modified very little by environmental conditions. Different populations of the same species can have completely different "programs." The seasonal termination of the life cycle sometimes appears "smart," since the organisms seem to know in advance what the environmental conditions will be like, although that is, of course, not possible. Stream insects with a 1-year life cycle will never experience a spring flood in a brook. Nevertheless, they must lay their eggs so that their larvae will not be in a sensitive stage during the next flood condition. The destructive power of the spring high water is such a strong selective factor that the individuals with an incorrectly programmed life cycle would quickly disappear.

Competition and predation have a similar effect on the temporal allocation of the life cycle. Organisms must be adapted so that their offspring are not born during a period of extremely low food supply or high mortality. This is only possible when such events are somewhat predictable and when there is a "trigger" that occurs at a fixed time in relation to the event. Day length and water temperature are excellent triggers, for both are closely correlated with the seasons.

Many insects that have their larval development in water hatch in a brief period and form swarms. This simplifies the problem of finding a mate, especially in populations with low densities. The swarms may also provide high prey densities that satiate predators (cf. Section 4.3.2), thereby reducing the predation risk for individuals. The development time of the individuals must also be synchronized by some external stimulus. Thus many insect larvae cease their development for a short period and wait for the appropriate time to hatch.

Diapause

Many animals go into a diapause, which is a stage of arrested development and reduced metabolic rate. The eggs in the ephippia of cladocera are such a diapause stage, for the eggs cease development after a few divisions and require a stimulus before they resume development. Diapause stages of copepods are especially common. Some cyclopoid copepods bury themselves in the sediment in the fourth copepodid stage and can rest there for many months. The onset of diapause is controlled by day length. It is still not known what factor(s) regulate the cessation of the diapause. This phenomenon was covered in the first chapter (Section 1.3).

Some species, such as *Cyclops vicinus* and *Cyclops kolensis*, have their diapause in the summer. This is usually interpreted as a strategy for avoidance of predation pressure by fish, which is high in the summer. There is also experimental evidence that the summer diapause apparently also serves to avoid competition. Cyclopoid nauplii are inefficient filter feeders and require high concentrations of algae to grow and develop. They would be very poor competitors against cladocerans with their highly efficient filtering mechanism. It can be argued that it would be a disadvantage if the nauplii had to develop during a period when cladocerans were in high densities in the lake. If one collects lake sediments during the summer, the resting copepodids of both species can be "awakened" by aerating the water. If after they become adults and produce nauplius larvae one places the larvae in the lake water with its natural algae at times of the year when the animals normally do not appear, the nauplii will starve. There is insufficient food present. The nauplii will develop further, however, if the water is enriched with cultured flagellate algae. Clearly, the nauplii are food-limited in the summer (Santer and Lampert 1995).

The calanoid copepod *Diaptomus sanguineus* produces two types of eggs, subitaneous eggs, which begin to develop immediately, and resting eggs, which can remain in the sediments for several months before they hatch. In March, all the females in a pond with fish begin to switch from subitaneous to resting eggs, using day length as a cue (Hairston and Olds 1987). This permits the population to disappear from the water column before the fish begin actively feeding. The switching is not directly caused by the fish, for the copepods begin to produce resting eggs at the same time even when they are moved to a pond without fish. There is some individual genetic variation in the precise time that the females begin to switch reproductive modes. Fish maintain the synchrony in the population by eliminating the females that have too great a delay in resting egg production. Hairston examined two ponds, one of which had lost its fish population. Before the fish had died, the females switched at the identical time in both ponds, but once the fish were gone, the time began to shift in the fishless pond and after 2 years occurred 1 month later. Apparently, in the absence of predators, females that produce subitaneous eggs for a longer time simply contribute more offspring to the next generation. This is a case where microevolution could have proceeded even faster if there had not been large numbers of resting eggs buried in the sediments with delayed hatching that contributed individuals to later generations, thereby slowing down the rate of change (Hairston and DeStasio 1988).

Constant selection is necessary to produce such exact synchronization of the time of reproductive switching. The fact that the majority of the copepods produce resting eggs before the fish become active results from the highly predictable predation pressure. Ponds that are not so predictable lead to an entirely different life cycle strategy. *Diaptomus sanguineus* also lives in ponds that occasionally dry out and therefore are without fish. Such drying out events are caused by climatic variation and are thus not predictable. In these ponds the females produce both types of eggs irregularly throughout the summer and do not respond to the environmental factors that act as cues for resting egg production. This ensures that an individual's resting eggs are present in the "resting egg pool" in the sediments at all times. Diapausing eggs buried in the sediments can remain viable for very long periods. The oldest eggs of *D. sanguineus* that hatched came from a depth of about 30 cm in the lake sediments and were estimated to be 330 years old (Hairston et al. 1995). This cache of eggs in the lake sediments is analogous to the "seed bank" in soils, providing recruitment to the lake copepod population for several years and "buffering" the effects of environmental fluctuations.

Spreading of the risk

Environmental conditions are not always so predictable that it is an advantage to terminate the life cycle so precisely. A species with a very narrow reaction norm may miss the most favorable period if that year has unusual climatic conditions or there is a random shift in the structure of the food web. This could lead to a catastrophe for a population with a fixed life cycle. An example of this was observed in Lake Constance. A large calanoid copepod *Heterocope borealis* lived in the oligotrophic Lake Constance, where it overwintered in resting eggs in the sediments. The nauplii hatched in March and then moved into the surface water. In 1956, the predatory copepod *Cyclops vicinus* immigrated into the lake as it became more eutrophic. It had a different life cycle with a maximum of adults at precisely the time when the *Heterocope* nauplii migrated to the surface. Within 2 years *Cyclops vicinus*, which prefers eating nauplii, had eradicated the *Heterocope* population, whose life cycle had suddenly become ill adapted (Einsle 1988).

Just as a stockbroker does not invest in only one type of stock, it can be advantageous to "spread" the risk to avoid a disaster. Such a strategy does not allow one to have maximum gains, but, in the worst case, everything is not lost.

The drying out of some small ponds represents an example of a risk that is not predictable. The cladocerans that live in the pond would have the highest reproductive rate if they produced only subitaneous eggs, since resting eggs do not contribute to the population growth rate. Occasionally, they produce resting eggs, nevertheless, that lie in the sediments as an "insurance" that allows for repopulation if there is a disaster.

The diapause serves to overcome unfavorable periods. Some years, however, the conditions may not be so bad as to prevent further development of the population. There may be years during which the animals can survive without resting eggs, but such years are not predictable. Under these conditions, it would be beneficial if some animals could "try their luck" and not go into the resting stage. These animals would have a considerable advantage during such favorable periods. This potential advantage allows a population to maintain a certain proportion of animals that have a different life cycle.

Nilssen (1980) described examples of such risk spreading. Cyclopoid copepods in Norway can complete their life cycle in 1 or 2 years. If they develop rapidly, reaching sexual maturity in 1 year, they are relatively small and have few eggs. As an alternative, they can develop more slowly, becoming sexually mature after 2 years, but then they are larger and have more eggs. Both alternatives can coexist in a single lake. The proportion of "1-years" increases with increasing predation pressure. The large copepods are especially vulnerable to predation, since the fish are size-selective predators. On the other hand, the fish population fluctuates greatly, so that it is sometimes advantageous to postpone the reproduction for 1 year on the chance that in the following year it may be less dangerous to be large and produce many offspring.

6.8.3 Cyclomorphosis

Many zooplankton regularly change their appearance during the course of the year. These changes are called cyclomorphosis because the morphological changes occur as seasonal cycles (Black and Slobodkin 1987). Summer animals may look entirely different from winter animals and may even be mistaken for different species (Fig. 6.36).

In some cases, such as for rotifers, the changes in shape may actually involve the re-placement of one species by another. In other cases, especially for planktonic clado-cerans, the changes that occur are actually within a single species. The formation of helmets by *Daphnia galeata mendotae* is a typical example in North America (Jacobs 1961). Closely related species can coexist in the same lake, if one species is cyclo-morphic and the other is not, such as *Daphnia galeata* and *D. hyalina* in Lake Con-stance.

The widespread *Daphnia galeata mendotae* is one of several cyclomorphic *Daph-nia* species in North American lakes (Jacobs 1987). A similar species typical of eu-trophic lakes in Middle Europe is *Daphnia cucullata* (Fig. 6.30). In the winter it has a rounded head and in early summer it develops a long "helmet." Already in June, the young *Daphnia* inside the brood pouch of round-headed females can be seen with pointed helmets. After the young hatch, their helmets grow at a faster rate than the rest of the body (positive allometry), so that the largest individuals develop the largest hel-mets. The length of the helmets increases relative to the body length until the end of July, then the relative helmet length slowly decreases. The length of the helmet can be up to 60% of the body length in mid-summer. Typical of cyclomorphosis, this follows a regular annual cycle (Lampert and Wolf 1986).

The most conspicuous cyclomorphosis involves the formation of body processes and spines or the development of crests. Cladocerans of the genus *Bosmina* can form a long rostrum and a high "hump back" (Fig. 6.36). Sometimes, however, the features are very small and difficult to recognize, such as tiny "neckteeth" in *Daphnia pulex* or longer antennal setae in *Daphnia galeata mendotae* (summary in Jacobs 1987).

Rotifers and cladocerans usually reproduce by parthenogenesis, which blurs the con-cept of species. A population consists of many clones, each the offspring from one

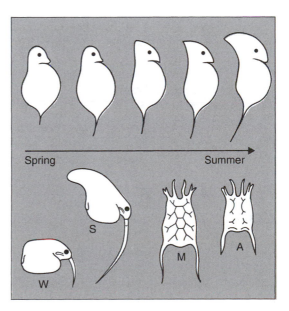

Figure 6.36 Cyclomorphosis in zooplankton. *(Above)* Seasonal changes in the shape of *Daphnia retrocurva* (from Riessen 1984). *(Be-low, left)* Winter form (W) and sum-mer form (S) of *Bosmina coregoni thersites* (after Lieder 1950). *(Below, right)* Morphology of the rotifer *Ker-atella quadrata* in May (M) and Au-gust (A) (from Hartmann 1920).

[243]

mother and genetically identical. The clones may differ morphologically, resulting in great variability within the population. In this case, there are two possible ways for a seasonal change in forms: a phenotypic or a genotypic change.

In the first case, the offspring from all of the animals would respond to the changes in the environmental conditions. In the second case, the clones that are best adapted and morphologically most divergent would separate; their proportion in the population would change. If, for example, the conditions favored *Daphnia* with helmets over those with round heads, a rare clone with a helmet could produce more offspring than a more abundant clone without. The rare clone would increase and the population would, on the average, develop longer helmets.

Evidence indicates *Daphnia* has a phenotypic change:

1. Helmets can be induced in isolated clones in the laboratory, although they are rarely so large as those found in nature.

2. Helmeted young can be found in the brood pouches of round-headed females.

3. There is no evidence from electrophoretic allozyme analysis (cf. Section 5.3) that there was a shift in the clonal structure of a population of *D. cucullata* during the period of helmet formation (Lampert and Wolf 1986).

This does not exclude the possibility that clonal shifts may be involved in the cyclomorphosis of other zooplankton such as *Bosmina* (Black 1980). It is also conceivable that there could be a combination of phenotypic and genotypic changes if there were an increase in clones that have greater phenotypic responses to the conditions.

Two lines of questions can help one explain cyclomorphosis:
1. *What induces the change in shape? Which factors regulate cyclomorphosis and make it predictable?* This is a question concerning *proximate factors*.

In laboratory experiments, high temperature, abundant food, and turbulence can induce helmet formation (Jacobs 1967). No one has yet found one factor alone that causes cyclomorphosis. There are many recent examples where morphological changes involved in cyclomorphosis can be induced by substances released into water by predatory invertebrates (*Chaoborus*, notonectids) (cf. Section 6.5.5). The relatively large helmet of *Daphnia cucullata* is positively correlated with water temperature in the field. The higher the temperature, the longer the helmet. When brought into the laboratory, the helmet stays relatively small with high temperature and abundant food. High helmets will be formed only if the animals are cultured in water that had previously contained *Chaoborus* larvae. The induction must involve a soluble chemical, since the experiment works with water that has been filtered free of particles (Tollrian 1990).

The role of such chemical interactions in cyclomorphosis has still not been completely explained, even if one can show that morphological changes can be induced in zooplankton by dissolved chemicals that signal the presence of a predator.
2. *Why do the plankton change their shape? What is the adaptive advantage of cyclomorphosis?* These are questions about *ultimate factors*.

There has been much speculation about the function of cyclomorphosis (see summary in Jacobs 1987). At first, the spines and helmets of zooplankton were interpreted to be "flotation organs" that reduce sinking in the summer when the viscosity is low because of the high temperatures. However, this hypothesis did not hold up. It appears that it is more likely a defense mechanism against predation (cf. Section 6.5.5).

If a mixture of *Daphnia* with and without helmets is offered to fish, the helmeted animals survive better (Jacobs 1987). Helmeted *Daphnia* appear to be better at escaping from fish. It is not clear whether their shape presents less form resistance or whether the helmet provides a better surface for the attachment of the antennal muscles.

Spines and helmets are a very effective defense against invertebrate predators. Experiments have demonstrated that plankton with exaggerated features of cyclomorphosis have a better chance of surviving in the presence of *Chaoborus* larvae, predatory copepods, and notonectids. Invertebrate predators have difficulty capturing large prey (cf. Section 6.5.4), since their feeding is limited by the size of their mouthparts. It has been demonstrated, however, that *Daphnia* with "neckteeth" or helmets are not only more difficult to handle, but are also captured less frequently by *Chaoborus*, indicating that the protection afforded is not only interference with handling (Havel and Dodson 1984).

One hypothesis (Dodson 1974a) assumes that the formation of a transparent helmet and a long tail spine makes *Daphnia* more difficult for invertebrate predators to capture, without increasing its visibility to fish. Cyclomorphosis does not result in an increase in the visible parts of the *Daphnia*, such as the eye, gut, and eggs. The zooplankton increase only their transparent structures, thereby avoiding making them more visible to fish that feed visually and at the same time making them less vulnerable to invertebrate predators.

There is no proof that these factors are important in the field. Conspicuously, however, the greatest morphological changes occur during the periods of the highest predation pressure. The cyclomorphic induction response of the prey to a chemical signal from the predator is the most convincing evidence for the predator defense advantage of cyclomorphosis, since in this case the proximate and ultimate factors are the same (predator). Cyclomorphosis could also easily be regulated without chemical induction. Temperature is a reliable "zeitgeber" or indicator, for it is closely correlated with the appearance of young fish and insect larvae and is highly predictable.

There is still the question why the zooplankton do not always have such protective structures. Why do they only form them when they expect predators? It is not easy to assess the "costs" of defense that could be saved when there is no danger. One hypothesis proposed for *Daphnia catawba* states that helmeted animals are overall more slender and thus have smaller brood pouches (Riessen 1984). The number of eggs (reproductive capacity) depends on the amount of food available. The highest food availability is during the spring maximum of small algae, before the young fish are abundant. The brood pouch must be large, meaning the animals must have a rounded shape in order to take advantage of the supply of food. Later, when the number of eggs produced is limited by the lack of food, it is not an advantage to have a large brood pouch with few eggs. Then it is better to invest in defense. Cyclomorphosis could be seen as the optimization of fitness by maximizing reproduction and minimizing mortality at the appropriate times of the year.

This does not seem to apply to *D. galeata mendotae* (1987) and *D. cucullata* (Tollrian 1990). There is no difference in the number of eggs produced by helmeted and round-headed animals when they are given optimal food conditions. In this case, the costs must be energetic (e.g., less efficient swimming, slowed development), for there does not seem to be a morphological cost for cyclomorphosis, such as a smaller brood pouch.

It appears that cyclomorphosis has arisen independently in many groups, since there is such a wide range of morphological features involved and closely related types differ dramatically. It is therefore unlikely that there is a single proximate or ultimate factor for all cases. It is also unlikely that such a pronounced change in shape is "neutral" and without some adaptive value (Jacobs 1987). However, there is not yet been a documented field study clearly demonstrating the gain in fitness due to cyclomorphosis.

6.8.4 Vertical Migration

Both marine and freshwater zooplankton exhibit an interesting daily (diel) migration behavior. Usually, the zooplankton are in the deep water during the day and swim up to the surface at night (Fig. 6.37). There are also cases of "reverse" migration, where the animals swim into the deep water at night and up to the surface in the day. Diel vertical migrations have been described ranging from a few centimeters to over 100 m (Hutchinson 1957).

The amplitude of the migration and the vertical distribution pattern varies from species to species and with the developmental stage for a given species. It can also be influenced by the transparency of the water and the food conditions. Some zooplank-

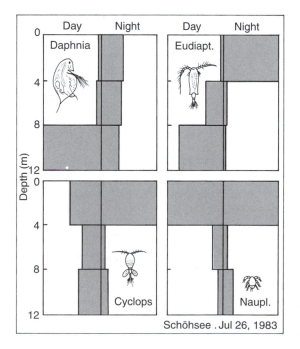

Figure 6.37 Relative distribution of zooplankton captured at various depths during the day and night in Schöhsee (Holstein) on July 26, 1983.

ton migrate up and down as a narrow band, and others disperse at night throughout the entire water column.

We cannot get a true picture of how an individual migrates from samples of the plankton. A vertical profile of plankton samples only allows us to see the changes in the density of organisms at particular depths. We then estimate the vertical movement of the population from the changes in the distributions. The day–night differences in the mean depth or median of the population provides a measure of the amplitude of migration. Such observations of the mean behavior of the population can lead to false interpretations of the behavior of the individual. The distance between the mean values of the distributions is a valid measure of the actual distance traveled only if all animals move synchronously. This is not always so. It is possible that there are always some animals moving upwards, while others are moving downwards. We can only measure the net effect, which can seriously underestimate the swimming activity of the individual.

During the 1960s there was much interest in the neurophysiological basis of vertical migration behavior. It was found that phototaxis and geotaxis are both involved in the regulation of the behavior. Light is the primary controlling factor (*proximate factor*). Since zooplankton constantly adapt to the ambient light conditions, the stimulus for the onset and termination of the migration is the relative light change (dI/I) and not the change in absolute light intensity (I). The highest light intensities occur around midday, but the greatest relative change per unit time takes place shortly after sunset and before sunrise. Below a certain threshold in the rate of relative light change zooplankton will not respond. When the light changes more rapidly and exceeds the threshold, however, the zooplankton attempt to correct for the altered light intensity. If the light increases, the zooplankton swim downward into the darker waters, and if the light decreases, they swim toward the surface. The migration stops when the light changes more slowly than the threshold value, that is, when it is slower than the adaptation time of the eye (Ringelberg et al. 1967).

The widespread occurrence of vertical migration by many different taxa strongly suggests that it has some adaptive value. What is then the *ultimate factor* that provides migrating animals with a fitness advantage? There is no reason to believe a priori that the same ultimate factor applies to all taxa, but all filter-feeding zooplankton should derive comparable benefits from vertical migration. Migrating animals spend the night in warm, food-rich water and the day in the deep, cold water, where both the quantity and quality of the food is low (primary production is restricted to the euphotic zone). Time spent in the cold water is a disadvantage in terms of reproduction of the zooplankton. Meager food conditions in the deep water limit the energy available for egg production, and the swimming involved in migration also uses energy reserves. The eggs in the brood pouch are then exposed to cold temperatures that slow their development. Longer egg development means lower reproductive rates. Thus there is a strong selective pressure to stay in the warm, food-rich surface water. Migrating genotypes should be quickly eliminated by nonmigrating genotypes. This is not what happens, however. There must be a selective advantage for migrating.

There are a number of hypotheses that attempt to address this problem. They can be divided into two categories:

1. The disadvantages of migrating are exaggerated. Actually, changing conditions encountered during migration lead to a more efficient use of energy or to faster population growth (higher fitness by increasing the production of offspring).

2. Avoidance of the surface water during the day reduces a source of mortality that is light dependent (higher fitness through decreased mortality).

Some hypotheses in the first category assume that there is a metabolic advantage derived from switching between feeding at high temperature to resting at lower temperatures or that animals grow larger at lower temperatures and thus can produce more offspring (demographic advantage). It is interesting that the algae in the surface waters have a higher nutritional quality in the evening. At the end of the light period they contain more food reserves, which then are metabolized during the night. Furthermore, it has been shown experimentally that zooplankton have higher feeding rates when they are hungry than when their guts are full. It is theorized that animals can obtain more energy by moving into the surface waters to feed on these energetically valuable algae after a period of starvation than they if they simply fed the entire day at the surface. All experimental tests of this "energetic bonus" hypothesis have concluded that it is energetically better to stay near the surface (see summary by Lampert 1993). Even if migration provided an energy bonus, it would not compensate for the negative effects of prolonged development time (Kerfoot 1985).

The second possibility that migration increases fitness by reducing mortality seems more likely, since there is essentially no support for the first group of hypotheses based on an increased reproductive rate. There are some excellent examples to support the decreased-mortality hypotheses.

Daphnia galeata and *D. hyalina* are two closely related species that coexist in Lake Constance. Although they are very similar and even form interspecific hybrids, they differ greatly in their migration behavior (Stich and Lampert 1981). During the summer *D. hyalina* migrates extensively, whereas *D. galeata* stays in the surface waters, migrating very little (Fig. 6.38). The different strategies can be observed directly in this example (Table 6.1).

Daphnia hyalina has fewer eggs because it lives under poorer food conditions during the day. Their eggs also require a longer time to develop since they live at lower temperatures. Both of these conditions cause the migrating *D. hyalina* to have much lower birth rates than the nonmigrating *D. galeata*. If there were no losses, each *D. galeata* would produce 86 offspring after 30 days, compared to 5 young produced by *D. hyalina*. This would lead to the rapid exclusion of *D. hyalina* by *D. galeata*. In spite of this, population densities of *D. hyalina* during the summer migration period are actually greater than the *D. galeata* densities. The nonmigrating species must have a considerable mortality rate that compensates for its reproductive superiority (Stich and Lampert 1981). It is likely that this mortality factor is related to light, since the migrating zooplankton generally avoid the surface waters during the day.

Light, especially UV light, can actually be a direct cause of mortality. It cannot, however, account for the great amplitudes of vertical migration, since short-wavelength light does not penetrate very deep in the lake (Kirk 1994). The most convincing hypothesis (Kozhov 1963) states that vertical migration is a strategy to avoid predators

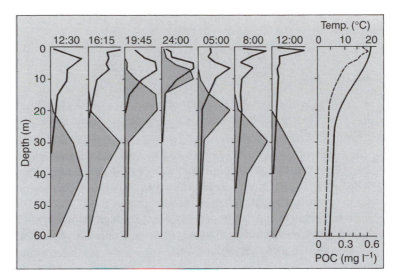

Figure 6.38 A comparison of the different vertical migration patterns of *Daphnia galeata* (plain) and *D. hyalina* (shaded) in Lake Constance (August). Depth profiles of temperature (solid line) and particulate carbon (particles <35 μm, possible food for *Daphnia*) are shown in the right-hand graph (redrawn from Stich and Lampert 1981).

that feed visually, which are mainly fish (cf. Section 6.5.3). Fish can see at very low light intensities, but a zooplankter has a much smaller likelihood of being seen at night or in deep water. The predator-avoidance hypothesis thus states that zooplankton retreat into the darker depths during the day and come to the surface only under the protection of night.

TABLE 6.1
Daily Average Environmental Conditions and Population Growth Parameters for Migrating *Daphnia hyalina* and Non Migrating *D. galeata* in Lake Constance in August 1977 (cf. Fig. 6.30)

	D. galeata	D. hyalina
Mean concentration of particulate carbon<		
30 μm (mg/l)	0.34	0.16
Mean temperature (°C)	14.0	7.1
Eggs/adult *Daphnia*	7.1	3.7
Development time (d)	8.8	14.5
Abundance (individuals/m²)		
In August	38,200	91,500
In September	154,800	272,000
Birth rate (b) (d^{-1})	0.147	0.055
Population growth rate (r) (d^{-1})	0.047	0.036
Death rate (d) (d^{-1})	0.100	0.019

Several predictions can be derived from this hypothesis (Lampert 1993):

1. *Zooplankton must migrate upwards in the evening and downwards in the morning.* This is, as a matter of fact, the "normal" behavior. "Reverse" migrations can sometimes be explained by indirect effects (see below).

2. *Vertical migration should occur mainly in zooplankton that are most visible to fish.* It can be seen that large, adult zooplankton, especially females with eggs, migrate more than young animals.

3. *The amplitude of the migration should vary as a function of the activity of the fish.* In Lake Constance, *D. hyalina* migrates only in the summer, when the fish in the lake are active. Recently, there is much evidence that zooplankton migrate only at times when fish are abundant (Ringelberg et al. 1991) and will cease migrating after a reduction in the fish population (e.g., through biomanipulation; Dini and Carpenter 1988) and that the amplitude of the migration is correlated with the strength of the year class of fish (Frost 1988).

These examples show that the pattern of vertical migration can change quite rapidly in response to changes in the predation pressure. This raises the question whether the migration response is due to strong selection or to the ability of the zooplankton to detect the predators. The selection hypothesis would require that there is a variety of migrating and nonmigrating genotypes present in the population. When fish move into the pelagic region, they would eat the nonmigrating animals, leaving only the migrating forms. Selection experiments indicate that the phototactic response of zooplankton has a genetic basis (DeMeester and Dumont 1988). It is therefore possible that nonmigrating genotypes could be eliminated at certain times of the year in a lake.

New experiments raise doubts about this hypothesis. Two of these studies were conducted in plastic bags. Bollens and Frost (1989) examined the migration behavior of the marine copepod *Acartia hudsonica* in Dabob Bay (Seattle, Washington). Copepods in the plastic bags without fish stayed near the surface. The copepods began to migrate normally, however, as soon as fish were added. Such rapid initiation of migration essentially precludes the possibility that the change was caused by selection. The authors suspected that the zooplankton detected the predators by means of a mechanical stimulus. Other experiments (Ringelberg 1991, Dawidowicz et al. 1990) have shown that fish can release a chemical stimulus that influences the migration behavior. As described above, zooplankton respond to changes in light intensity, but they become more highly motivated to respond to light by the "smell" of fish (Loose et al. 1993).

Figure 6.39 provides evidence for a chemical cue that causes *Daphnia hyalina* to migrate. This experiment was performed with a *Daphnia* clone obtained from the migrating population in Lake Constance (see Fig. 6.30), using the Plankton Towers at the Max-Planck Institute for Limnology at Plön. Surface water from each tower was circulated through an external aquarium. If the aquarium was empty, daphniids remained in the epilimnion during the day and night. One small planktivorous fish in the aquarium was sufficient to change the diel pattern of *Daphnia*'s depth distribution. Although the daphniids had no physical or visual contact with the fish, these zooplankton mi-

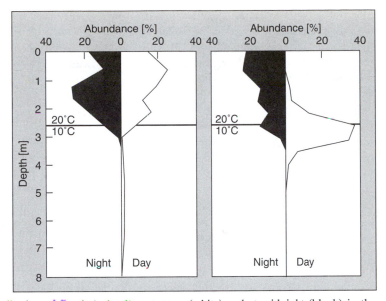

Figure 6.39 Distribution of *Daphnia hyalina* at noon (white) and at midnight (black) in the Plankton Towers in the absence (left) and in the presence (right) of a chemical cue from fish. With "fish smell" zooplankton are motivated to react to the light and migrate across the thermocline (horizontal line) (from Loose 1993).

grated downwards during the day. This experiment also confirms the predictions mentioned above: (1) migrations are controlled by light; (2) large individuals migrate deeper; (3) the mean depth of the *Daphnia* population increases with increasing abundance of fish in the aquarium (Loose 1993).

In another plastic bag experiment, Neill (1990) was able to explain the cause of a "reverse" migration. For many years there were no fish in Gwendolyne Lake (British Columbia, Canada). During the day the copepod *Diaptomus* was in the surface water and at night would migrate into the deep water. By doing this, the copepod apparently avoided the predatory *Chaoborus* larvae, which had a "normal" migration (in the deep water during the day). *Chaoborus* were eradicated from the lake by the accidental introduction of trout from a nearby hatchery. When the lake was sampled again 2 years later, the copepods had stopped migrating and always stayed in the surface water. Although there were enough fish to keep *Chaoborus* out of the lake, apparently they were not abundant enough to induce a normal migration in the copepods. As soon as Neill placed *Chaoborus* larvae in plastic bags with the copepods, the copepods began to have a reverse migration. A chemical stimulus was involved, since the same migration response was caused by adding to the enclosures water that had previously held *Chaoborus*. The copepods responded within 4 hours after the addition of *Chaoborus* water. The chemical induction and the speed of the response show that it was not caused by a selection of certain genotypes and that the animals were programmed for the migration behavior.

Even if the direct costs of swimming are low (Dawidowicz and Loose 1992), diel vertical migration can produce considerable demographic costs by displacing zooplankton from their optimal habitat (Loose and Dawidowicz 1994). On might expect, therefore, that patterns of vertical migration must be variable in order to minimize the costs and maximize the net gain for the animals. Amplitude and timing of migration should not only be determined by the presence of predators, but also by temperature and food conditions or the presence of inhibitory cyanobacteria in the epilimnion (Haney 1993). For example, if the food abundance is very low, animals may be forced to remain in the epilimnion and take the risk of being detected by a fish, as their reproductive rate may otherwise be too low under the combined effects of low food and low temperature. On the other hand, if the thermal gradient is shallow and the hypolimnion contains sufficient food, as in a hypertrophic lake, animals may remain in the hypolimnion all the time and not take the risk of migrating upwards (Gliwicz and Pijanowska 1988). As a result, the observed patterns may be very different, despite the underlying driving force of predator avoidance.

Vertical migration is a good example of optimization of a behavior by natural selection. It also illustrates how the optimization of the fitness of an individual can affect the entire ecosystem. If zooplankton stay in the deep water, they reduce the grazing pressure on the phytoplankton, allowing the algae to grow uninhibited during the day. This allows for an increase in the primary production and also changes the competitive interactions between the different algal species. As some zooplankton specialize in certain species of algae, the different migration patterns of the various zooplankton species (cf. Fig. 6.37) means that each algal species is affected differently by the migration (Lampert 1992). This produces a quantitative as well as qualitative "cascading effect" (cf. Section 7.3.4) of fish–zooplankton–algae.

REVIEW QUESTIONS FOR CHAPTER 6

1. Does it follow from the "competitive exclusion principle" that all existing species are the "winners" of competition?

2. In a chemostat, two phytoplankton species (a chrysophyte and a chlorophyte) compete for nitrate nitrogen. The chrysophyte has a low μ_{max} (0.7 d^{-1}) and a low k_s (0.1 μmol l^{-1}). The chlorophyte has a high μ_{max} (1.4 d^{-1}) and a high k_s (1 μmol l^{-1}). Which species would you predict would win the competition at dilution rates of 0.1, 0.3, 0.5, and 0.65?

3. The diatoms *Fragilaria* and *Stephanodiscus* compete for silica and phosphorus in a chemostat. Both have identical maximal growth rates (1.2 d^{-1}). The half-saturation constants of Si-limited growth are 2 μmol l^{-1} for *Fragilaria* and 0.6 μmol l^{-1} for *Stephanodiscus*. The half-saturation constants for P-limited growth are 0.03 μmol l^{-1} for *Fragilaria* and 0.1 μmol l^{-1} for *Stephanodiscus*. Construct ZNGIs

for a dilution rate of 0.5 d^{-1} and predict the outcome of competition (winner–loser or coexistence) for the following Si and P concentrations in the inflow medium (in μmol l^{-1}):

Si: 200P: 1

Si: 800P: 4

Si: 100P: 1

Si: 40P: 1

Si: 80P: 4

Si: 200P: 1

Si: 60P: 4

Si: 3P: 1

Si: 12P: 4

4. Keddy's competition experiments with the phytometer *Lythrum salicaria* did not detect reversals in the rank order of competitive strength of different macrophyte species. Tilman's chemostat experiments with phytoplankton did indicate such reversals when experimental conditions such as resource ratios were changed. Can you explain possible reasons for the differences in the results of these two experiments?

5. Planktonic bacteria are frequently better competitors for dissolved P than phytoplankton. Nevertheless, phytoplankton are not excluded from lakes where P is the primary limiting factor. There are several possible explanations for the persistence of phytoplankton. Could you mention a few?

6. Why are allelopathic interactions more commonly observed between benthic than between planktonic algae?

7. The negative correlation between *Daphnia* and rotifer abundance could be explained either by exploitative or by interference competition. Which observations support the latter?

8. Which observations and experiments could support the hypothesis that the minimum of phytoplankton biomass (clear-water phase) commonly observed in early summer is caused by grazing and not by sedimentation, nutrient limitation, or light inhibition?

9. Under which conditions can filter-feeding zooplankton cause an increase of the net growth rate of a phytoplankton species?

10. Organisms in a lake experience predictable and unpredictable changes in the environment. List some examples of predictable and unpredictable events that occur in lakes. What adaptations have organisms evolved in response to these events? Identify the environmental cue used by the organism in each of these cases.

CHAPTER 7

Communities

7.1 CHARACTERISTICS OF COMMUNITIES

The sum of all of the interacting populations in a habitat is called a "community" (bio-coenosis). Practically all modern ecology textbooks use this definition. At second glance the definition is not so clear, for, strictly speaking, it would be difficult completely to exclude interactions between populations that are far apart. An atom of oxygen respired by a fish in a stream may have been released into the atmosphere by the photosynthesis of an alga in a distant lake and been dissolved back into the habitat of the fish. However, no one would think of assigning both organisms to the same community. It is more useful, then, to view a community as a system of populations that are bound to one another by *strong interactions* and surrounded by a surface of weaker interactions with populations outside the community.

Even then it is not easy to define the community either in terms of its spatial limits or its species association. A major comprehensive research effort would be needed actually to determine the actual strength of the interactions. In reality, communities are defined a priori, without measuring the strength of the interactions. For example, when we distinguish between a planktonic and a benthic community, we assume that there are strong interactions within each of these communities, and that the shading of the benthos by the plankton and the effect of bottom organisms on the flow of matter between the sediment and open water can be characterized as "weak interactions."

Ecologists often define communities by common functional characteristics rather than by assumed strong interactions. For example, the open-water communities "plankton" and "nekton" are distinguished from one another by the ability to move. It is quite possible, however, that there are stronger interactions between zooplankton and fish populations than between many of the populations within the plankton.

7.2 "SUPERORGANISM" OR "SIEVE"

There is much debate about the strength and type of interactions between populations and the degree of integration of the communities. The "superorganism" concept is an extreme position originated by the terrestrial plant ecologist Clements and introduced into limnology by Thienemann. Communities and ecosystems (Chapter 8) are viewed as highly integrated, natural entities, or "organisms of a higher order." Features (*"emergent properties"*) are ascribed to the whole (communities, ecosystems) that cannot be explained by the combined effects or interactions of the individual components (populations). It is assumed that that "the whole is greater than the sum of the parts."

It is important to make the distinction between "emergent properties" and "collective properties" when discussing the superorganism concept. The existence of collective properties is widely accepted, whereas emergent properties are hotly debated.

A classic example of a collective property is the statistical correlation between the concentration of a limiting nutrient and the potential biomass of phytoplankton. This correlation can be explained by the limited variation in the chemical composition of the different algal species and the fact that algae cannot use more nutrients for biomass formation than are present. Interactions between algal populations cannot result in greater production of biomass. Thus this phenomenon can be explained as the sum of all component processes. It is not necessary to look for "system" properties that go beyond the activities of the individual organisms.

"Self-regulation" of communities is a typical emergent feature proposed by proponents of the superorganism concept. This suggests that the community as a whole can hold certain properties constant (e.g., the total energy flow) when there are external disturbances, analogous to the ability of a homeothermic organism to keep its body temperature constant.

The greatest problem with the superorganism concept is that communities do not have a centralized genome. Thus a community has no means, therefore, of reproducing to perpetuate its identity. The system as an entity cannot then be subject to the forces of natural selection and cannot evolve adaptations to the environment.

The *"individualistic concept"* is an extreme position opposing the superorganism concept that can be traced back to Gleason (1926). It states that populations respond to the external environment completely independently of one another (Harris 1986). In this concept the local environmental conditions can be visualized as a sieve that regulates the species composition. A supply of organisms is transported to the system (active migration, passive transport) and the best-suited species are admitted and those poorly suited are excluded. The species composition is primarily the net result of stochastic transport processes and autecological requirements. In its most extreme form, the individualistic concept negates the importance of interactions between populations. The absence of interactions between populations would be found, however, only during the early phase of colonization when the habitat is essentially empty or in habitats that are extremely disturbed, where the populations are held well below their upper limit by external disturbances.

[255]

A third concept is needed to describe communities in most habitats. We will refer to it as the *"Darwinistic"* concept, in accordance with Harper (1967). The Darwinistic concept refutes the idea that entire communities and ecosystems result from evolutionary selection and reproduction of selected features and that community characteristics can be optimized at the expense of the populations. It recognizes, however, the fact that organisms modify their environment and become the "environment" for one another. Thus populations that have passed through the "sieve" of lethal limits have interactions that become factors that select for the evolutionary adaptations of the individual components of the community. The community is more than a random collection of individuals, even though it is not a superorganism.

7.3 INTERNAL STRUCTURE OF COMMUNITIES

7.3.1 Food Chains and Food Webs

Populations within a community are bound by a network of interactions. The most important interactions are of a trophic nature, that is, "eat" and "be eaten". In diagrams predator–prey interactions are usually portrayed vertically and the competition relationships are portrayed horizontally. The simplest presentation of the vertical connections within a community is the *food chain*. Plants are eaten by herbivorous animals (*primary consumers* or *secondary producers*). These animals are in turn eaten by carnivorous animals (*secondary consumers*). Bacteria are placed in this concept as decomposers, that is, the remineralization ("destruction") of the dead organic matter. The individual links of the chain correspond to trophic levels (see Chapter 8) in the transport of energy and matter. Traditionally, terrestrial food chains have three links (grass–zebra–lion or grass–cattle–humans); aquatic food chains usually have four links (phytoplankton–zooplankton–planktivorous fish–piscivorous fish or microphytobenthos–benthic invertebrates–benthivorous fish–piscivorous fish).

The food chain concept is too simple for natural systems, except for extremely species-poor habitats. For example, filter-feeding organisms (e.g., many "herbivorous" zooplankton) select their food much more on the basis of size than according to its trophic role. This makes it difficult to assign organisms to a particular level in the food chain. When a *Daphnia* eats a phytoplankter, it is a "primary consumer"; however, if it eats a phytoplankton-eating zooflagellate or a ciliate, it becomes a "secondary consumer." Planktivorous fish eat "herbivorous" as well as "carnivorous" zooplankton; they can be a third, fourth, or fifth link in the food chain, depending on what their food had eaten. The more complex *food web* has replaced the simple food chain. The pelagic food web shown in Fig. 7.1 is extremely simplified. One important simplification is the grouping of populations into functional categories (e.g., "herbivorous" zooplankton). Such functional categories are best referred to as "guilds."

A further simplification seen in Fig. 7.1 is that it ignores the changes in trophic roles that may occur during the ontogeny of a species. For example, the early copepodid stages of the "carnivorous" copepods are usually "herbivorous" and the juvenile stages of piscivorous fish are usually planktivorous.

[256]

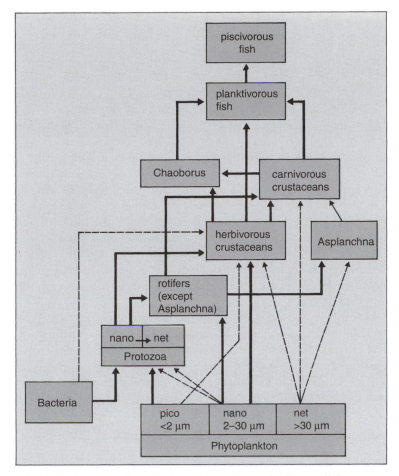

Figure 7.1 A simplified food web for the pelagic region. Only the feeding on living organisms is shown. (*Solid lines*) All or most of the species of predators and prey categories are involved. (*Dashed line*) Only a few of the species of predators and prey are involved.

According to Fig. 7.1, organic matter flows from all components to the bacteria, which use excretory products and dead matter (detritus), but do not feed on or kill living organisms. The feeding of heterotrophic bacteria on the excretions and remains of other organisms makes this a circular component in the food web, since the bacteria themselves also serve as food for the protozoans and herbivorous zooplankton. This transfer of matter and energy has been recently given the name *microbial loop* (cf. Section 8.2.4) and has become an important topic in marine and freshwater microbial research (Azam et al. 1983).

The study of microbial interactions became possible only recently with the development of new techniques for the identification of microorganisms in field samples. More progress can be expected with further development of molecular techniques, such as low-molecular-weight RNA analysis (Höfle 1993) and specific DNA and RNA probes (Lim et al. 1993) that allow the identification of species or functional groups. Growing knowledge about individual species and groups or microorganisms and their interactions has led to the concept of *microbial food webs*. Microbial food webs consist of bacteria, picoplanktonic cyanobacteria, and various types of protozoa (heterotrophic and mixotrophic flagellates, ciliates).

The organization of microbial food webs is very similar to metazoan food webs (Riemann and Christoffersen 1993). The basic difference is that microbial food webs start with dissolved organic matter. Otherwise, processes and interactions in both food webs are similar. In microbial webs, heterotrophic nanoflagellates (HNF) and ciliates fill the role of predators. Their functional responses are similar to those of metazoans; they compete with one another and they feed selectively. Feeding HNFs release nutrients derived from their prey, stimulating further prey production (cf. Section 6.4.4). Flagellate grazing also leads to the selective removal of prey species. For example, the size structure of bacterial populations can shift to small particles due to removal of the preferred large bacteria by protozoans (Bird and Kalff 1993). It is not difficult to see the analogy to the effect of fish predation in metazoan food webs. As in algal communities, intense grazing leads to the development of grazing-resistant forms (Jürgens and Güde 1994).

Microbial food webs are coupled to metazoan food webs as protozoans are a good food for many zooplankton. In particular, *Daphnia* has a strong impact on flagellates and ciliates. This results in a very different structure of the microbial food web depending on the presence or absence of large cladocerans (*Daphnia*) (Jürgens 1994). Figure 7.2 contrasts the situations in a *Daphnia*-dominated system and in a system dominated by small zooplankton (e.g., under strong pressure by planktivorous fish, cf.

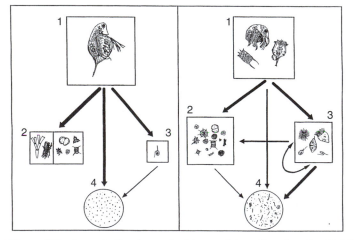

Figure 7.2 Impact of metazooplankton on interactions in microbial food webs, when the zooplankton is dominated by *Daphnia* (left) or small crustaceans and rotifers (right). Organisms are grouped into four categories: (1) metazooplankton, (2) primary producers, (3) phagotrophic protozoans, (4) bacteria. Sizes of boxes represent the relative amount of biomass. Width of arrows indicates the interaction strength (from Jürgens 1994).

Fig. 6.31). In the presence of *Daphnia*, edible phytoplankton and all protozoans are suppressed. This results in the dominance of inedible phytoplankton and the low abundance and diversity of protozoan grazers. There is little impact of protozoans on the bacteria. *Daphnia* has fine filters, enabling it to graze on bacteria, although small bacteria are at the lower limit of its food size spectrum (cf. Section 4.3.11). Thus *Daphnia* grazing results in the dominance by small bacteria. Grazing is less important in systems lacking *Daphnia,* since the small grazers are less efficient. As a result, phytoplankton in these systems are small and mostly edible. Protozoans are abundant and diverse. There are many predator–prey interactions among protozoans. Some protozoans also graze on phytoplankton. The strong impact of protozoans on the bacteria populations results in the development of filamentous bacteria and bacterial aggregates that are resistant to protozoan grazing, but not to grazing by *Daphnia*. Thus *Daphnia* has a key role (cf. Section 7.3.3) in structuring microbial food webs.

When the flow of matter and energy is considered quantitatively, as presented in Chapter 8, the food web becomes much more simplified, since only a few of the strands are important. Few, if any, investigations have considered all potential components of a food web in sufficient detail. In the pelagic zone of many lakes it appears, however, that only two strands of the food web are very important: The four-link chain nanophytoplankton–"herbivorous" crustaceans–planktivorous fish–piscivorous fish and the five-link chain picophytoplankton–nanoprotozoans (zooflagellates)–"herbivorous" crustaceans–planktivorous fish–piscivorous fish. Because of this simplification, the use of the food chain concept and the related *trophic level* concept can still be justified.

7.3.2 Problems of Aggregation

Communities consist of populations. Strictly speaking, an accurate analysis of the structure of a community should consider each population separately. In most community studies, however, populations are placed into collective categories (*"aggregations"*) for the following reasons:

1. The definition of populations is by no means clear for organisms with little or no sexuality (almost all phytoplankton, many rotifers, and cladocerans) (cf. Section 5.1).

2. Species determination may require so much effort (e.g., setting up of cultures for certain algal species) that it is not practical to include this as part of a community study that involves large numbers of samples in space and time.

3. Most important, an intelligent aggregation can result in a greatly improved ability to forecast changes and make generalizations. The various species are often so similar in their requirements and activities that the differences in conditions leading to the dominance of one or the other species are too slight to be detected. One would have to attribute the appearance of such species viewed separately to "random variation," whereas, dealt with as a group, their occurrence would be more predictable. A well known example of this is the spring maximum of phytoplankton in Lake Windemere, which has been studied for over 3 decades (Reynolds 1984). The pre-

diction that *Asterionella formosa* will be the dominant species during the spring maximum holds true for about 80% of all the years in the study, but the same prediction made for diatoms would be true for all years. Similarly, the biomass of individual algal species contributing to the spring peak is variable, but the total biomass of diatoms is relatively constant.

Populations can be grouped according to taxonomic or functional criteria. *Taxonomic aggregation* (taxa above the species level) has the advantage that it is not biased by the persuasion of the ecologist; however, it does not always produce a useful aggregate. The plankton dinoflagellate *Gymnodinium* is an extreme example of a misleading aggregation. It includes species that are autotrophic, some that use dissolved organic carbon (DOC), and still others that eat other organisms. A more meaningful and functional taxonomic aggregation is the genus *Daphnia*. Within this genus, there is a greater effect of body size on feeding rates and food spectra than between different daphniid species. Likewise, the cyanobacterial family Nostocacae is a unified and meaningful grouping. All representatives can fix N_2, form large colonies, are highly inedible for the zooplankton, and are capable of migrating vertically, by means of gas vacuoles.

A *functional aggregation* is generally more meaningful than a taxonomic aggregation. First, it can use the guilds defined for the food web. It can then be further differentiated, as, for example, according to size. Other environmental requirements, such as temperature, water chemistry, or competitive ability, can be used for additional subdivisions. Within the net phytoplankton, for example, the diatoms form a useful functional grouping based on their high sinking rate, high silica demand, low light requirement, and ability to compete for phosphorus. One should keep in mind that a functional aggregation is highly subjective and tentative.

7.3.3 Keystone Species

Populations within a food web directly affect their "neighbors" and indirectly influence other populations with which they are integrated. Effects of population interactions can travel through the food web, although the effects may be dampened. Species with especially far- reaching effects are called keystone species (*keystone predators*; Paine 1969). Just as one cannot remove a keystone in an arch without causing its collapse, a keystone species cannot be removed without causing a dramatic change in the community.

Such keystone species need not be associated with major pathways of matter and energy in the food web. For example, a disease vector with negligible energy flow could alter the composition of a community by eradicating a dominant population within a guild, thereby allowing its competitors to become dominant.

Piscivorous fish are frequently keystone species in lakes. They eat planktivorous fish that, in turn, eat zooplankton (cf. Section 6.5.3). If there are few piscivorous fish (e.g., because of overfishing), the population of planktivorous fish will increase, eliminating the large herbivorous zooplankton and invertebrate predators, so that the small zooplankton become dominant (cf. Fig. 6.31). The reduced zooplankton biomass and the shift to smaller zooplankton results in lower grazing pressure, increased growth of the small algae, and lower water clarity. The macrophytes and their associated fauna may also change in response to the poorer underwater light conditions. In this case, one guild, or even one species, can completely restructure a community, even though its biomass is very small.

Cormorants have taken on the role of keystone species in the shallow lakes of the Norfolk Broads in England (Leah et al. 1980). Of two neighboring lakes, only one was connected to a nearby stream. Cormorants drastically reduced the population of planktivorous fish in the isolated lake, whereas there were frequent immigrations of fish into the other lake from the stream. Both lakes had similar nutrient concentrations. The water in the isolated lake became clear after the large zooplankton became abundant. Macrophytes then increased in response to the higher water transparency and the diversity of the benthic animals also increased.

Predatory fish can also be an important link in the benthic food chains. Small prairie streams provide a striking example (Power et al. 1985). Prairie streams have extremely variable flow. In the summer the stream may consist of isolated pools connected by small rivulets, but after a rainstorm the brook can become a gigantic river.

During the dry period, one can recognize from a distance which pools have a large predatory fish (smallmouth bass, *Micropterus*). The bottoms of these pools are covered with a brown layer of algae, with only occasional small strips of bare gravel. Pools without smallmouth bass are clean and have algae growing only in the very shallow water. Small, algal-grazing minnows (*Campostoma*) are normally numerous enough to keep the growth of algae in the pools under control. Hence, the pools are clean. The predatory smallmouth bass eat most of the *Campostoma* or chase them out of the pools. Those that remain stay in the very shallow regions of the pool, where they create clean "edge strips," while the deeper parts of the pool become covered with benthic algae. Clearly, the contrasting algal growth creates very different living conditions for the benthic animals (e.g., insect larvae and other periphyton grazers). The algivorous minnows had a negative effect on crayfish, probably because of resource monopolization. On the other hand, the production of snails (*Physella*) was indirectly enhanced by the minnows. By removing the overstory algae, the minnows increased the growth of the algae closely attached to the substrate, which are fed upon by the snails (Vaughn et al. 1993). In this case, the predatory fish can be viewed as a keystone species.

The impact of a predator on a community depends on the strength of its interaction with other species. Brönmark (1994) studied the effects of Eurasian perch (*Perca fluviatilis*) and tench (*Tinca tinca*) in benthic enclosures. Perch and tench both feed on macroinvertebrates and thus belong to the same trophic level. Only tench eats large numbers of snails, however. Snails appeared to have the strongest interactions within the macroinvertebrate community. As tench reduced the number and biomass of grazing snails in the enclosures, there was an increase in the periphyton biomass on the macrophytes (*Elodea canadensis*) and a decrease in the biomass of the macrophytes, probably due to shading by periphyton. Enclosures with perch that do not feed on snails had no such "cascading" effects.

The idea that the effect of fish can be translated through the zooplankton to the phytoplankton has led to its practical application as a form of *biomanipulation* in eutrophic lakes. It is based on the premise that a reduction in the population of planktivorous fish will lead to larger zooplankton, increased grazing, and clearer water.

When many lakes are compared, there is a significant logarithmic relationship between phosphorus concentration and algal biomass (cf. Fig. 8.11). The deviation of some of the individual lakes from the regression line is quite striking. Many of the lakes that are below the regression line, having less chlorophyll than predicted for their phosphorus concentration, are lakes that had small populations of planktivorous fish. J. Shapiro concluded that one should be able to improve the water quality of a eutrophic lake by manipulating the fish populations. He referred to this as "biomanipulation," although *"food- chain manipulation"* may be more appropriate. In the first experiments fish populations were eliminated with rotenone, a fish poison that also kills many of the zooplankton, although they quickly repopulate the lake as the poison is rapidly decomposed (Anderson 1970). The species composition of the zooplankton changes within a short time. As expected, the large filter-feeding zooplankton appear and suppress the algae (Fig. 7.3).

Lakes always respond very rapidly to the elimination of fish by rotenone. Poisoning of lakes is not only a questionable method, it also has short-lived results. Migrating planktivorous fish usually reenter the lakes and reestablish the dominance of small zooplankton. Also, the composition of the algae frequently shifts toward highstanding crops of grazing-resistant species. Both of these effects can be seen in Fig. 7.3. Thus efforts are being made to develop "less harsh" methods that produce stable, long-term effects. One method is to remove planktivorous fish by intensive netting and then stock the lake with large numbers of piscivorous fish. Many north temperate lakes in Europe and North America are suited to the use of walleye, pike perch, northern pike, and largemouth bass. After a period of repeated stocking, the population of piscivorous fish may reproduce naturally, and the altered food web structure can be maintained at a stable level. Even if the total algal biomass is not reduced, the phytoplankton are often dominated by colonial forms that scatter the light less than the small algae, producing clearer water (Benndorf et al. 1988).

The piscivorous fish are keystone species, since their effect transcends several trophic levels, ultimately changing not only the plankton, but the entire lake. This is especially apparent from the biomanipulation efforts in The Netherlands. In one case, fish were removed by draining the lake. The following year the water stayed very clear due to large *Daphnia* that kept the algae under control, despite very high concentrations of nutrients in the lake. Following the manipulation, light penetrated to the bottom of the lake and massive amounts of macrophytes began to grow immediately. These aquatic plants stabilized the sediments and bound the nutrients. The following year, the zooplankton biomass decreased, but the lake stayed clear. Then they were able to place fish back into the lake and establish high densities of piscivorous fish, since especially pike use the macrophytes as a preferred habitat and hiding place. At this point, the lake was beautiful and clean, but of limited use for swimming. Sufficient clear space for swimmers was made by periodically removing the plants, without negatively impacting the restoration effects (Van Donk et al. 1989).

Results from lake biomanipulation efforts are contradictory. Even where the manipulation has succeeded, a single theory based on predation cannot explain all the ef-

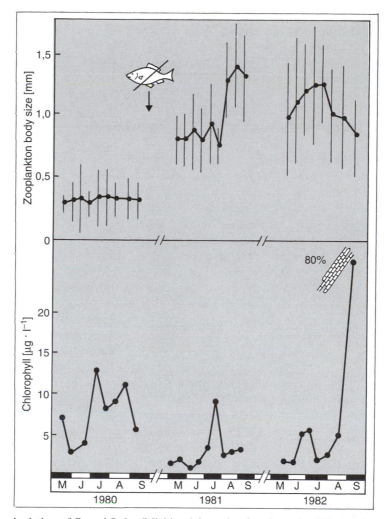

Figure 7.3 Biomanipulation of Round Lake (Michigan) by poisoning the entire fish population with rotenone. (*Above*) Body size (mean and standard deviation) of the zooplankton in the summer before (1980) and in the summer following (1980, 1982) biomanipulation. (*Below*) Biomass of the phytoplankton (measured as chlorophyll) before and after biomanipulation. The maximal value after biomanipulation (September 1982) was caused by the massive development of the cyanobacteria *Aphanizomenon flos-aquae*, which is largely resistant to grazing by zooplankton (data from Shapiro and Wright 1984).

fects produced. It can be shown in the laboratory that some of the large colonial algae (especially cyanobacteria) cannot be eaten, even by the largest zooplankton (e.g., *Daphnia magna*). These are the algal species that usually dominate when a biomanipulation fails to reduce the algal biomass. Sometimes, however, they do not appear after biomanipulation. One possible answer is that the large colonies begin as small aggregates

[263]

made up of only a few cells that can be grazed. The success of the biomanipulation may depend on having high densities of zooplankton present when the small colonies are first developing and are still edible. If the zooplankton appear too late, they are unable to gain control of the algae (Gliwicz 1990b). Another possibility is that zooplankton release large quantities of CO_2, lowering the pH and making an unfavorable chemical environment for the development of cyanobacteria (Shapiro 1984).

7.3.4 The "Bottom-up/Top-down" Controversy

The keystone species concept, the size-efficiency hypothesis, and experience from biomanipulations have led to the assumption that, in contrast to the flows of energy and matter, the controlling influence ("controls") in communities flows from the top to the bottom (*"top-down"*). This idea of control flowing downward in the food chain has been termed *trophic cascade* (Carpenter et al. 1985). The biomass and species composition of each guild is determined as those species left over by the predators. The traditional *"bottom- up"* concept maintains that "many prey can feed many predators," whereas the top-down hypothesis proposes "where there are many predators, few prey will remain." The bottom- up hypothesis requires that the biomass of all the trophic levels are positively correlated and depend on the fertility (limiting resources) of the habitat: more available nutrients → more algae → more zooplankton → more planktivorous fish → more piscivorous fish. The top- down hypothesis predicts, however, that the adjacent trophic levels will be negatively correlated: more piscivorous fish → fewer planktivorous fish → more zooplankton → fewer algae → more available nutrients.

Proponents of the top-down hypothesis typically use the results from mesocosm enclosures ("limnocorrals"; cf. Fig. 2.1b). Pelagic communities are often manipulated in these experiments by adding fish, reducing zooplankton populations, and adding nutrients. Stocking of fish in the mesocosms usually produces the results predicted by the top-down hypothesis: reduction in the zooplankton biomass, shift to small zooplankton species, and increase in phytoplankton. Alternatively, an increase in phytoplankton can also be caused by fertilizing the enclosure, producing a bottom-up effect. Time and space limitations of this type of experiment do not allow one to see whether this effect is also carried through the higher trophic levels to the fish. In a typical enclosure experiment (Fig. 7.4) in which fish and nutrients were manipulated, the presence of the sunfish (7 cm long) suppressed the larger zooplankton, the cladoceran *Diaphanosoma birgei*, only when the system did not receive nutrient additions. Populations of the small cladoceran *Bosmina longirostris* did not react to the presence of the sunfish. However, both zooplankton responded positively to higher algal concentrations following nutrient enrichment. Larger zooplankton such as *Daphnia* were not present in this lake due to high fish densities. From other experiments one could conclude that *Daphnia* would have been excluded entirely in the enclosures with fish.

The relative importance of "bottom-up" or "top-down" controls has been debated mainly as an "either-or" question. There are few quantitative comparisons of these two effects. To do this, one would have to make dose-effect-type comparisons of the effects of nutrient and fish manipulations on all trophic levels. The question would be "How many fish produce how many algae?" This type of investigation would be dif-

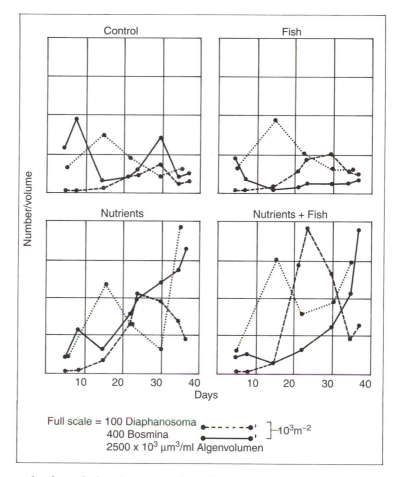

Figure 7.4 An example of a typical enclosure experiment for manipulation of plankton communities. Four groups of 3 plastic bags (1 m³) were treated differently: the controls were untreated; one group received one sunfish (7 cm) per bag; the third group was enriched with nutrients (N and P); the fourth group received nutrients and a fish. The changes through time in the most important components of the plankton are shown. (*Solid line*) The intermediate-sized cladoceran *Diaphanosoma birgei* (full scale = 10^5 individuals/m²). (*Dotted line*) The small cladoceran *Bosmina longirostris* (full scale = 4×10^5 individuals/m²). (*Dashed line*) Volume of "edible" phytoplankton (full scale = 2.5×10^6 μm³/ml) (from Vanni 1987).

ficult to conduct because of the tremendous technical effort that is needed to deal with even a few enclosures. Thus most manipulations make comparisons in terms of "all or nothing" or "much and little."

Whole lake observations have similar contradictory results. On the one hand, the top-down hypothesis is frequently confirmed by responses to biomanipulations and changes in plankton communities after fish immigrations or a fish dieoff (Vanni et al. 1990). On the other hand, investigations comparing lakes with differing trophic conditions show a positive correlation of phosphorus not only with the phytoplankton biomass, but also with the biomass of the total zooplankton, the crustacean zooplankton, the fish populations, and the fish harvested (cf. Section 8.5.3). These studies support the bottom-up hypothesis. Year-to-year comparisons within the same lake with essentially constant nutrient conditions usually support the top-down hypothesis, whereas comparisons of different lakes tend to support the bottom-up hypothesis.

Manipulated streams provide results similar to the enclosures. The manipulation of top predators results in cascading top-down effects. Bechara et al. (1992) compared experimental streams with and without the top predator, brook trout (*Salvelinus fontinalis*). When brook trout were present, large invertebrate grazers (mayfly nymphs) were suppressed. This resulted in a short-term doubling of the periphyton biomass. Small grazers (chironomid larvae), however, profited from the abundance of periphyton. Thus there was a strong top-down effect without changing the availability of nutrients. On the other hand, streams that receive nutrient additions have higher algal biomass than nutrient-poor streams. As shown earlier for lakes, both top-down and bottom-up mechanisms also are important in streams.

Rosemond et al. (1994) examined the relative effects of herbivory and nutrients on stream algal communities. They manipulated nutrient levels (N and P) and grazing by snails in experimental streams with a simple two-trophic-level community structure. The addition of nutrients and removal of snails produced the greatest effect on the community structure and growth of the periphyton. The effects were less pronounced when grazers were removed at low nutrient levels or when nutrients were added under grazing pressure, indicating both mechanisms are important. Snails grew better at high nutrient levels. They were controlled bottom-up as the system had no predator on snails. The relative strength of bottom-up and top-down control varied for different parameters, but clearly the combined bottom-up and top-down control effects were stronger than either one of the two controls.

Species composition, as described earlier for biomass, can be examined for controls from below (e.g., influence of the Si:P ratio on the phytoplankton composition, Fig. 6.10) or from above (e.g., shift toward poorly edible phytoplankton species with increasing influence by zooplankton). Clearly, although the two controls represent independent mechanisms, they are not mutually exclusive. It is easy to include species-specific mortality rates in the competition model (cf. Section 6.1.3), where they shift the outcome of the competition by changing the R^*. Grazing pressure does not necessarily eliminate the competition for resources; it only changes the conditions under which it takes place.

7.3.5 Attempts at Synthesis in the "Bottom-up/Top-down" Controversy

Dampening of the effects
McQueen et al. (1989) proposed a synthesized bottom-up/top-down concept that suggests the effects of both nutrients and fish are increasingly dampened as they pass through the food web. Thus, being at the bottom trophic level, phytoplankton show the greatest bottom-up effects, whereas the higher trophic levels (from zooplankton on upwards) respond more to the top-down effects. This concept is illustrated by a 7-year

study of Lake George, Canada (Fig. 7.5). In the beginning (1980–1982) the lake had a large population of piscivorous fish and an moderate population of planktivorous fish. A winter fish kill caused by low oxygen under the ice (1981-1982) reduced both fish guilds. The population of planktivorous fish quickly recovered with the low predation pressure and reached a maximum in 1984. The piscivorous fish slowly recovered and the population of planktivorous fish slowly declined. The zooplankton reached maximum total biomass as well as maximum body size during the minimum of the planktivorous fish. This effect did not show up in the phytoplankton, for their biomass (measured as chlorophyll) was relatively high the same year as the zooplankton maximum. The gradual decrease in the phytoplankton during the entire period of the observations seems best explained by the simultaneous decrease in available nutrients (measured as total phosphorus). The between-year comparison shows a strong and negative correlation between piscivorous and planktivorous fish (top-down), a weaker, but likewise negative correlation between the population of planktivorous fish and zooplankton biomass (top-down), no significant correlation between zooplankton biomass and phytoplankton biomass (although there may have been a shift in species composition), and a significant positive correlation between phytoplankton biomass and total phosphorus (bottom-up).

Vertically alternating controls

Persson et al. (1988) developed a model in which there is a switching of controls from above and below that is related to the number of trophic levels in the food web. The original version of this model was developed outside limnology by Oksanen et al. (1981). At this time there are few empirical data available to verify the model. The number of trophic levels possible in a given habitat depends on the availability of resources (bottom-up), since there are large losses (> 80%) between trophic levels. There is a linear correlation between the potential primary production and the biomass of the plants in extremely infertile habitats that only have one trophic level (primary producers). When there is enough primary production to support a second trophic level, further increase in the potential production leads to an increase in the herbivore biomass (bottom-up), but to no further increase in the plant biomass, since they will be controlled by the herbivores (top-down). Further increase in production can support a third trophic level (primary carnivores) that prey on the herbivores (top-down). With increasing productivity the biomass of the plants will then increase (bottom- up), as they are released from the control by the herbivores. A fourth trophic level (secondary carnivores) can exist if the productivity increases further. The primary carnivores are then preyed upon by the secondary carnivores (top-down), releasing the herbivores from predation pressure to be controlled once again from the "bottom-up" and the plants are controlled by the herbivores (top-down). Summarizing, one can say: The number of trophic levels is controlled bottom-up; the highest trophic level is likewise controlled bottom-up; the next trophic level down is under top-down control; bottom-up and top-down control alternate between the trophic levels (Fig. 7.6). Deviating from the original model of Oksanen, Persson et al. postulate that a further increase in productivity will cause a decrease in the number of trophic levels. This is based on empirical observations and does not follow from the logic of the Oksanen model. In ex-

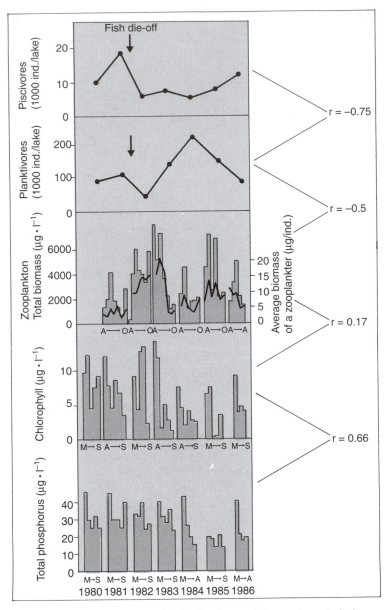

Figure 7.5 The case study of Lake St. George. The number of piscivorous fish and planktivorous fish (population census in the fall). Total biomass of zooplankton (bar graph) and mean biomass per individual (solid line), monthly average from April to October (1986 only to August); phytoplankton: chlorophyll concentration, monthly average from April/May to September; total phosphorus: monthly average from May to August/September. On the right-hand side are the correlation coefficients between the different trophic levels (from McQueen et al. 1989).

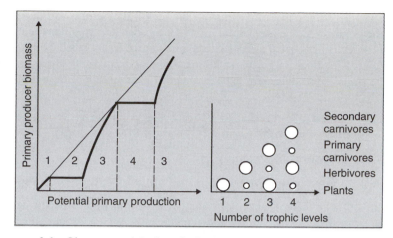

Figure 7.6 Diagram of the Oksanen model of vertically alternating control. (*Left*) The relationship between the plant biomass and the potential primary production and number of trophic levels. (*Right*) The effect of the number of trophic levels on the type of control of the individual trophic levels (large circles: bottom-up; small circles: top- down).

tremely eutrophic lakes, for example, cyprinids often dominate the fish populations or a fish die-off may lead to the elimination of all fish.

A limitation of the Oksanen model is that it makes the simplified assumption that there are discrete trophic levels. The advantage of this model over the cascading trophic level theory and the bottom-up/top-down hypothesis is that it does not view fish as an independent variable that externally controls the interactions within the system, but rather as a product of the boundary conditions that determine the development of the community within a habitat.

Is the bottom-up/top-down controversy a matter of scaling?

A comparison of many lakes with contrasting nutrient concentrations usually supports the bottom-up hypothesis. However, studies of lakes with similar nutrient levels (or different years for the same lake) and different types of fisheries management often verify the top-down hypothesis. It is premature, however, to conclude that a "large-scale" view (lake comparisons, eutrophication models) generally supports the bottom-up hypothesis, and a "small-scale" view (investigations within a lake, mesocosm experiments) usually supports the top-down hypothesis.

Even the small-scale approach can result in rapid shifts in controls from above and below, depending on the time period that is observed. An example is the spring succession in nutrient-rich lakes, when the algae first have their maximum, followed by the zooplankton maximum and the clear-water phase (cf. Section 6.4, Fig. 6.15). At the beginning of the season the growth rate of the zooplankton depends on the amount of algae available (bottom-up). Eventually, the grazing rate becomes so high that the clear-water phase develops (top-down).

The apparent change between controls from below and above is caused by the different time lags in the bottom-up and top-down mechanisms. Individuals that are eaten disappear immediately, but it takes time to translate an improved nutritional state into an increase in population density. The length of the time lag depends on the size of the organisms. The normal duration of a mesocosm experiment is usually sufficient to observe the effects of a fish manipulation on the lower trophic levels, but it is too short to see the effects of nutrient additions on the size of the fish population. This can only be observed in long-term experiments or by comparing different lakes.

7.4 SPECIES RICHNESS AND DIVERSITY

7.4.1 Measures of Diversity

Species richness and *diversity* are among the most intensively discussed concepts in the fields of theoretical ecology and conservation. Both concepts can be applied to entire communities as well as to subcomponents, such as trophic levels, guilds, and taxonomic aggregates. Species richness or total number of species is certainly the easier measure to understand, but in practice, it is difficult to determine. The discovery of rare species is very strongly dependent on the sample size and how intensively one searches; so it is practically impossible to determine the actual species richness for a community or subunit. In contrast, most diversity indices that are used stabilize after a relatively few, frequent species have been found. Thus rare species have little effect on these indices.

A diversity consists of two components: the number of species and the *equitability*, that is, the uniformity of abundance in the species found. If, for the same number of species, one species is dominant and all others are rare, there will be a lower diversity of the community, guild, etc. than if all species were represented equally. *Hurlbert's PIE* or "probability of interspecific encounters" is one of the most biologically meaningful concepts, of the many proposed *diversity indices* (Washington 1984). It shows the probability that two individuals randomly encountering one another belong to different species:

$$PIE = N/(N + 1)(1 - \Sigma p^2_i)$$

where N is the total number of individuals and p_i the proportion of species i in the total number (N_i/N).

At higher values of N, PIE approaches the simplified version of the *Simpson index* (D):

$$D = 1 - \Sigma p^2_i$$

Both indices express the equitability of the frequent species and are extremely insensitive to the effects of rare species. Indices based on information theory indicate the *"information content"* of an individual. These indices also rely primarily on the dominant species, but are somewhat more sensitive to rare species. Although there is much debate about the ecological relevance of the concept of "information," they are still the most widely used indices, especially the index H' of Shannon and Weaver:

$$H' = -\Sigma(p_i \log p_i)$$

[270]

It is difficult to calculate diversity indices when individuals cannot be easily defined, such as colonial protists. There are also problems within communities due to extreme size differences, where the smallest forms, often difficult to determine as to species, make up a trivial amount of the biomass, but may entirely dominate in numbers. For example, this is true for the "picoplankton," which in most cases can only be enumerated as a collective category. In such cases, it is usually better to calculate the diversity indices on the basis of biomass, rather than numbers of individuals.

7.4.2 Causes and Maintenance of Diversity

Thienemann's *"biocoenotic laws"* represent an early attempt to explain the diversity of different communities:

1. The more diverse the environmental conditions (i.e., the more species that can live near their optimum), the greater the number of species that will occur, although there will be only a few individuals of each species.

2. The more harsh the environmental conditions (i.e., the further they are from the optima of most species), the more a few species will dominate the total picture. These will be represented in very large numbers, however.

There are examples from many extreme biotopes, many of which are a result of anthropogenic changes. For example, over 100 tubificid worms per cm^2, more than one animal for each mm^2, were found in the highly polluted sludge in the Hamburg Harbor, Germany.

Quite early there were attempts to hypothesize large-scale trends in diversity: increase in diversity from the poles to the tropics; lower diversity on islands than on continents. These global trends were primarily based on macroscopic organisms in terrestrial and in littoral, marine communities. Lewis (1987) argues against a latitudinal trend in diversity for phytoplankton and zooplankton in lakes; it may hold, however, for fish, although there has been no systematic investigation of this. Even less is known about benthic communities in this regard. The extremely high species richness of the fish populations in Lake Tanganyika cannot be viewed as representative of the tropics, since this lake belongs to the small group of tertiary lakes that are much older than the large majority of lakes that originated in the Ice Age.

If speciation and immigration are the processes that contribute to the formation and maintenance of diversity in a community (cf. Sections 6.1.1 and 6.1.3), then the exclusion of inferior competitors is a force that acts to decrease diversity. Various forms of abiotic stress, to which only a few species are resistant, also tend to reduce diversity (e.g., toxic pollution, extreme temperatures, oxygen depletion). This idea corresponds to Thienemann's second biocoenotic law. Extreme conditions for one group of organisms are not necessarily extreme conditions for another group. The diversity of animals would certainly decrease in response to an oxygen deficit, whereas the diversity of microaerobic and anaerobic microorganisms would increase.

The effects of grazing on diversity are not so clear. A nonselective grazing pressure that is constant through time has the same effect on a guild of prey organisms as an increase in the flowthrough rate of a chemostat. Other species may be selected, but the number of species capable of coexisting will not change. This can change if the predator alters its feeding behavior. If, for example, it concentrates on the most abundant prey, this would work against competitive exclusion and increase the likelihood of coexistence. Temporal changes in grazing pressure (e.g., due to predator–prey oscillations) also contribute, independent of selectivity, to an increase in the diversity of the food organisms, for it creates changes in the conditions of competition.

Next to recruitment of new species, temporal and spatial changes in the conditions of competition are the most important mechanism of maintaining diversity against the pressure of competitive exclusion. As described earlier in the "intermediate disturbance hypothesis" in Section 6.1.4, intermediate frequencies of disturbance should lead to maintenance of the highest diversity. Intervals of about three generation times between disturbances produced the maximum diversity in experiments with phytoplankton (Gaedeke and Sommer 1986).

Cycles of very different lengths can occur in lakes (Fig. 7.7). The daily cycle is too short and the yearly cycle is too long to have an important effect on the diversity of phytoplankton. Predator–prey cycles between *Daphnia* and edible algae usually have a period length of 30 to 50 days (McCauley and Murdoch 1987). This is longer than the optimal time interval for keeping diversity, but compared to constant conditions, it still can promote diversity. Bad weather fronts can cause aseasonal increases in the mixing depth. Such fronts are highly irregular, but at temperate latitudes tend to occur most often at intervals of 5 to 15 days. This is very near the interval of optimal diversity for phytoplankton. The effects of the lack of aseasonal mixing events can be illustrated with the extremely stable summer in 1989 in Plußsee, a wind-protected lake in Northern Germany (Fig. 7.8). The thermocline stayed between 4 and 6 m throughout the period from the middle of May until well into September. Following the clear-water phase, there was a continuous increase in the total phytoplankton biomass, reaching a plateau in August and September. The proportion of the dinoflagellate *Ceratium hirundinella* steadily increased during this period and finally achieved one-sided dominance of 98% of the total biomass. This produced a minimum diversity index of $H' = 0.11$ (natural log base), an extremely low diversity compared to values of about 2 during the spring bloom and shortly following the clear water phase.

7.5 STABILITY

The concept of stability is often used uncritically in ecology, especially in the *diversity-stability hypothesis*. This hypothesis states that the stability of a community increases with its diversity (complexity). This hypothesis has been discussed mainly in the field of terrestrial ecology. It will be dealt with here primarily because of its importance in popularized ecology, and as an example of *myth creation* in science. The following arguments are used to support the diversity–stability hypothesis:

1. Island communities that are species-poor are more sensitive to the introduction of new species than species-rich mainland communities.

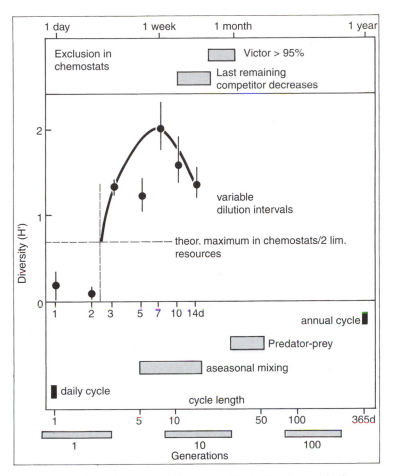

Figure 7.7 The diversity of the phytoplankton and time scale of potential changes in the competition conditions. (*Above*) The time course of exclusion in chemostat experiments. (*Middle*) Diversity of phytoplankton (*H'*, calculated as natural logarithm) in competition experiments with periodic dilution and variable intervals (from Gaedeke and Sommer 1986). (*Below*) Tthe time scale of naturally occurring changes in the competition conditions.

2. Simple experimental systems and mathematical models (one predator–one prey) show strong oscillations and often go to extinction.

3. Agricultural monocultures are sensitive to pests.

4. Populations in species-poor arctic and boreal communities have drastic fluctuations more frequently than populations in species-rich tropical communities.

With the exception of the first argument, none of the other arguments has been substantiated (summarized in Goodman 1976). Even complex model systems are often unstable, often even more so than simpler systems. In contrast to agricultural species, nat-

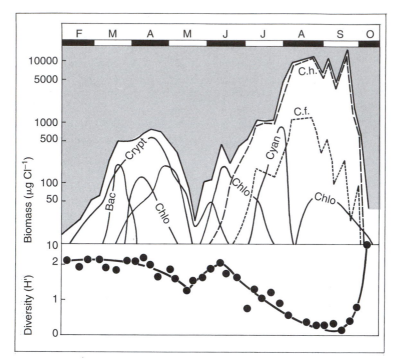

Figure 7.8 The development of phytoplankton populations in Plußsee 1989. (*Above*) Total biomass and biomass of individual taxa (C.h., *Ceratium hirundinella*; C.f., *Ceratium furcoides*, Bac, Bacillariophyceae; Crypt, Cryptophyceae; Chlo, Chlorophyceae; cyan, Cyanophyceae). (*Below*) Diversity index (*H'*, calculated as natural logarithm) of the phytoplankton.

ural monocultures, such as reeds, are not especially susceptible to pests. In the case of agricultural species, it is not the diversity, but rather their lack of evolution of resistance against parasites and their being placed in the wrong place, that makes them vulnerable. Finally, the constancy of the tropical rain forest is overestimated. Sufficiently long-term observations also shows strong population fluctuations in the tropics. The tropical rain forest, especially noted for its great diversity, is especially sensitive to disturbances caused by humans. After destruction for agricultural development, the vegetation of the temperate zone recovers much more quickly than that of the tropical rain forest. Possibly, the differences in constancy of the communities simply reflect differences in the constancy of the climatic conditions.

The concept of "stability" in the diversity–stability hypothesis is apparently inadequately defined. We must differentiate at least three classes of stability (Fig. 7.9):

1. *Constancy:* This usually refers to the lack of change in the number of individuals, biomass, number of species, etc. Constancy need not only mean a fixed set of conditions, for one can speak of consistent cycles and consistent trends. The observa-

tion that there were no unpredictable changes does not explain whether the constancy was caused by the external conditions (absence of disturbances) or to the ability of the populations to resistance disturbances.

2. *Resistance (inertia):* This refers to the ability to keep a stationary condition, a cycle, or a trend despite a disturbance. Resistance leads to constancy, but it is not the only possible cause.

3. *Elasticity:* Elasticity is the ability to return to the stationary condition, cycle, or trend after a deviation from the normal condition. Elasticity only leads to constancy over a long period.

An investigation of the relationship between stability and diversity must first clarify which type of stability is meant. The diversity–stability hypothesis is often used in connection with resistance. This is not justified, based on the correlative and experimental investigations to date (see Goodman 1976).

Lakes in the temperate zone and especially streams often show a high degree of elasticity (cf. Section 5.7). The organisms are adapted to strong fluctuations and disasters and therefore have resting stages and/or specialized strategies for colonizing. This enables the system to regenerate itself very quickly following a short disturbance, even if it was a strong disturbance. It becomes more difficult if the disturbance lasts longer so that the system slowly transforms into another state. A good example of this

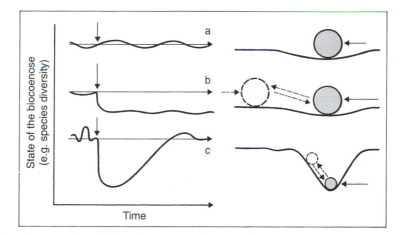

Figure 7.9   A visual presentation of the various types of community "stability." The condition of the system is represented as a ball in a valley. In the adjacent graphs a measurable parameter (e.g., diversity) is shown. (a) The system has a high resistance. A minor disturbance (arrow) cannot alter its course. The community either does not change or changes only within the boundaries of normal seasonal variation. (b) A more powerful disturbance can change the course of the system and elevate it to a new stable level. It will remain there until it is reverted back to its previous condition by a new disturbance. (c) The small ball represents a system with little resistance, but a high degree of elasticity. Even a small disturbance causes large changes, but the system rapidly returns to its initial condition (from Lampert 1978).

is the eutrophication of the lakes through the addition of phosphorus. A one-time addition of phosphorus (e.g., as fertilizer) has no lasting effect on the lake. The phosphorus is quickly incorporated into the biomass and then settles out of the water. As long as there is oxygen in the deep water, the phosphorus will stay in the sediments as insoluble ferric iron complexes (Section 8.3.4). This disturbance is so short that the food web structure does not change.

The problem of eutrophication of lakes is that the disturbance occurs continually over long periods. Phosphorus is added to the lakes with sewage, erosion of land, and from the air. This causes a continually elevated production that slowly changes the food web. The ball in Fig. 7.9b is slowly rolled to a new level. If the disturbance is then stopped (e.g., by diversion of the sewerage), the lake will not "get better" immediately. In the meantime, the reduction in nutrient addition now becomes a disturbance to the new system. The feedback of phosphorus from the anaerobic sediments and the internal phosphorus cycling determine the productivity. Long-lived keystone species have changed; for example, there are more planktivorous, carplike fish (cyprinids). Filamentous cyanobacteria inhibit the reappearance of large, filter-feeding zooplankton. These mechanisms work to keep the lake in its new, eutrophic condition. This is painfully obvious when one attempts lake restoration (Sas 1989). Sometimes artificial measures must be taken to speed up the process of oligotrophication. In the picture analogy: One must give the ball a shove to roll it back to its old position.

Shallow eutrophic lakes can exhibit two alternative stable states, either a clear-water state dominated by aquatic vegetation or a turbid state, characterized by high algal biomass. There is no intermediate state. It is either one or the other. Phytoplankton biomass in clear-water shallow eutrophic lakes can be two orders of magnitude lower than in turbid lakes. It is not clear how dense beds of aquatic macrophytes stabilize the clear-water state. Competition and allelopathy have been suggested as mechanisms. However, once a system has switched to the turbid state, it takes considerable nutrient reduction to achieve recolonization by macrophytes, since algae absorb too much light to allow the rooted plants to grow on the bottom. Top-down manipulations are more effective in this case. One season of enhanced grazing due to reduced fish stocks may be sufficient to reestablish the macrophytes, which subsequently suppress the algae. There are numerous examples of neighboring shallow lakes that persist for many years in two different states (Scheffer et al. 1993).

7.6 LAKE COMMUNITIES

7.6.1 Communities in the Pelagic Zone

The concept of the pelagic zone refers to the open water zone in lakes and oceans. Two communities are traditionally included in this zone: the *plankton* and the *nekton*. The distinction is based on swimming ability. Plankton are suspended in the water, whereas nektonic organisms are active swimmers. Certainly there are many plankton that are capable of actively swimming (vertical migrations of zooplankton and flagellates), but they generally are not capable of swimming against strong water currents. The view of plankton and nekton as separate communities is probably not justified, considering the discussion of the interactions between fish and zooplankton presented in Section 7.3. It is more useful to speak about an integrated community in the pelagic zone.

The phytoplankton as primary producer level contains cyanobacteria and algae in the size range of about 0.5 μm to 1 mm (colonies to more than 1 mm). Higher plants, red algae, and brown algae are not found in the phytoplankton. Conventionally, three size categories are defined: *picoplankton* (<2μm), *nanoplankton* (2–30 μm), and *net plankton* (>30μm).

The zooplankton in freshwater consists primarily of protozoans (flagellates and ciliates; a few to a few hundreds of micrometers), rotifers (30 μm to 1 mm), and crustaceans (copepods and cladocerans; some 100 μm to 1 cm). In addition, there are a few insect larvae (*Chaoborus*), the larvae of the zebra mussel (*Dreissena*), water mites, and fish larvae. There are several trophic levels within the zooplankton, including herbivores, bacterivores, and zooplanktivores. The nekton of lakes is made up almost exclusively of fish species. Freshwater seals are limited to a very few of the large lakes (e.g., Lake Baikal). Pelagic fish may be either planktivores (usually zooplanktivores) or piscivores.

The *bacterioplankton* (usually less than 1 μm; autotrophic bacteria may be larger) have the most diverse trophic roles (see Table 4.1). Most of the bacteria in the aerobic zone are decomposers and to a lesser degree primary producers that are chemolithoautotrophs. In the anaerobic zone there are also photolithoautotrophic primary producers. Some planktonic fungi also decompose organic matter, and others are parasites. Recent evidence indicates an astonishingly high number of planktonic viruses (10^8/ml) in lakes (Bergh et al. 1989). The ecological role of these viruses has not yet been examined.

A unique peculiarity of pelagic communities is that body size increases progressively in going up the food chain. The primary producers are microscopic sized in the pelagic zone, whereas in terrestrial communities the dominant primary producers are often large and long-lived trees. This has some interesting effects on the functioning of pelagic communities:

1. Plants can form light and chemical gradients in their habitat, but they are too small physically to structure their habitat physically.

2. Herbivory almost always consists of eating entire plants, rather than "nibbling." This causes immediate mortality for the plants involved and not just a reduction in vitality.

3. Across major taxonomic groups, reproductive rates decrease with increasing body size, whereas generation times and time lags in the population responses increase with body size.

The ability of a pelagic community to resist disturbances must therefore be based on top-down effects of fish. The resistance of a forest, in contrast, is usually related to bottom- up effects of the long-lived trees.

A most unusual community known as *neuston* lives at the air–water interface in the surface film of lakes (cf. Section 4.2.7). Numerous types of algae and bacteria attach themselves to the surface by taking advantage of the surface tension. A few animal specialists can utilize this food source. The cladoceran *Scapholeberis mucronata* filter feeds on the small particles while hanging attached to the surface film, and the larger insect water striders (Gerridae) prey on other small animals while "walking" on the upper water surface. The tension of the surface film is also used by insect larvae and snails that come to the surface for short periods to breathe.

7.6.2 Benthos

The community that lives on the bottom of lakes and streams is called the *benthos*. Benthic organisms can live in the substrate (in mud and sand), move on the substrate surface, grow attached to the substrate, or move about freely in the vicinity of the bottom (e.g., benthic fish). The composition of the substrate can have an important influence on the type of communities that develop. Communities living on the surface of substrates can be divided into *epipelic* (on mud), *epipsammic* (on sand), *epilithic* (on rocks), and *epiphytic* (on submersed plants). The prefix "endo" is used to describe the analogous communities that live within the substrate (*endopelic, endopsammic*). The benthic habitat above the compensation level (cf. Section 4.3.5) is called the *littoral,* zone and the habitat below the compensation level is called the *profundal* zone.

Benthic primary producers include cyanobacteria, all higher taxa of eucaryotic algae, and flowering plants, except for gymnosperms. Plants are categorized according to size as macrophytes and periphyton (aufwuchs, microphytobenthos). Higher plants and charophyte algae (Characea) are *macrophytes*.

As a rule macrophytes tend to be larger than their predators and are thus rarely entirely consumed; Jacobsen and Sand-Jensen (1992), for example, found on average only 4% loss of leaf mass for *Potamogeton* species in Danish lakes and even less for other macrophytes. Herbivores that eat macrophytes actually are more like parasites. Many emergent macrophytes have a siliceous epidermis that protects them from being eaten (reeds, rushes). Death and decomposition following the growing season is a more important loss for macrophytes than herbivory. The importance of macrophytes for littoral communities goes beyond its trophic role: They provide substrate for aufwuchs, attachment surfaces for the eggs of fish and amphibians, and hiding places for fish and other animals. They also compete for light with the aufwuchs and littoral plankton (shading). Macrophytes reduce the water current speed, thereby increasing the sedimentation rate and inhibiting the chemical exchange between the littoral and pelagic zones.

Periphyton is usually smaller than its predators and can be eaten whole. This makes the process of herbivory more important in regulating periphyton populations than for macrophytes. There are two different types of food chains emanating from the periphyton. One type involves the eating of aufwuchs algae by protozoans and very small metazoans (up to a few millimeters) that also belong to the aufwuchs community. The second type involves the grazing of the entire aufwuchs community (algae, animals, bacteria, fungi) by a variety of larger and very mobile animals (snails, insect larvae, etc.) (cf. Section 6.4.5).

The highly diversified littoral region offers many *microhabitats*, creating a zooben-thos that is very diverse. Many animals that live in the transition region between wa-ter and land either breathe atmospheric air, such as insects and certain snails, or carry out a portion of their life cycle outside the water.

Except for a few chemoautotrophic bacteria, the dark *profundal* zone is occupied by a community solely consisting of consumers. Animals that live there must utilize the organic matter that rains down from the epilimnion or is carried out from the lit-toral region. Much of this material has already been partially decomposed and there-fore has a low nutritional value. The biomass of benthic consumers is dependent on the amount of organic matter that reaches the bottom of the lake. Organic matter may also come from allochthonous sources. For example, in Lake Constance there is a clear relationship between the number of tubificid worms in the lake bottom near the mouth of the lake tributaries and the pollution load of these streams (Probst 1987). Oxygen depletion and the build up of H_2S create chemical stress in the profundal zone of eu-trophic lakes during the summer stagnation. During this period the deep-water fauna may become impoverished.

7.6.3 Coupling of Habitats

Benthic and pelagic habitats are not completely separate from one another. There are numerous connections between their component parts. As previously noted, the deep-water fauna is very dependent on the production in the euphotic zone. Animals can greatly accelerate the vertical transport of organic matter. For example, the feces of zooplankton, especially copepod fecal pellets, are an important source of food for the benthos. These fecal pellets often contain large amounts of partially digested algae. Due to their large size, the pellets sink rapidly to the bottom before the organic mat-ter is decomposed.

Small lakes have a large number of connections between the littoral and pelagic zones. Phytoplankton that is carried into the littoral region provides food for filter feed-ers such as mussels and littoral cladocera. Some animals are planktonic in one part of their life cycle and benthic in another. An example of this is the zebra mussel (*Dreis-sena polymorpha*). Adults live in the littoral region of lakes, in large rivers, and in boat canals, where they attach themselves to a substrate by means of byssal threads. Their veliger larvae are free-swimming and can make up an large portion of the summer plank-ton in eutrophic lakes. Also, some cyclopoid copepods are planktonic as juveniles and adults, but spend a part of their development as a resting stage in the sediments.

During their ontogenetic development, fish change their habitat according to the lo-cation of their food. As juveniles, many cyprinids, for example, eat plankton and there-fore stay in the pelagic zone. When they become larger, they shift to feeding on the bot-tom and move into the littoral regions of the lake. There may also be daily cycles in habitat shifts. Some planktivorous fish spend the day in schools near the shore and move into the open water at night to feed on pelagic plankton. The larvae of the phantom midge *Chaoborus* may be buried in the sediment during the day, where they can live under anaerobic conditions, and at night they migrate up into the pelagic zone to feed on zoo-plankton (Fig. 7.10a–e). By day they can be considered benthos; by night, plankton.

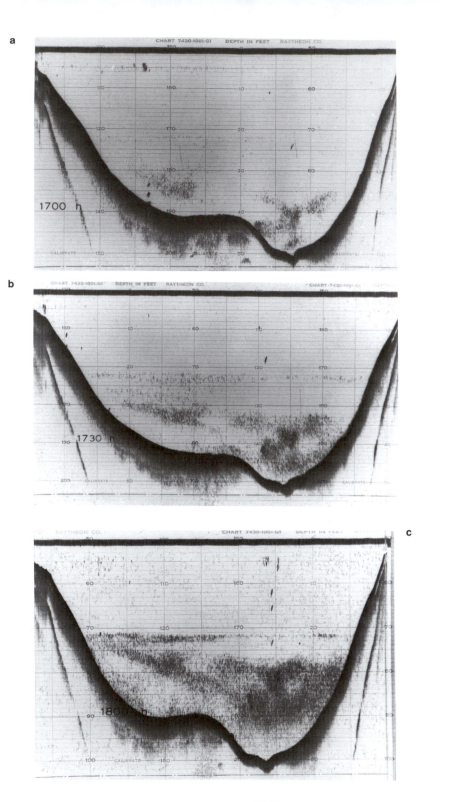

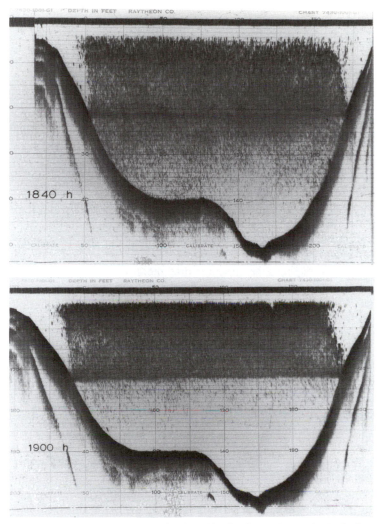

Figure 7.10(a–e) Echo graphs showing the distribution of *Chaoborus* larvae in Barbadoes Pond, a small lake in New Hampshire. The animals spend the day in the sediments. In the evening they emerge from the sediments and migrate upward to feed in the epilimnion. At 17:00 h only a few animals are visible in the deep water; at 19:00 h they are all in the epilimnion. The lower boundary of their distribution marks the oxicline (with permission of J. F. Haney).

7.7 STREAM COMMUNITIES

7.7.1 Types of Streams

Streams have different selection factors than lakes (cf. Section 3.4). The driving force is the directed current. To avoid being swept downstream, organisms must either be attached or be good swimmers that can occasionally rest in dead water regions. For this reason, by far the dominant communities in streams are benthic. Plankton also occurs in the upper reaches of rivers that are in the outflow of lakes, but such plankton is often rapidly consumed by filter-feeding benthic organisms or lost downstream in rapidly flowing rivers. *River plankton* is only found in large rivers with stretches of quiet water and in the river backwaters (Reynolds et al. 1991). Much as a chemostat, when the retention time of the water is longer than the generation time of the plankton, the river is continuously inoculated with plankton as they are transported downstream.

Streams are much more dependent on their watersheds than lakes. This is especially true for the allochthonous input of energy in the stream. Very often such inputs from outside are larger than the stream's own primary production; this leads to the domination of stream communities by consumers and decomposers. Streams are not divided into vertical zones, since there are no distinct trophogenic and tropholytic zones that are defined by light as in lakes. At best, a stream cross section can be divided into its water mass, the stream bed (*benthic zone*), and the subsurface region below the stream bed (*hyporheic zone*). All these zones are in close contact with one another in small streams. Throughout the stream course there are changes in the external conditions, stream flow, and all the related abiotic parameters as well as the effects of the watershed. Corresponding to these physical-chemical shifts are changes in the stream communities. This suggests streams can be divided horizontally according to their position downstream.

There have been attempts to classify different parts of streams and rivers by defining zones with similar environmental conditions. The zones have been characterized by indicator organisms. The classical system used in western Europe is based on fishes and was developed for the River Rhine at the beginning of the twentieth century by the German forester Robert Lauterborn. It can be used to explain the principle of many classification schemes. Each zone is named after a characteristic fish species. The first is called the "trout zone" after the brown trout (*Salmo trutta*). This is the region of the stream (mostly at higher altitudes) where the summer temperature does not exceed 10°C, the water is completely saturated with oxygen, and the bottom substrate consists of coarse gravel to provide interstitial refuge for the fish larvae. The trout zone is followed by the "grayling zone" with the grayling (*Thymalus thymalus*). Here the stream is already wider and the current velocity is lower. Occasionally plants grow along the banks. Summer temperature is higher, but rarely exceeds 15°C. The water is still saturated with oxygen, and the bottom is covered with gravel. In the middle reaches of rivers the water velocity is still high enough to prevent mud deposition on the river bed, but summer temperatures may exceed 15°C, and minor oxygen deficits may occur. This zone is named the "barbel zone" after the barbel (*Barbus barbus*). In the lowlands, where the river flows more slowly, mud is deposited and macrophytes grow in the quiet regions. Temperatures may exceed 20°C, and the oxygen saturation can drop considerably. Characteristic fishes of this zone are temperature- and low-oxygen-tolerant cyprinids. It is named the "bream zone" after the bream (*Abramis brama*).

This classification scheme demonstrates the principal shortcoming of a system that is based on indicator species, for such species are usually restricted to a certain biogeographic area. The fish fauna of North America, for example, is much richer than the European fish fauna. Moreover, there are large regional differences in species composition. The Western European classification system cannot be used in North America even under similar climatic conditions, for the indicator species may be unimportant or absent. There have been attempts to replace them with American fish species, but this can only be applied for local river systems such as New Hampshire, or for particular streams such as the Ohio River (for a review see Hynes 1970).

The same objections are valid for other schemes based, for example, on insects. Illies (1961) tried to overcome these difficulties by concentrating more on types than on particular species so as to create a general classification scheme that can be used worldwide. His system is based on stoneflies (Plecoptera). Their distribution is determined mainly by temperature and structure of the bottom substrate. The basic premise is that similar types of stream insects have evolved convergently everywhere in response to similar environmental conditions. Even if the species in different areas are not identical, one can find types characteristic for the particular conditions. Illies' zonation scheme results in three broad categories that can be subdivided (each habitat is denoted by the suffix "on":

1. Krenon—headwater region

2. Rhithron—mountain stream region (salmonid zone, cool, mean monthly temp. $\leq 20°$ C):

   epirhithron—upper MSR (upper trout zone);
   metarhithron—middle MSR (lower trout zone);
   hyporhithron—lower MSR (grayling zone).

3. Potamon—lowland river region (warm, mean monthly temp. $\geq 20°C$):

   epipotamon—upper LRR (barbel zone);
   metapotamon—middle LRR (bream zone);
   hypopotamon—lower LRR (brackish water zone).

This system can be applied to streams at all geographic latitudes, although there are shifts in the relative proportions of the rhithron and potamon. The rhithron is limited to extremely high altitudes as one approaches the equator, whereas the potamon becomes insignificant in polar regions.

7.7.2 The River Continuum Concept

The river continuum concept (Vannote et al. 1980) takes the classification of streams a step further in a predictive model that integrates the geomorphological features of streams with the composition and function of the biological communities (cf. Section 2.1). Streams are physically open systems. From the source of a stream to its mouth, there is a continual change in physical factors such as width, depth, flow velocity, discharge, temperature, and entropy gain. This hypothesis proposes that the biological organization of a stream is adapted to these gradients, forming a *continuum*. Producers and consumer populations achieve a steady state; over major sections of a stream, communities develop that are in equilibrium with the physical features of the stream. The river continuum concept is based on streams that originate in forested regions. Clearly, it must be modified when applied to other river systems, such as those involving alpine and arid regions where low-order streams have little or no allochthonous input from trees and are thus more dependent on autochthonous primary production.

Communities can be divided into three general categories according to the size of the stream: small streams (first–third order), medium-sized streams (fourth–sixth order), and large streams (> sixth orders) (Fig. 7.11).

Processes of synthesis (production) and decomposition (respiration) change along the course of a stream. Small streams near their source are often lined with vegetation. This reduces available light and the amount of photosynthesis, but supplies a large quantity of allochthonous organic matter. This results in a production to respiration ratio ($P{:}R$) in the stream of <1. As the stream becomes larger the direct effects of introduced organic matter diminish and autochthonous production and downstream transport become more important. Thus the $P{:}R$ ratio shifts toward a value >1. The vegetation along the stream becomes unimportant in the higher-order streams. These large streams have a considerable quantity of fine particulate organic matter that is transported downstream. Once again there is a light limitation of photosynthesis for organisms within these streams, this time caused by high turbidity. Accordingly, the $P{:}R$ ratio shifts toward a value <1.

There is a great diversity of dissolved organic substances in low order streams. Such small streams are most closely connected with their watershed and therefore collect, transform, and transport a variety of organic compounds. *Labile* substances that are easily decomposed are rapidly taken up by microorganisms or adsorbed physically. *Refractile* substances are left over and transported downstream. Along with a decreasing diversity of dissolved substances, there is also a decrease in the importance of coarse particulate organic matter (CPOM) as one moves downstream, although the more resistant particulate matter may reach further downstream.

One can therefore view a stream system as a continuous gradient that shifts from its source to its mouth from a system that is highly heterotrophic to one that is primarily autotrophic, with daily and seasonal variation, and finally back again to a system that is dominated by heterotrophic activities (Fig. 7.12). This continuum is disturbed where tributaries of lower order converge with the main stream.

There is a downstream decrease in the size of organic particles, resulting in an increase in the importance of species that can most effectively utilize fine particles. Morphological and behavioral adaptations of stream organisms reflect these food relationships. The invertebrates can be divided into four *functional feeding groups*: shredders, collectors, scrapers, and predators. Shredders utilize coarse organic particulate matter (>1 mm), such as leaves. Collectors filter fine or ultrafine particulate organic matter (FPOM, 50 μm–1 mm; UPOM, 0.5–50 μm) or they ingest it with the sediments. The microbial aufwuchs community (e.g., fungi) on these particles also is important for the nutrition of collectors. Scrapers graze on the aufwuchs algae. Since there is a dominance of CPOM in low-order streams, shredders and collectors tend to predominate and scrapers are poorly represented. In intermediate-order streams scrapers and collectors become important and the shredders diminish. Finally, in the large, high-order streams collectors dominate, since particles in this region are already small. There is

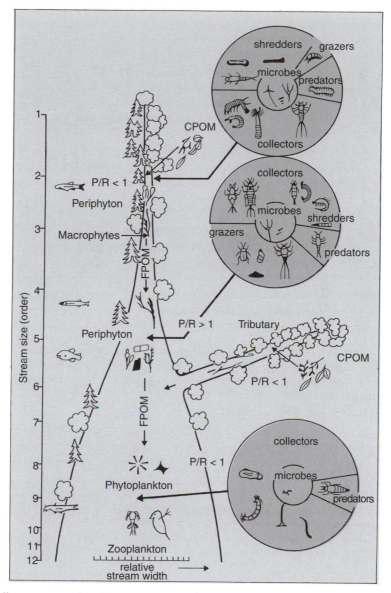

Figure 7.11 A diagram presenting the stream as a continuum of abiotic factors and communities. The relative proportions of the various feeding groups are shown in the circles (redrawn from Vannote et al. 1980).

approximately the same proportion of predators in all stream orders, as they feed on all types of primary consumers. Fish populations also shift from species-poor communities of cold-water fish to species-rich communities of warm-water fish.

There is a biological steady-state equilibrium at every location along a stream. Energy is carried in, utilized, stored, and transported downstream as partially utilized or

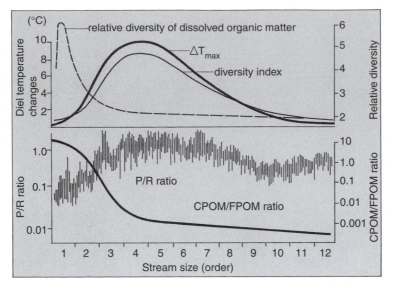

Figure 7.12 Gradients along a stream course. CPOM, coarse particulate organic matter; FPOM, fine particulate organic matter (from Vannote et al. 1980).

unused matter. This matter then serves as an input to the next location. One must assume that if species are resource limited in a system that there is a tendency at that location for most efficient use of the resources, that is, that there is as little loss of resources as possible. Additional organisms can occupy the habitat as soon as resources become available. If a dominant species disappears due to environmental conditions (e.g., temperature), it will be replaced by another species; thus, for example, there are species shifts throughout the year. Also, if one species utilizes a part of the available energy, since it specializes in its feeding, other species can use the remaining energy. Although the same principle applies to all ecosystems, in streams, the resources must be used immediately or they will be transported downstream. Vannote et al. (1980) postulate that stream communities have a tendency to maximize the use of resources and to utilize resources uniformly over time.

The tendency toward efficient use of resources holds for the predictable short-term changes in abiotic factors (e.g., light, temperature). Let us assume that a species is threatened by predators during the day and must therefore hide. If this were the only species present in the system, then the energy that enters that part of the stream during the day would be lost as it would immediately leave. This represents available resources during the day that could be utilized to establish another species, if it were small or camouflaged. One could describe the same scenario, by assuming that certain species best utilize a resource within a certain temperature range during the daily temperature cycle. The largest daily temperature changes occur in the intermediate-order streams (cf. Section 3.4.2); the largest diversity of species also occurs in these streams.

Despite different effects of specific daily and seasonal rhythms, this process leads to an increase in the number of species and to a reduction in the fluctuations in the use of energy; that is, it leads to a degree of stability (constancy). This, however, is not a "system feature" (cf. Section 7.2), even if there is a maximization of resources "within the system." It is simply a result of optimal resource utilization by each individual population.

There is no ecological succession in a flow-through equilibrium, in contrast to a closed system (cf. Section 8.7.1). The destiny of a lake is prescribed; depending on its size, sooner or later it will become land due to the accumulation of inorganic and organic matter. In contrast, flow-through equilibrium changes take place in evolutionary, rather than ecological, time realms. This does not mean that streams do not change. Geological events may modify the stream course, and changes in the watershed will alter the input of matter. However, a new flow-through equilibrium will quickly become established. This is a prerequisite for the regeneration ability of streams following disturbances (cf. Section 5.7).

REVIEW QUESTIONS FOR CHAPTER 7

1. In eutrophic lakes oxygen shortage under the ice cover can kill entire populations of fish. Sometimes the water appears clearer during the year following the winter fish kill than before, even though the concentration of nutrients is unchanged. What might account for the clear water?

2. Describe the type of correlation (positive or negative) you expect with (a) dominant top-down control and (b) bottom-up control between: piscivorous fish and herbivorous zooplankton; piscivorous fish and phytoplankton; planktivorous fish and herbivorous zooplankton; planktivorous fish and phytoplankton.

3. Evidence for top-down control of pelagic food webs can sometimes be found by comparing interactions in different lakes. Would you expect to find such evidence more easily when comparing lakes with similar nutrient concentrations or when comparing lakes with very different nutrient status? Explain.

4. Calculate the Shannon diversity index H' and the Simpson index D for hypothetical communities of 3, 10, and 100 species with different equitability. Compare the sensitivity of both indices against equitability and species richness for conditions of: (a) high equitability: all species equally abundant; (b) medium equitability: dominant species 50%, all others equally abundant; (c) low equitability: dominant species 90%, all others equally abundant.

5. How can disturbances promote diversity? Give some examples of natural and anthropogenic disturbances in a lake and in a stream. Explain how the magnitude and frequency of these disturbances would affect the diversity of aquatic communities in each of these systems.

CHAPTER 8

Ecosystem Perspectives

8.1 THE ECOSYSTEM AS AN ECOLOGICAL UNIT

The activities of organisms set in motion the transformation of energy and matter and have important effects on the quality of their abiotic environment. One of the most impressive examples of this on a geological time scale was the change from an anoxic to the present-day oxic atmosphere on earth, brought about by photosynthesis by cyanobacteria and green plants and also the accumulation of calcium carbonate in the shells of marine organisms. We have already discussed mechanisms by which organisms can cause changes in the environment in a much shorter time period, such as the depletion of nutrients by planktonic algae that occurs during the summer growth period. Thus organisms not only interact with one another, but also with their abiotic environment. Communities consist of interacting populations. The comprehensive unit of communities and their interactions with the abiotic environment is the *ecosystem*.

The ecosystem concept is not so easily defined as the term "system" may suggest. Ecosystems are more than incidental aggregations of organisms in a particular habitat, for the organisms also modify the environment. Jones et al. (1994) describe the role of organisms as "ecosystem engineers." The activities of beavers provide a good example. By creating a dam beavers change a lotic system (stream) into a lentic system (pond). Beavers thus have a major impact on other communities, such as the phytoplankton and zooplankton. Plankters could not exist there without the beavers, but this is a nonreciprocal interaction, for the plankters have no impact on the beavers. The beavers must be considered as part of the "ecosystem beaver pond," for they not only construct and maintain the pond; they contribute nutrients and debris to the system. Beavers also utilize and modify the forest ecosystem by cutting and removing trees. While the ecosystem for the plankton has clear boundaries, this is not so for the beavers. Hence, the problems of setting boundaries for ecosystems are similar to those for communities (cf. Section 7.1).

[288]

Some standard books in ecology, such as that by Odum (1959), present an analysis of ecosystems at the beginning, rather than at the end, as presented here. This underlies more than a difference in the teaching approach of the authors. Authors such as Odum begin with either an explicit or implied "superorganism" concept (cf. Section 7.2). The analogy with organisms stems from the fact that energy flows through ecosystems and matter circulates within the ecosystem. According to "holistic" ecologists, populations and organisms sort themselves into this flow of energy and matter, thereby searching, so to speak, for their place in the ecosystem. "Mechanistic"-oriented ecologists, such as the authors of this book, believe that energy and matter do not actually "flow" in ecosystems, but rather that flows of energy and matter are simply abstractions representing the sum of the activities of individual organisms. It is the scientist who, by summarizing the activities of organisms into categories such as guilds and trophic levels, defines the pathways for transfer of energy and matter that create a picture of energy and matter flowing through an ecosystem. It is quite natural that the general laws of physics and chemistry apply to ecosystems. Especially important in this regard are the first two laws of thermodynamics, the law of conservation of matter, the principle of electron neutrality, and the boundaries of variability in biomass stoichiometrics. These general laws do not account, however, for the diversity of flows of energy and matter.

As discussed earlier (Section 7.2), the superorganism concept of ecosystems is weakened by the lack of a central genome in the system. Therefore, an entire ecosystem cannot evolve, cannot be "optimized," and cannot possess a defined fitness (Calow 1992). There is a debate in the environmental sciences as to whether an optimum status of an ecosystem can be defined as a measure of "ecosystem health." It seems unlikely that general rules for the functioning of an ecosystem will be found. The actual flow of matter and energy in an ecosystem depends on the activities of the individual organisms, which in turn are regulated by the principle of fitness. For example, the direction of the flow of energy and matter captured and produced by phytoplankton depends on how well the algae are eaten and how rapidly they sink (Fig. 8.1). One can find similar examples for other branches of the flows of energy and matter. The distribution of the flows is the result of natural selection conditions that affect the guilds.

8.2 FLOW OF ENERGY

8.2.1 Sources and Carriers of Energy

Thermodynamically, both organisms and ecosystems are open systems. They can be maintained only by a continuous flow of energy through the system. Light or the energy released by an exergonic chemical reaction can serve as an energy source (cf. Section 4.3.1). Chemoorganotrophic organisms, which use organic matter as a source of energy, are dependent on the products of other organisms; only phototrophic and chemolithotrophic organisms do not require the synthesis of organic compounds by other organisms. Possible energy sources for entire ecosystems include sunlight, allochthonous input of organic matter, and reduced inorganic chemical bonds. There are

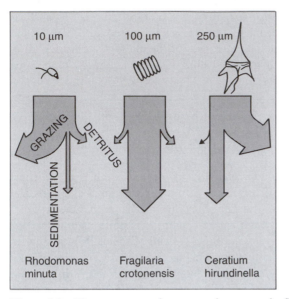

Figure 8.1 The movement of energy and compounds, fixed but not respired by phytoplankton through the processes of zooplankton grazing (left arrow), sedimentation (middle arrow), and detritus formation in the epilimnion (right arrow): The small flagellate *Rhodomonas* is eaten efficiently and almost none sinks out of the epilimnion; cell death and formation of detritus in the epilimnion are relatively unimportant. The colonial diatom *Fragilaria* is poorly grazed and has a high sinking velocity. Epilimnetic detritus formation is not important, since the dead colonies sink rapidly. The large flagellate *Ceratium* is almost completely inedible for most zooplankton and only its cysts and cells in poor physiological condition sink out of the epilimnion. Thus detritus formation is the most important sink.

few ecosystems where chemolithotrophy is the primary initial process of energy flow, such as hot springs in the deep ocean bottom (hydrothermal vents) that supply H_2S. Ecosystems that use allochthonous organic compounds as an energy source must depend on other ecosystems. On a global scale, sunlight is by far the most important energy source and oxic photosynthesis is by far the most important initial process in the flow of energy.

Organic matter is therefore the universal carrier of potential energy within an ecosystem. Its various forms include living organisms (biomass), dead particulate matter (detritus), and dissolved organic matter (DOM). Some organic compounds cannot be utilized or can only be utilized by a few specialized organisms and therefore, regardless of their energy content, provide little useful energy (cf. Section 4.3.11). Such compounds are referred to as *refractory*.

Both the *pool size* and flow (*transfer rate*) affect the transfer of energy. The size of pools (biomass, detritus, dissolved organic matter) has energy as a dimension (= work, force times distance, 1 N m = 1 J = 1 W s). Flows have the dimension of energy per unit time (= activity, 1 N m s^{-1} = 1 W). By dividing the size of the pool either by the sum of all incoming flows or by the sum of all outgoing flows, one can calculate the theoretical residence time of energy in a particular pool, that is, the time that would be needed to fill up a pool that has been emptied (turnover time).

8.2.2 Efficiency of Energy Transfer

Theoretically, all forms of energy except for heat are interchangeable. Each time there is a *transfer* of energy a portion is lost as heat (Second Law of Thermodynamics). There are other losses that make some of the energy converted by primary producers unavailable to consumers. Metabolic processes and locomotion require energy, which in the energy budget of an organism comes under the category of respiration (cf. Section 4.4.3). Incompletely oxidized organic matter is excreted and egested. This material in turn can be used as an energy source by detritivores, but not predators.

The *efficiency* (*effectiveness*) of energy transfer is expressed by several energy flow quotients. The efficiencies derived from energy balance of an individual has already been dealt with in Section 4.4.3; the *assimilation efficiency* is the ratio of assimilation to ingestion, the *net production efficiency* (K2) is the ratio of production to assimilation, and the *gross production efficiency* (K1) is the ratio of production to ingestion. There are also efficiencies that are defined at the population level. For predator–prey relationships, the *predation efficiency* is the ratio of ingestion rate of the predator and the production rate of the prey, that is, the proportion of the prey production that is eaten by the predator. The *ecological efficiency* refers to the ratio of predator production to prey production. This is a measure of the amount of energy that is passed from one link of the food chain to the next. Ecological efficiencies are usually from 0.05 to 0.2; that is, 80–95% of the energy is lost at each transfer in the food chain. Because of these losses, unlike the flow of matter, energy must always flow in one direction in ecosystems.

8.2.3 Trophic Levels and Pyramids

The concept of "trophic level" is closely related to the food chain concept (cf. Section 7.3.1). Organisms within a food chain that can be assigned to the same position are collectively referred to as a trophic level. The trophic level concept has proved to be useful in many studies, although it is a simplification, since in nature one usually encounters food webs rather than food chains.

Because the ecological efficiency is always low, there is a rapid decrease in the productivity of subsequent trophic levels. With an ecological efficiency of 0.1, the percentage of the primary production in the second level is 10%, the third level 1%, the fourth level 0.1%, etc. The number of possible trophic levels is therefore dependent on the amount of primary production. This is also the basis of the Oksanen model of vertically alternating controls in communities (cf. Section 7.3.5). The restriction of the number of trophic levels by the primary production is not a fixed number. A low primary production can be compensated in a large habitat by the large hunting area available to the predator. The presence of sharks in extremely oligotrophic tropical oceans is an example of this.

"Trophic pyramids" allow one to visualize the decreasing production with increasing height of the trophic level. The blocks that represent the production rates become smaller as one goes up to higher trophic levels, giving rise to the appearance of a pyramid. Such presentations are mainly a heuristic tool, since it is usually difficult to assign a population within a food web to a single trophic level, and there are difficulties in estimating both primary production (Box 4.3) and secondary production (Box 4.4). Originally it was assumed that there must be pyramids of biomass and numbers that correspond to the production pyramids. Sometimes such pyramids have been found, but it is not necessarily so if organisms at a lower trophic level are significantly smaller than the organisms at the next highest level, for the lower trophic level would then have a higher specific production rate (cf. Section 4.5). In this case, a smaller biomass could produce enough to support the next level.

8.2.4 The Detrital Food Chain and the Microbial Loop

The greater the resistance of the primary producers to herbivory, the more important the *detritus food chain* becomes as a part of the energy flow through the ecosystem (cf. Fig. 8.1). An extreme example of this in aquatic systems is found in shallow reed areas of a lake. The ecological efficiency is lower for detritivores than in the direct food chain, since detritus has a large percentage of refractory compounds. Much of the use of detritus is by microorganisms (bacteria and fungi). There is much controversy over whether animal detritivores actually get their nutrition directly from the detritus or from the attached microorganisms. This is also likely to differ depending on the specific detritivore. If most of the energy in the detritus food chain first flows through microorganisms, there must be a larger number of links in the chain. This leads to further efficiency losses.

The *microbial loop* is a special type of detritus chain that has recently become known because of its importance in the pelagic region (cf. Section 7.2.2). It begins with dissolved organic excretions from pelagic organisms (primarily phytoplankton). Much of this dissolved organic matter is not refractory and can be efficiently utilized by bacteria, which in turn are eaten by heterotrophic protozoans (see food web in Fig. 7.1). Since these protozoans also eat picosized phytoplankton, this transfer step is usually included in the microbial loop, even though it is really a case of herbivory. The heterotrophic protozoa (mostly flagellates) are in turn eaten by metazoan zooplankton (cf. Fig. 7.2), thereby channeling the energy from the microbial loop into the "classic" food chain. At present there is much discussion over whether the microbial loop is a "link" or a "sink" in the energy flow of the pelagic ecosystem. The answer depends mainly on whether one views the microbial loop from its base (production by the bacteria and picoplankton) or its top (transfer to metazooplankton). The total production by the bacteria and the picophytoplankton may be larger than the production by the nanophytoplankton, especially in oligotrophic lakes and in lakes with a large allochthonous input of dissolved organic matter. Because of their small size these particles cannot be used by most zooplankton and are therefore lost from the food chain. If the particles are eaten, however, by flagellates (5–20 μm), they then become available to filter-feeding zooplankton. With an ecological efficiency of 0.1, 90% of the energy is lost as heat at the flagellate step. Thus, to provide the metazooplankton and further links in the food chain with the same amount of energy, the bacteria and picoplankon must produce ten times as much as the nanoplankton. Some of the bacteria can also be eaten directly by the metazoopolankton. Therefore, one could also argue that the flagellate step only represents a gain of energy if the predatory efficiency of the zooplankton on the small particles is less than 10%. The ecological importance of the microbial loop is probably more due to its effect on the mortality of the bacteria and picoplankton and in the regeneration of nutrients that is tied up in these small particles than its effect on the energy flow through the food chain (Riemann and Christoffersen 1993).

8.2.5 Diagrams of Energy Flow

Many field investigations were begun during the International Biological Program (IBP) of the 1970s to determine the total flow of energy or matter through an ecosystem. The final goal of these ecosystem studies was to construct a diagram of the energy flow or carbon circulation through an ecosystem. Much information was gathered and brought together here, but in most cases only a small part of the flow rates were actually measured, and such measurements had serious inaccuracies. The numerical values of individual flow rates should therefore be viewed as first-order estimates of the size of energy flow in ecosystems rather than precise results. Considering the inaccuracies in estimating production rates in the field, transfer efficiencies should probably not be estimated to decimal places, as they are often given. Published energy-flow diagrams are also qualitatively incomplete, since the importance of the microbial loop has only been recognized in the past few years.

We present here the classic energy flow diagram of the Silver Springs ecosystem as one of the most instructive and complete examples (Fig. 8.2). This is the result of a very complete ecosystem study (Odum 1957). Silver Springs is a large spring in Florida with very clear-water. The important primary producers are aquatic macrophytes (*Vallisneria*) and the periphyton that grows on its surface. Snails and turtles are the dominant herbivores. Fish are the carnivores. As in all similar diagrams, the energy flow through the higher trophic levels is depicted as extremely small, compared to the primary production. This characterization of animals as quantitatively insignificant differs considerably from their description as having important regulatory effects as "keystone species" (cf. Section 7.3.3).

One must also realize that a particular energy flow diagram only applies to a specific situation in a particular body of water. The explosion of ecological data gathered during the IBP still did not lead to the formulation of general laws, except for the rule of thumb that ecological efficiencies are generally from 0.05 to 0.2. One could question whether even the most complete and comprehensive measurement of all energy flows within an ecosystem would allow for a causal explanation for the direction of the energy flow, especially when regulatory components, such as keystone species, would be ignored as "quantitatively unimportant."

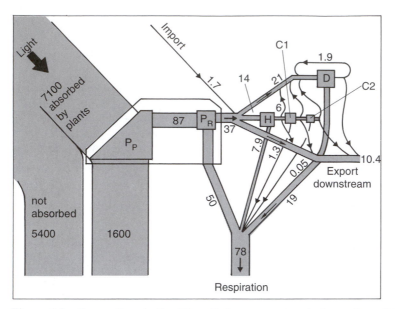

Figure 8.2 Energy flow in the Silver Springs ecosystem (redrawn from Odum 1957). Flow rates in 10^6 J m^{-2} year^{-1}: P_p, photosynthetic biomass of the plants; P_R, respiratory biomass of the plants (e.g. roots); H, herbivores; C1, first-order carnivores; C2, second-order carnivores; D, detritivores.

8.3 CYCLING OF MATTER

8.3.1 General Features

The transfer of energy in ecosystems is always coupled with the transfer of matter. There are, however, some important differences. The most important difference is that *matter can circulate in an ecosystem, whereas energy can only flow through the system*. The reason for this is that even completely remineralized decomposition products (e.g., CO_2) can be reused as a resource. Also, biological and chemical reactions and transfers of matter follow only the law of conservation of matter, and not the second law of thermodynamics.

In contrast to the popular belief in the concept of *"balance of nature,"* natural ecosystems are by no means completely closed. Matter is constantly removed from every ecosystem by physical transport processes such as outflow, erosion, and the evolution of dissolved gases. A portion of the organic matter produced is deposited at the same location within the system and is not recycled or exported. The *sediments* in many water bodies act as a sink, since the anoxic decomposition of organic matter is much slower than the oxic decomposition. There is a long-term accumulation of substances that are chemically resistant to decomposition or become buried in the sediments. Paleolimnology attempts to reconstruct the history of a lake from remains of organisms and organic molecules (e.g., chlorophyll derivatives) found in the sediments. Fossilized energy carriers such as coal, oil, and natural gas are the result of long-term accumulation of undecomposed organic matter.

In addition to biological activity, abiotic reactions such as precipitation, dissolution, dissociation, and redox reactions are also important in the cycling of matter. All of the reactions that serve as a basis for chemolithotrophic production (cf. Section 4.3.8) can also occur spontaneously without the involvement of organisms. On the other hand, there are also inorganic reactions that are caused by the activity of organisms, such as biogenic marl precipitation. This close interaction of biological and geochemical processes is reflected in the expression *biogeochemical cycles*.

A critical step in all biogeochemical cycling is the transformation of matter from the dissolved to the particulate phase. Primary production is generally the most important mechanism for achieving this transition. Often there is an inverse relationship between the dissolved (available) and the incorporated pools of an element within an ecosystem; that is, the maxima of the dissolved concentrations coincide with the minima of the particulate concentrations and conversely (Fig. 8.3a and b). Of course, heavy importation of allochthonous matter may obscure such patterns.

The biogeochemical cycles of individual elements are coupled, since by and large all organisms are made up of the same elements and the stoichometric composition varies within fairly narrow limits. Silica in diatoms is an exception (cf. Section 6.4.4). However, the composition of potentially limiting elements is so variable (especially phosphorus; see Droop's formula, Section 4.3.3) that one cannot apply a general formula to calculate the cycling of an element from one organism to the next. Also, there are differences in the resolubilization of various elements from dead organisms and animal feces. Phosphorus goes into the dissolved phase more rapidly than nitrogen and carbon, and these latter elements more rapidly than silica. Mobile elements such as phosphorus cycle between the dissolved and particulate phase many times within the epilimnion ("short-circuited cycle") before it sinks, whereas immobile elements, such as silica, sink out of the epilimnion after a single incorporation into biomass and may redissolve only after reaching the sediments. The following cycles of carbon, nitrogen, phosphorus, and silica are presented as examples of some of the biologically relevant elements in lake ecosystems.

8.3.2 The Integration of Lakes into Larger-Scale Biogeochemical Cycles

As stated earlier, biochemical cycles in lakes are not entirely closed. Lakes receive substances from the watershed and from the atmosphere; they export substances via the outflow and atmosphere and they bury substances in the sediments. Depending on the balance of these processes, water passing through a lake can be either enriched or impoverished of particular substances.

Modifications of water in the watershed
Before entering a lake, water is chemically altered while passing through the lake's *watershed*. The geology of the watershed and the chemical composition of the precipitation are extremely important. Average rainwater is roughly a 5000-fold dilution of

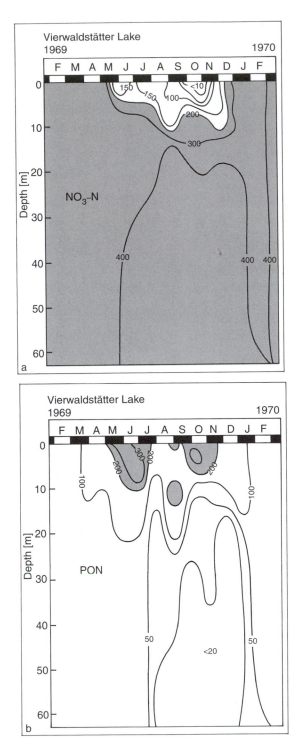

Figure 8.3 (a) and (b) Complementary time–space distributions of nitrate- nitrogen and particulate organic nitrogen (PON) in Vierwaldstätter Lake (Switzerland). Concentrations in μg N/l; concentrations over 200 μg N/l are shaded (from Stadelmann 1971).

seawater, with addition from the atmosphere of nitrous oxides, sulfur oxides, and ammonium. The hierarchy of cation concentrations is $Na^+ > Mg^{2+} > Ca^{2+} > K^+$, and the hierarchy for anions is $Cl^- > SO_4^{2-} > HCO_3^-$.

On its way to a lake, water is modified mainly by rock *weathering*. This leads to an enrichment of bivalent cations, especially calcium. The increased content of calcium permits a higher uptake of carbon dioxide from the atmosphere (see Section 3.1.6). Depending on the pH, CO_2 is then converted into bicarbonate and carbonate.

Soft waters are formed if the rocks in the watershed are resistant to weathering (granite and other acidic siliceous rocks) and if high precipitation minimizes the contact time on and in the ground. Soft waters have a low salinity and low alkalinity The rank order among cations is $Ca^{2+} > Na^+ > Mg^{2+} > K^+$. In extremely soft waters and especially in lakes along coastal regions, Na^+ may dominate over Ca^{2+}. The rank order of anions is similar to rainwater: $Cl^- > SO_4^{2-} > HCO_3^-$.

Hard waters develop if rocks weather easily (e.g., limestone) and precipitation is low or moderate. This results in water with higher salinity and alkalinity. The rank order of cations is $Ca^{2+} > Mg^{2+} > Na^+ > K^+$, and the rank order of anions is $HCO_3^- > SO_4^{2-} > Cl^-$. Carbonate rocks were deposited by organisms in the geologic past, either as calcium from skeletons or by biogenic decalcification (see Section 3.1.6). Therefore, most of the calcium in present-day fresh waters is recycled calcium from organisms and very little originates from primary weathering of calcium silicate rock.

Nutrient loading of present fresh waters is also dominated by recycled matter from dead organisms. There are fundamental differences in the geological origin of different nutrients, such as nitrogen and phosphorus. The primary geological source of phosphorus was the mineral hydroxyl-apatite, whereas in geologic times, nitrogen was brought into the global water cycle through the activities of organisms, mainly through their uptake of ammonium (the major nitrogen compound of the early, anoxic atmosphere) and the microbial fixation of nitrogen. Today, however, most of the N and P entering lakes are derived from soils. Most soils tend to retain phosphorus, but easily release nitrate and ammonium. This is probably the main reason why in lakes phosphorus tends to limit algal growth more than nitrogen. Human activities can accelerate the release of nutrients into lakes and streams. Agricultural activities tend to increase the N:P ratio in the nutrient loading of waters, while domestic wastewater tends to be rich in P, lowering the ratio of N:P. The topic of nutrient loading is further developed in Sections 8.5.4 and 8.6.1.

Modification of the water within lakes

Water from the watershed enters a lake from surface and ground water. Deposition of substances in the lake sediments leads to a decrease in their concentration. If exchange with the atmosphere is negligible, the retention of a particular element may be calculated as the difference between the total amount of the element entering the lake via surface and groundwater inflows minus the total amount leaving the lake. At times the sediment can release substances instead of trapping them and thus cause a negative retention. The most conspicuous example is the release of phosphorus from anoxic sediments into the overlying water. If the sediments are oxic, bioturbation can lead to a flux of materials from the sediments to the overlying water. For example, benthic animals feeding within the sediments may excrete substances onto the surface of the sediments and into the water. However, negative retention is only possible on a short time scale, for only those substances that have been previously deposited can be released from the sediments. In the long run, the sediment must act as a sink for substances in a lake.

Balances are more complicated for elements that are *exchanged with the atmosphere*, such as carbon and nitrogen. Exchange of gases with the atmosphere tends to reduce the degree of undersaturation and supersaturation in the surface waters. The net balance of the exchange depends on the biological processes with the lake. For example, losses of CO_2 caused by photosynthesis and the deposition of marl (calcium carbonate) and undecomposed substances will result in a net import of CO_2. A lake heavily loaded with organic matter and high rates of respiration by heterotrophic bacteria may have a net export of CO_2. The total balance for nitrogen in a lake depends on the balance between nitrogen fixation and denitrification. Nitrogen fixation lowers the concentration of N_2 in the water, allowing for an influx of this gas from the atmosphere. Denitrification increases N_2 in the water and may result in export from the surface water to the atmosphere.

8.3.3 The Carbon Cycle

Dissolved inorganic carbon (DIC), dissolved organic carbon (DOC), and particulate organic carbon (POC) are the most important pools of carbon in aquatic ecosystems. They usually occur as DIC > DOC > POC, in order of the relative size of the pools. Usually most of the POC is bound in detritus and the remainder in living organisms.

Figure 8.4 is a diagram of the carbon cycle in a hardwater lake (Lawrence Lake, Michigan). This diagram illustrates the complexity of the processes involved in the carbon cycle, but it cannot simply be applied to all other types of lakes. The plant, microbial, and detrital components are much more detailed than the animal components, for they appear relatively unimportant, from the standpoint of mass flux.

The DIC pool consists of CO_2, HCO_3^-, and CO_3^{2-}. Shifts between the three forms follow the laws of the carbon dioxide, bicarbonate, carbonate equilibrium (cf. Section 3.1.6) and are indirectly affected by the pH and the photosynthetic and respiratory activity of the aquatic organisms. The most important input of DIC is from the atmospheric CO_2, and the dominant abiotic outputs are evolution of dissolved CO_2 into the atmosphere and precipitation of marl (calcite). Respiration is the biological input, and photosynthesis and chemosynthesis are the biological outputs. The DOC pool consists of a mixture of various substances. The biologically reactive, low-molecular-weight compounds are less abundant than the refractory, high-molecular-weight humic substances (cf. Section 4.3.10). In addition to allochthonous inputs, the most important sources of DOC are secretion and excretion by organisms from all trophic levels and autolysis of detritus. Uptake by heterotrophic microorganisms, especially bacteria, is the most important sink for DOC. Excretion products are rapidly utilized by bacteria and therefore do not accumulate as the refractory substances do. In hard-water lakes some of the humic substances may be lost from the DOC pool through coprecipitation with marl.

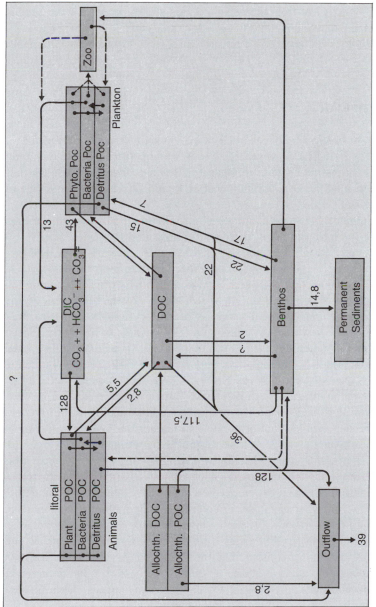

Figure 8.4 The carbon cycle in Lawrence Lake (Michigan). Transfer rates in g C m^{-2} year^{-1}; - - - -: transfer rates that are negligible in the authors' opinion (from Wetzel 1983).

POC consists of carbon bound in organisms and detritus. Primary production is the source of POC. Shifts in the POC pool occur because of death, feeding, and parasitism. POC can be transformed into DIC by respiration and into DOC via secretion, excretion, and autolysis. POC is lost from the pelagic region and imported into the benthic zone through sedimentation. Sedimented POC may be resuspended into the pelagic region by wind-driven water currents and by occasional benthivory by pelagic animals.

8.3.4 The Nitrogen Cycle

The most abundant nitrogen component in lakes is dissolved elemental nitrogen (N_2; solubility approx. 15–20 mg/L); however, relatively few organisms can use this form. Nitrogen fixation requires the enzyme nitrogenase, which is found only in the procaryotes (cf. Section 4.3.7). The contribution of N fixation to the nitrogen cycle of lakes is highly seasonal.

For autotrophic organisms that do not produce nitrogenase, dissolved nitrate, nitrite, and ammonium are the most important inorganic nitrogen forms. These forms of inorganic nitrogen are carried into lakes through surface water, ground water, and precipitation. All three nitrogen compounds can be utilized by autotrophic organisms. On the other hand, the decomposition of N-rich organic compounds releases only ammonium ("ammonification"). In contrast to vertebrates, which excrete either urea or uric acid, zooplankton excrete ammonium. Thus it has become customary in marine biology to refer to nitrate as "new" and to ammonium as "regenerated" nitrogen. To go from nitrate to nitrite to ammonium a series of microbial transformations are required that depend on the presence or absence of oxygen (Fig. 8.5). The nitrate respiration that takes place in an anoxic condition (cf. Section 4.3.9) transforms nitrate to either *ammonium* ("nitrate ammonification") or elemental nitrogen ("denitrification"). *Nitrification*, which takes place under oxic conditions (cf. Section 4.3.8) transforms ammonium through the intermediate step of nitrate into nitrate. Because this process requires oxygen, it results in a characteristic depth profile in eutrophic lakes (Fig. 8.6). Since the three forms of nitrogen are taken up by phytoplankton, all three have a minimum concentration at the depth of maximum photosynthesis (0–6 m). Nitrate increases rapidly with depth until it begins to be removed by nitrate respiration. Ammonium begins to increase below 7 m, whereas nitrite, as a relatively temporary intermediate stage in nitrification, has a limited maximum near the lower boundary of the oxidized zone.

Dissolved organic nitrogen (DON) comes from the excretion from organisms and the decomposition of detritus. It consists mainly of polypeptides and other complex amino groups. Simple amino acids are taken up quickly by bacteria and are therefore present in low concentrations. Microorganisms can also excrete peptidases (exoenzymes) that assist in the splitting of peptides into smaller units.

A diagram of the most important components of the nitrogen cycle in a lake with an anoxic hypolimnion is presented in Fig. 8.7. When there is oxygen in the hypolimnion the anoxic conditions and associated transformations occur only in the sediments.

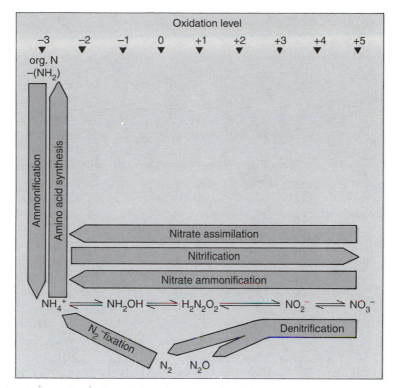

Figure 8.5 Biochemical transformations in the oxidation levels of nitrogen (from Wetzel 1983).

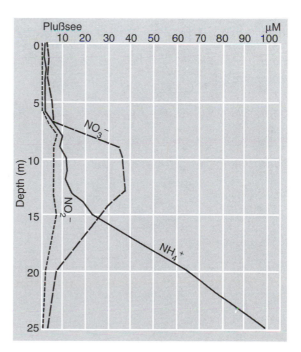

Figure 8.6 Vertical profiles of nitrate, nitrite, and ammonium in Plußsee.

[301]

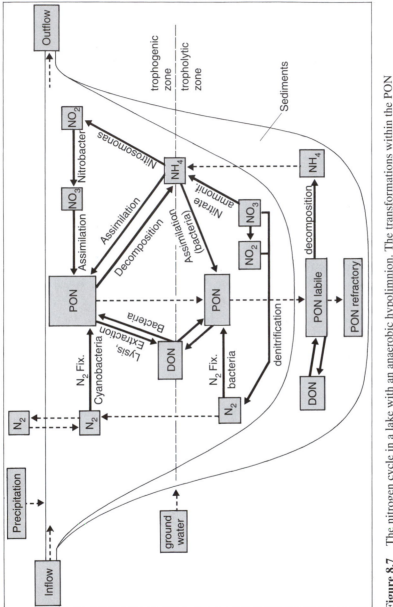

Figure 8.7 The nitrogen cycle in a lake with an anaerobic hypolimnion. The transformations within the PON pool (grazing) are not indicated. (*Dashed lines*) Transport processes; (*Solid lines*) Biological and chemical transformations.

8.3.5 The Phosphorus Cycle

In contrast to nitrogen, phosphorus in ecosystems occurs in only one form as PO_4 (*orthophosphate*). Similarly, phosphorus is bound to organic compounds by ester bonds between phosphoric acid derivatives and carbon chains. Orthophosphate is much less mobile than nitrogen in the sediments and precipitates more easily, since it combines with several different cations (Al^{3+}, Fe^{3+}, Ca^{2+}) to form highly insoluble products and also it readily adsorbs to clay. If the load of phosphorus is not increased artificially by anthropogenic influences (waste water, detergents), lakes usually receive more nitrogen than phosphorus in the surface and ground water. In unpolluted lakes phosphorus is more often the limiting factor for plant growth than nitrogen.

The fractions of phosphorus that are differentiated in limnological investigations of phosphorus cycling are primarily determined by the methodology and do not correspond precisely with the chemical species (Strickland and Parsons 1968).

Soluble reactive phosphorus (SRP) passes through a filter with 0.1 or 0.2 μm pores and is measured directly with the molybdenum blue method. In the older literature this fraction is often referred to as orthophosphate, but it consists of this free ion as well as many other labile phosphorus compounds. Actually, the SRP fraction closely approximates the amount of phosphorus available for use by phytoplankton, since some of these other compounds can be broken down and transported across the cell membrane in addition to the free phosphorus ion.

Total dissolved phosphorus (TDP) passes through a filter with 0.1 or 0.2 μm pores and is measured with the molybdenum blue method after an acidification and oxidation step. The portion that does not react without the acid oxidation consists of colloidal phosphorus as well as the dissolved organic phosphorus with low molecular weight (ca. 250). Both of these fractions continually contribute to the SRP pool.

Total phosphorus (P_{tot} or TP) is determined from an acidified and oxidized unfiltered sample and thus contains both the dissolved and particulate components (P_{part} or PP). The PP is either determined by the difference between the filtered and unfiltered samples or by measuring the particulate matter collected on the filter.

The most important internal sink of soluble reactive phosphorus is uptake by algae and bacteria. It is noteworthy that the C-heterotrophic bacteria can be P-autotrophs; that is, in taking up inorganic phosphorus they compete with the phytoplankton for SRP. Despite the fact that phosphorus is often the growth-limiting factor, algae and bacteria lose SRP, and to a lesser degree, low-molecular-weight, nonreactive phosphate (Lean and Nalewajko 1976). This *back and forth flow* of phosphorus is the most rapid of the interacting cycles in lakes. The turnover time of the SRP pool, estimated from the gross uptake of radioactive phosphorus (^{32}P), may be less than 10 minutes under P-limited conditions. Many researchers have taken interest in this rapid cycling, but it is of little ecological consequence, since the growth of organisms and transfer of phosphorus through the food chain depends on the net uptake of phosphorus. Important internal supplies of phosphorus occur through animal excretion and the decomposition of nonreactive components by microbial exoenzymes (phosphatases).

During the growth period, SRP in the epilimnion may be reduced to its limit of detection (0.03 μM, ca. 1 μg/l). Periods in which zooplankton grazing exceeds phytoplankton production (clear-water phase) can usually be recognized by an increase in the SRP concentration. SRP and particulate phosphorus have complementary seasonal changes; due to sedimentation, the total phosphorus in the epilimnion usually decreases during the stratified period unless there is a major external addition. As the thermocline deepens in the fall there is a mixing of P-rich water into the epilimnion, causing another increase in the P_{tot} concentration in the epilimnion (Fig. 8.8).

Figure 8.9 shows a diagram of the P cycle in a stratified lake. Phosphorus moves into the sediments through sinking of organisms and adsorption on sinking clay minerals and marl particles. The redox conditions in the sediment–water boundary layer are of critical importance in determining the fate of the phosphorus in the sediments. The concentration of phosphorus in the anoxic interstitial water of the sediment is always an order of magnitude or more higher than in the open water. According to the concentration gradient, then, there must always be a diffusion of P into the open water. When the deep water and the sediment–water boundary are oxidized (oxidized microzone), iron occurs as the oxidized Fe^{3+} form. Phosphorus that diffuses up from the

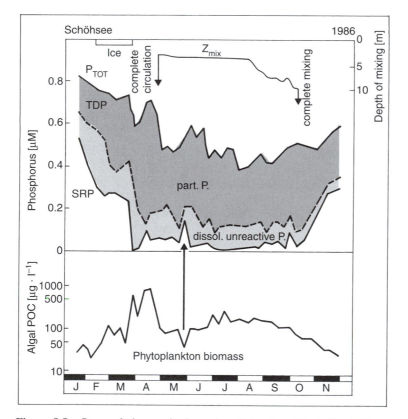

Figure 8.8 Seasonal changes in the various P fractions (P_{tot}, TDP, SRP; concentrations in μg) in Schöhsee related to the mixing depth and the development of the phytoplankton biomass. The arrow indicates the clear-water phase.

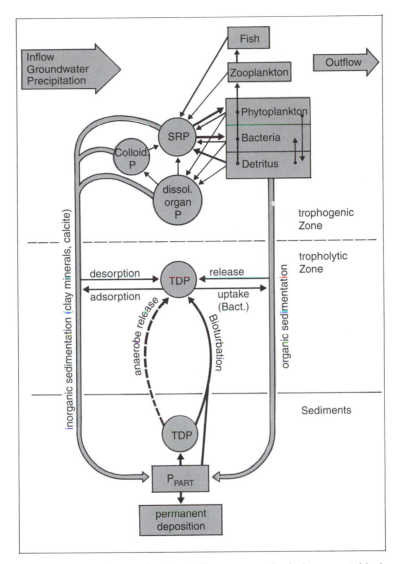

Figure 8.9 Diagram of the P cycle in a stratified lake. Gray arrows: physical transport; black arrows: biological transformations.

sediments forms an insoluble complex with $Fe(OH)_3$, that precipitates. If the Fe^{3+} is reduced to Fe^{2+}, the complex dissolves and the phosphorus can go back into solution. The reduction of iron occurs at a redox potential of 0.2–0.3 mV, which may be found in the hypolimnion of productive lakes at an oxygen concentration of about 0.1 mg/l (cf. Section 3.3.1). Thus, phosphorus accumulates in the anoxic hypolimnion, since it remains soluble there and does not precipitate back to sediments. Occasionally, turbulence-induced mixing of the thermocline causes phosphorus to move into the epilimnion, stimulating the growth of the algae. This, in turn, leads to further oxygen consumption and greater release of phosphorus from the sediments ("galloping eutrophication"). In this way the sediment acts as a P trap, as long as its surface is oxidized and as a P source when it becomes reduced and the oxidized microzone disappears. There is also some release of phosphorus into the hypolimnetic water from oxidized sediments, through excretion from benthic animals and by their mechanical disturbance and mixing of the upper few centimeters (bioturbation).

8.3.6 The Silica Cycle

Silica is somewhat peculiar, since in freshwater there is only one group of organisms, the diatoms, that use it in significant amounts. The effect on the silica cycle by chrysophytes with siliceous scales is negligible. Diatoms, on the other hand, require large quantities of silica.

Soluble silica occurs as ortho-silicic acid, which does not dissociate at pH values < 9. It is derived from the weathering of silicate minerals. The only significant sink in lakes is the uptake by diatoms. Although the shells of diatoms dissolve in water, the process is very slow, requiring on the average 50 days for one-half of the particulate silica of a dead diatom to dissolve. Also, silica does not dissolve during passage through the gut of herbivorous zooplankton; in contrast to other algal nutrients, silica is excreted in the particulate form rather than dissolved. This and the rapid sinking of dead and detrital diatoms (several meters per day) means that there is essentially no short-circuited cycling of silica in the epilimnion (Sommer 1988b). This is also evidenced by the lack of recovery of soluble Si in the epilimnion following the collapse of a diatom maximum (Fig. 8.10). Increased Si concentrations in the epilimnion occur only after allochthonous additions or mixing from Si-rich deep water. Particulate silica in the sediments may partially dissolve and return to the open water by diffusion and bioturbation. Diatom shells are resistant and accumulate with time, making up some of the most important microfossils in paleolimnology.

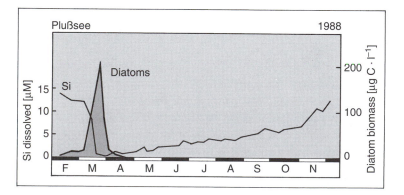

Figure 8.10 Concentration of dissolved Si (μm) and biomass of the diatoms (μg C/l) in the epilimnion of Plußsee.

8.4 ARE STREAMS ECOSYSTEMS?

The definition of an ecosystem implies that it must have some type of boundary that separates it from other systems, regardless of whether it is a "microcosmos" that has characteristics greater that the sum of the parts or it is simply an aggregation of integrated communities and abiotic factors (cf. Section 8.1). Two major ecosystem features are that energy flows through the system and materials cycle within it. Generally it is assumed that outside energy is fixed within the system (e.g., light energy is fixed by green plants) and is then converted back to heat by heterotrophic organisms. Heterotrophic activity also releases essential substances (e.g., oxidized carbon), making them once again available for energy fixation.

The ecosystem concept is sometimes applied to streams and rivers, referring to them as flowing water ecosystems. Application of the holistic concept of ecosystem ("superorganism") to streams creates major difficulties.

First of all, streams are primarily heterotrophic functioning systems. The energy for streams is fixed mainly in terrestrial systems in the watershed, rather than in the stream itself. Energy is always transported as allochthonous organic matter into the water, where it is then utilized. This means that the processes of energy fixation and decomposition are separated. A stream system could not exist without the energy supplied by its watershed; it is energetically dependent on its watershed. On the other hand, decomposition can take place quite well outside of the stream. Such a dependency is one-sided. It does not make sense to treat a one-sided, dependent structure as an independent system. Streams can therefore be better understood by treating them as a part of a larger system, such as the valley (Hynes 1975).

There is a similar problem in considering the cycling of matter in streams. Substances released by decomposition are not available for the producers, for they are carried downstream in a flowing wave and, at best, may be utilized by downstream organisms. Materials "spiral" downstream, rather than cycle. Compared to the quantity of nutrients that are transported from the watershed, the amount regenerated in the stream is negligible. Also, with regard to inorganic matter, the stream system is largely controlled by external influences. Certainly there are situations where the stream exports matter into the watershed (e.g., the role of the River Nile water for the fertility of the Egyptian agriculture), but even then it is actually a matter of moving the nutrients from the land (upper reaches) by the stream (means of transport) back to the land (lower reaches).

Streams are thus dependent on the watershed in terms of energy flow as well as the movement of matter and are best understood as part of the entire watershed system. This explains the great elasticity of streams following disturbances (cf. Section 5.7). Even when considerable stretches of a stream are disturbed, it represents only a localized effect. For example, the large watershed is not affected by an accidental poisoning in a brook. The watershed represents a buffer that remains intact. Impacts on the watershed (e.g., deforestation), however, can cause dramatic effects on the stream. The effects of deforestation of an entire watershed on its stream were carefully studied in a famous large-scale study at Hubbard Brook in Northeastern United States (Likens et al. 1977).

Thus the concept of an ecosystem as a self-regulating system cannot be correctly applied to streams. Reference to streams as ecosystems should be restricted to describing the "system" of interrelationships of biotic and abiotic factors. Used in this way, the phrase *stream ecosystem* has heuristic value, for it draws attention to the importance of abiotic factors that are especially important in streams.

It may be surprising that one of the classic ecosystem studies that first quantified energy flow through all trophic levels was conducted on a stream, Silver Springs, Florida (Fig. 8.2). Closer examination reveals that Silver Springs is a very atypical stream. Most of the energy in Silver Springs comes from photosynthesis by macrophytes and their associated periphyton. Allochthonous energy inputs come mainly from the feeding of fish by tourists. There is actually no cycling of matter, for the system is supplied entirely by spring water that supplies an excess of nutrients. The absence of a cycling of matter is not critical in this case, for the study was restricted to examining the flow of energy. Silver Springs is energetically independent of its watershed, but entirely dependent on the spring water for matter. It can be viewed as a special type of system in steady state. Of course, the fact that it is in steady state makes it ideal for the study of energy flow.

8.5 COMPARISONS OF ECOSYSTEM PRODUCTIVITY

8.5.1 "Empirical Models"

The concept of "productivity of an ecosystem" is similar to the concept of "fertility" in agriculture. It refers to the ability of any trophic level to produce biomass. The productivity of the various trophic levels are usually positively correlated, since the lower trophic levels form the food base for the higher trophic levels. In ecosystems with little allochthonous organic input, the primary production is representative of the productivity of the entire ecosystem.

One of the main goals of comparative ecosystem research, as carried out in the International Biological Program of the 1970s, was the measurement of productivity of different ecosystems and the discovery of global trends in the control of productivity by climatic and geochemical conditions. In addition, there was an attempt to find statistical correlations between input parameters and the productivity and biomass of individual trophic levels. Using comparative studies, one can develop "empirical models" that allow for probabilistic predictions of aggregated parameters, such as production and biomass for specific trophic levels. Such empirical models are simply *regression analyses* between input and output parameters, such as primary production and secondary production. The input parameter need not be the production or biomass of the next lower trophic level; for example, it is possible to determine the statistical correlation between fish production (output) and primary production (input). In many cases empirical models also use *surrogate parameters* (e.g., the geographical latitude) as a substitute for the radiation of light energy or chlorophyll for the biomass of the phytoplankton. Of course, unless the surrogate parameter and the actual parameter of interest are closely correlated, much variability will be introduced into the analysis.

Significant correlations are found through regression analysis only when there are large differences in the dependent and independent variables in the different lakes. This large-scale approach almost always favors the "bottom-up" hypothesis; that is, there are positive correlations between production and biomass parameters in the different trophic levels. "Top-down" effects simply appear as overlapping variability between the individual lakes (cf. Section 7.3.5).

8.5.2 Primary Production and Biomass of the Phytoplankton

Comparative investigations of the IBP have shown that the size of the primary production in lakes declines as one goes from the tropics to the poles. This is especially true for the most productive lakes in the various zones, and less so for the unproductive lakes. There are several potential causes for this effect that may have additive effects. The amount of light energy is higher in the tropics. The longer growth period in the summer appears to have a greater effect than the longer summer day length at higher latitudes. Nutrients are remineralized more rapidly at higher temperatures and are therefore more available in the epilimnion for primary production, before they eventually sediment out. The polymictic mixing regime in tropical lakes continually cancels out the spatial separation between availability of light and nutrients.

There is a general trend with geographical latitude, but there are large differences between individual lakes at any particular latitude. Much of this variation is caused by differences in the availability of limiting nutrients. Thus within the temperate zone of North America and Europe the average daily primary production in the growth period ($mg \, C \, m^{-3} \, d^{-1}$) is closely correlated with the concentration of total phosphorus ($\mu g/l$):

$$PPR = 10.4 P_{tot} - 79, \qquad r^2 = 0.94$$

(Smith 1979). This empirical model can be used to predict with a high degree of statistical certainty the primary production from the total phosphorus concentration.

Practical needs associated with the eutrophication of lakes have led to the development of a large number of empirical models to predict the biomass of the phytoplankton from the supply of nutrients. Chlorophyll is usually used as a surrogate parameter for biomass, although occasionally microscopically determined biovolume or POC (detritus is included) is used as the dependent variable. The conceptual basis for the model is the assumption that only a given amount of biomass can be formed from a certain amount of a limiting nutrient. One should note, however, how the temporal and spatial average values have been derived when comparisons are made of different phosphorus-biomass models. Temporally, there are models that use yearly averages, averages for the growth period, or maximum values; spatially, vertical integral values are used to calculate mean values for the euphotic zone, the epilimnion, or per unit lake volume or lake surface area.

The best-known empirical model is the comparative eutrophication study of the OECD (Vollenweider and Kerekes 1982). This model can be used for a discussion of the problems of all phosphorus-biomass models and all similar regression models. The OECD model includes two steps: The first step describes the relationship between the phosphorus loading and the phosphorus concentration in a lake (cf. Section 8.5.4). The second step conveys the relationship between phosphorus concentration and phyto-plankton biomass. When chlorophyll is used as a surrogate parameter for biomass, the regression equation obtained is:

$$\text{Chl a} = 0.28P_{tot}^{0.96}, \qquad r^2 = 0.77, \qquad n = 77$$

where Chl a is the yearly average in the euphotic zone, in $\mu g/l$, and P_{tot} is the yearly average for the lake, in $\mu g/l$. Despite the high significance of the regression equation, there is considerable variance between the individual data points (Fig. 8.11). The 95% confidence interval for the dependent variable covers almost an order of magnitude.

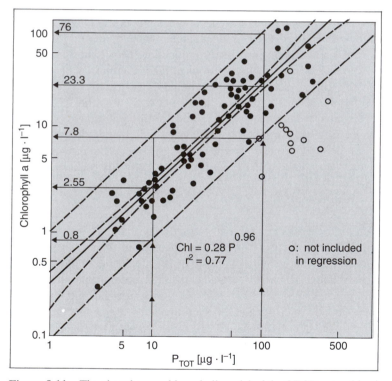

Figure 8.11 The phosphorus–chlorophyll model of the OECD eutrophication study (from Vollenweider and Kerekes 1982). Each data point represents one lake. Log-log regression of chlorophyll (yearly average in the euphotic zone) and total phosphorus (yearly average for the entire water column). Inner dashed line: 95% confidence interval around the regression line; outer dashed line: 95% confidence limit for the dependent variables; thin lines with arrows: predicted chlorophyll values for lakes with 10 and 100 μg P_{tot}/l.

At a P_{tot} content of 10 $\mu g/l$ the chlorophyll concentration lies with 95% confidence between 0.8 and 7.8 $\mu g/l$ and at 100 $\mu g/l$ between 7.8 and 76 $\mu g/l$. Thus, in order not to overlap in their 95% confidence intervals, lakes must differ in P_{tot} content by at least one order of magnitude. Most other empirical models that use double logarithmic regressions have similarly broad confidence intervals. Because of this variability such models are not useful as a calibration tool with which the measurement of one variable can eliminate the need for measuring another variable.

8.5.3 Secondary Production, Fish Yield, and Animal Biomass

If there is a constant ecological efficiency, secondary and primary production are directly proportional. Since lakes can vary about three orders of magnitude in primary production (Brylinsky 1980), there is a statistical correlation between primary production and secondary production despite differences in ecological efficiency (0.05–0.2). The data from the IBP showed there is a linear relationship between the annual gross primary production (PP) and the annual net secondary production (SP) (both in kJ m^{-2} year^{-1}):

$$SP = -36.05 + 0.128PP, \qquad r^2 = 0.82, \qquad n = 17$$

Fish yield, the catch of fish per unit surface area, is an economically important parameter. Although in fisheries management it is often referred to as "fish production," it is not a production value in the ecological sense, for the catch depends on the intensity of fishing as well as the productivity of the ecosystem. This error is removed if the fishing is conducted strictly according to the principle of sustainability; that is, the amount caught each year equals the amount grown. In this case the fish yield is less than the actual fish production, as not all fish species are caught. Fish production cannot be designated to a particular trophic level, since fish belong to several trophic levels. Nevertheless, statistical relationships have been derived for the productivity of a lake and its fish yield:

$$Y_F = 7.1P_{tot}^{1.0}, \qquad r^2 = 0.87, \qquad n = 21$$

$$Y_D = 0.012Chl^{1.17}, \qquad r^2 = 0.87, \qquad n = 19$$

$$Y_D = 10^{-6}PP^{2.0}, \qquad r^2 = 0.74, \qquad n = 15$$

where Y_F is the fish yield in g fresh weight m^{-2} year^{-1}, Y_D the fish yield in dry weight m^{-2} year^{-1}, P_{tot} the total phosphorus in mg/l (annual average), Chl the chlorophyll in mg/l (summer average), PP the primary production in g C m^{-2} year^{-1}. The equations in Section 8.5.3 are from Peters (1986), who summarizes a large number of "empirical" relationships for production and biomass of many different trophic levels.

[311]

8.5.4 Trophic Classification of Lakes

Early on in limnology it was known that large, deep lakes were usually less productive than those that are small and shallow. There are several reasons for the relationship between the size and productivity of lakes:

A lake receives its nutrients from its watershed mainly through erosion. Assuming the same supply of nutrients per unit surface area, the nutrient loading per unit lake volume depends on the ratio of lake volume: watershed area. Lakes that are large in relation to their watershed will have low nutrient loading.

In large, deep lakes only a small part of the epilimnion is in contact with the sediments during the summer, so that there is little opportunity for return of nutrients to the euphotic zone.

Larger lakes usually have a longer hydraulic retention time. During the time the water is retained a large portion of the nutrients is lost through sedimentation. The OECD eutrophication study found the following empirical relationship between the P_{tot} concentration in the lake (P_L), the P_{tot} concentration in the tributaries (P_{in}) and the theoretical hydraulic retention time (τ_w, in years) (Vollenweider and Kerekes 1982):

$$P_L = 1.55[P_{in}/(1 - \sqrt{\tau_w})]^{0.82}, \qquad r^2 = 0.86, \qquad n = 87$$

This equation shows that the difference between the P concentration in the lake and the tributaries is dependent on the τ_w.

The striking differences between very productive and very unproductive lakes inspired Einar Naumann and August Thienemann to develop a *lake typology* (Elster 1974). Lakes with little productivity and clear-water were referred to as *oligotrophic*, and lakes with high productivity and greenish water due to phytoplankton were called *eutrophic*. An additional *dystrophic* type described lakes with low conductivity and brown-colored water due to dissolved humic substances. Originally, lakes were assigned to a particular type according to qualitative criteria: Oligotrophic lakes have an orthograde oxygen profile in the summer, and eutrophic lakes have a clinograde oxygen curve (cf. Section 3.3.1). An additional criterion was the oxidized (oligotrophic) or reduced (eutrophic) sediment surface. A closely linked feature was the colonization of the sediments with different indicator organisms. In oligotrophic lakes, for example, the chironomid larvae *Tanytarus* spp. dominate the benthos, whereas *Chironomus* spp. is the dominant chironomid in eutrophic lakes. The latter are capable of anoxibiosis and thus can withstand long periods in the anoxic hypolimnion of eutrophic lakes.

As a result of the OECD study, there have been attempts to replace the qualitative lake typology with quantitative criteria. "Oligotrophic" and "eutrophic" are regions of a trophic continuum that can be separated from one another by established limits, such as by the P_{tot} concentration during the spring circulation:

ultraoligotrophic	$P_{tot} < 5 \ \mu g/l$
oligotrophic	P_{tot} 5–10 $\mu g/l$

mesotrophic	P_{tot} 10–30 $\mu g/l$
eutrophic	P_{tot} 30–100 $\mu g/l$
hypereutrophic	$P_{tot} > 100$ $\mu g/l$

These threshold values can also be expressed as production or biomass by using the empirical equations given in Sections 8.5.2 and 8.5.3. In this way the system can also be used for lakes in which phosphorus is not the factor limiting productivity.

Chemical data are more easily quantified and translated into technical procedures. However, the trophic condition of a lake is basically a biological phenomenon. Trophic condition cannot be determined with a single factor, as illustrated by comparing Schöhsee in Holstein (mean depth 13 m) and Lake Constance. Schöhsee is mesotrophic according to the OECD criteria (P_{tot} ca. 20 $\mu g/l$), but it has an anoxic hypolimnion and is therefore eutrophic in the traditional sense. Based on its high phosphorus concentration, Lake Constance (1980, $P_{tot} = 100$ $\mu g/l$) should be classed as eutrophic according to the OECD criteria, although the oxygen concentration near the bottom in the summer was about 50% saturation, which in the traditional scheme would be considered mesotrophic. These discrepancies can be explained by the lake morphometry: Lake Constance (mean depth ca. 100 m) has an enormous hypolimnion that brings in a large supply of dissolved oxygen during the spring circulation that can be used to decompose the sinking organic matter from the trophogenic zone. Schöhsee, in contrast, has a small hypolimnion with a correspondingly small supply of oxygen. Thienemann had already recognized the importance of morphometry for the trophic condition of lakes. He writes that lakes with a volume ratio of epilimnion to hypolimnion of <1 tend to be eutrophic, whereas lakes with a ratio of >1 are more likely to be oligotrophic. Clearly, several criteria are needed to characterize the trophic condition of a lake (Box 8.1).

8.6 ANTHROPOGENIC DISTURBANCES TO ECOSYSTEMS

8.6.1 Eutrophication: Causes and Consequences

The increase in trophic condition of a lake is referred to as *eutrophication*. It is a slow and natural process in the geological history of a lake. The lake basin gradually fills with sediments, reducing its volume (landing). This leads to an increase in the trophic condition, even if there is no change in the nutrient load. In the past decades the trophic condition of many of the inland waters has increased rapidly. Eutrophication has proved to be one of the most widespread and serious anthropogenic disturbances to aquatic ecosystems. There is considerable consensus today that the major cause for eutrophication is the increased loading of nutrients, especially phosphorus. Such *increasing anthropogenic nutrient loading* has even lead to the eutrophication of very large lakes, such as Lake Erie (25,800 km^2) and Lake Ontario (24,500 km^2). Increasing waste water, the introduction of phosphorus-containing washing detergents, increasing use of

BOX 8.1	SOME CHARACTERISTICS OF OLIGOTROPHIC AND EUTROPHIC LAKES	

	Oligotrophic	Eutrophic
Morphometry	deep	shallow
Volume ratio epi/hypolimnion	<1*	>1*
Primary production	low (50–300 mg C m^{-2} d^{-1})	high (>1000 mg C m^{-2} d^{-1})
Algal biomass	small 0.02–0.1 mg C/l 0.3–3 μg Chl a/l	large >0.3 mg C/l 10–500 μg Chl a/l
Nutrients	low P_{tot} after complete circulation < 10 μg/l	plentiful P_{tot} after complete circulation > 30 μg/l
Massive development of cyanobacteria	absent	present
O_2 depletion in the hypolimnion	little, >50%	strong, may be anoxic
O_2 profile	orthograde	clinograde
Profundal zoobenthos	diverse, O_2 requiring	species poor, tolerate low O_2
Chironomid larvae	*Tanytarsus* group	*Chironomus* group
Fish fauna	deep-water salmonids, coregonids (cold stenotherms)	cyprinids, centrarchids (sunfish) ("warm-water" fish)

* exceptions possible.

fertilizers, and increased erosion in the watershed are the major reasons for the increased loading of nutrients. The loading of nitrogen is increased by pollution of the atmosphere with nitrous oxides, where they combine with water vapor and are transported to lakes as precipitation, contributing to increased loading of nitrogen. Fertilization of farmland has its greatest effect on N loading, since, unlike nitrogen, phosphorus is relatively immobile once it is in the soils. Flooding, however, can transport to lakes large quantities of phosphorus derived from fertilizers. Domestic sewage results primarily in increased P loading. It contains on the average an N:P ratio of 4:1 and can lead to a shift from phosphorus to nitrogen limitation (cf. Section 4.3.7). Estimates of the various components of phosphorus loading in Lake Constance for 1976 show the importance of detergent phosphorus and fecal phosphorus in the sewage before the construction of modern waste water treatment (Fig. 8.12). The phosphorus content in detergents has now been reduced with the introduction of phosphate substitutes.

An especially undesirable consequence of eutrophication is the *massive development of cyanobacteria* (blue-green algae), which tend to form dense "surface blooms." Some strains are toxic or cause allergic reactions. Cyanobacteria excrete organic compounds that impart the water with bad odors and tastes, creating serious problems in drinking water. Some of the classic "problem algae" are the N_2-fixing *Anabaena* and *Aphanizomenon*, as well as the non-N_2-fixing genera *Microcystis*, *Limnothrix*, and *Planktothrix*. Most of the bloom-forming cyanobacteria appear in the late stages of eutrophication. The red-pigmented *Planktothrix* spp. are somewhat unique, for they can regulate their position in the water column and inhabit the metalimnion of deep lakes in the early stages of eutrophication (often already at 20 μg P_{tot}/l). *Planktothrix* is mixed throughout the entire water column during the periods of circulation and becomes layered in the metalimnion during the summer. When advanced eutrophication reduces light penetration, this stratum becomes too dark and *Planktothrix* disappears again (e.g., in Lake Zürich; cf. Section 4.3.5).

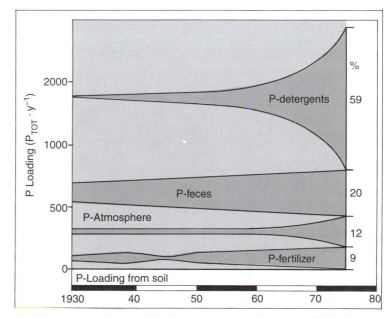

Figure 8.12 Annual phosphorus loading in Lake Constance before the successful implementation of sewage treatment (from model estimates of Wagner 1976).

One effect of eutrophication on the littoral zone is a worsening of the light climate for submersed macrophytes. Under oligotrophic conditions macrophytes that obtain their nutrients from the sediments have an advantage over the aufwuchs community, which gets its supply of nutrients from the water. As the supply of nutrients in the water increases, the macrophytes become increasingly overgrown with aufwuchs algae and clumps of filamentous forms such as *Cladophora* and *Spirogyra*. This, along with the poorer light conditions, leads to a suppression of the macrophytes and indirectly, of their associated fauna.

The main effect of eutrophication in the hypolimnion and sediments is the increased consumption of oxygen. As anoxic conditions develop at the sediment surface and in the hypolimnion the deep water fauna becomes impoverished. Even pelagic fish, which release their eggs in the open water to sink to the bottom (e.g., lake whitefish), cannot continue to reproduce naturally in a lake with an anoxic sediment surface. Anoxic conditions also lead to a series of chemical and microbial processes that otherwise would not take place: nitrate ammonification, denitrification, desulfuration (Section 4.3.9) and methane formation (Section 4.3.10). The reduced end products of these anoxic reactions can then be used as initial products for chemolithotrophic processes at the aerobic–anoxic boundary (Section 4.3.8). The release of phosphorus from the anoxic sediments (internal loading) is extremely important, as this sets in motion the *self-acceleration of eutrophication*.

Spectacular *fish kills* may result from very advanced eutrophication. Winter fish kills may occur when there is heavy oxygen consumption under an ice cover. Summer fish kills can be initiated by elevated pH caused by high photosynthetic rates (cf. Section 4.3.6) in lakes with high total ammonium concentration. The increased pH causes a shift from nontoxic ammonium ion to toxic ammonia (cf. Section 4.2.3).

8.6.2 Eutrophication: Nutrient Control and Lake Restoration

In the period between the First and Second World Wars the oligotrophic condition of large lakes was considered a liability, especially by fisheries specialists; there were even attempts to fertilize such lakes. Values have shifted with the rapid eutrophication of lakes in the past decades. The use of lakes for drinking water sources, swimming, and tourism has now become more important than fisheries. As a result, strategies have been developed to attempt to reverse the lakes toward their original condition.

The best method for fighting eutrophication is the prevention of anthropogenic addition of nutrients. As a first step, modern techniques can be used to reduce the impact of sewage. For example, fecal material can be composted instead of flushing it into sewage systems. Large-scale efforts have been made to reduce the phosphorus content of washing detergents or find phosphorus substitutes. The fertilization of agricultural land still presents serious problems, for much time is needed to change old practices, and, even then, improved techniques may have limited effects on the nutrient loading. Thus much of practical water management focuses on the removal of waste water and waste water treatment.

Waste water removal involves the collection of waste water from the watershed into a sewage canal system ringed around the lake and its release into the lake outlet. The logic is that increased autochthonous production is less damaging in a stream than in a lake (cf. Section 7.7.2).

Waste water treatment facilities are essentially technically intensified versions of the natural processes that occur in lakes. The most common waste water treatment has three stages. The mechanical stage first removes the coarse particles, the biological stage (activated sludge process) involves the microbial respiration of organic compounds, and the chemical stage precipitates out the phosphorus. Nitrogen can also be removed by creating alternating oxic and anoxic conditions, nitrifying the ammonium and denitrifying the nitrate (Box 8.2).

Restoration of a lake often requires measures in the entire catchment area, for pollutants are transported to the lake by tributaries. Lake Constance is an example of a joint international effort (Austria, Germany, Switzerland) to stop the eutrophication of a large lake. Lake Constance was originally oligotrophic, but showed signs of eutrophication with increasing populations and use of the lake after the turn of the twentieth century and by the 1970s had developed blooms of cyanobacteria. This restoration project required the construction of tertiary sewage treatment plants (removing phosphorus) for all towns adjacent to the lake as well as along the tributaries. The result was a reduction in the phosphorus concentration in the lake and biological signs of a reversal in the eutrophication.

Even after nutrient loading has been reduced, there is often a lag in the lake's response. It sometimes takes several years after the reduction in external nutrient loading until the phosphorus concentration in the lake decreases (Fig. 8.13; above). First, it takes time to exchange the water in the lake, depending on the type of basin and watershed. Second, phosphorus deposited in the sediments during the peak period of pollution may continue to be released from the sediments for several years. There may even be a temporary increase in internal loading as the phosphorus concentration in the overlying water decreases, increasing the diffusion gradient. After the phosphorus concentration in the lake decreases, it may take several years for a corresponding decline in the phytoplankton (Fig. 8.13; below). This type of lag can occur if at the height of eutrophication the phytoplankton becomes N or light limited, rather than controlled by phosphorus. Time is required to reduce the phosphorus concentration to the level where it becomes limiting. Time lags in the response of the algal biomass may also be caused by top-down effects. For example, populations of planktivorous fish that developed during the eutrophic phase may live for several years and continue to suppress the zooplankton grazers, maintaining low grazing pressure on the phytoplankton.

Lake Washington is one of the finest case studies of lake restoration that demonstrates how the structure of the food web and biotic interactions affect the success of nutrient reduction measures (Fig. 8.14). Earliest records from the 1940s indicate Lake Washington was an oligotrophic–mesotrophic lake, but with increased urbanization of the watershed and pollution from domestic and industrial sewage the lake rapidly became eutrophic. By the 1960s, there was a public concern caused by the lake's low water transparencies of only 1 m and noxious smells from blooms of cyanobacteria. After much debate over the question of whether phosphorus was actually the nutrient-limiting algal growth (Edmondson 1991), the public voted to build a pipeline to divert the sewage away from the lake and into the nearby Puget Sound. By 1968 the sewage

BOX 8.2   **OUTLINE OF A WASTE-WATER TREATMENT PLANT**

Waste-water treatment incorporates the same natural processes that occur in lakes under natural conditions. Technical procedures and added energy greatly intensify and accelerate these processes. The most important features are summarized below. The processes in the different basins of a treatment plant also occur in vertical strata of lakes and in the horizontal zones of streams. The processes above the dashed line take place in the oxic region (epilimnion) and the processes below the line in the anoxic sediments or hypolimnion of a lake.

Treatment process	Conditions	Equivalent process in a lake
Mechanical purification sedimentation basins	reduced turbulence	sedimentation
Biological step: oxidative decomposition of organic matter	aeration, turbulence	heterotrophic microbial decomposition
Simultaneous precipitation of phosphorus	addition of aluminum or iron salts	precipitation of phosphorus with iron hydroxide
Nitrification	strong aeration	nitrification
- - - - - - - - - - - - - - - - - - - -	- - - - - - - - - - - - - - - - - - - -	- - - - - - - - - - - - - - - - - - - -
denitrification	anoxic conditions, low-molecular C source	denitrification in anoxic sediments
methane formation in sludge digestion tanks	anoxic conditions, low-molecular C source	methanogenesis in sediments

diversion was completed and after only 2 years the lake phosphorus concentration dropped to a new low equilibrium. After a short time lag, summer water transparencies improved to 3 to 4 m and remained constant for about 5 years until 1975. However, when large filter-feeding *Daphnia* returned to the lake in 1976, average water transparencies suddenly doubled, increasing to values up to 10 m. Clearly, the second

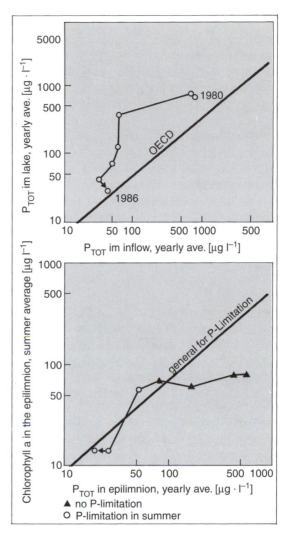

Figure 8.13 Delayed response of Schlachtensee (Berlin) to the reduction of external phosphorus loading. (*Above*) Total phosphorus in the lake in relation to the total phosphorus in the inflow and compared to the OECD model (cf. Section 8.5.4). (*Below*) Summer mean chlorophyll in the epilimnion compared to the summer mean total phosphorus in the epilimnion. Reference line: regression for years with P limitation from a lake restoration study of 18 European lakes (from Sas 1989).

increase in water clarity was not simply a direct response to the reduced phosphorus concentration. Several mechanisms appear to have contributed to the re-establishment of the *Daphnia* population in Lake Washington, including the disappearance of filamentous cyanobacteria that interfere with *Daphnia's* filtering apparatus (a bottom-up, biotic control) and reduced predation by mysid shrimp (a top-down, biotic control).

In situations where the nutrient sources for a lake are diffuse (groundwater, precipitation), waste water removal and treatment may not be effective. In such cases, or when the response to nutrient abatement is too slow, other lake restoration techniques can be applied. These measures may include increasing the export of nutrients from the lake, enhancing the immobilization of nutrients in the sediments or inhibiting the uptake of nutrients by the algae.

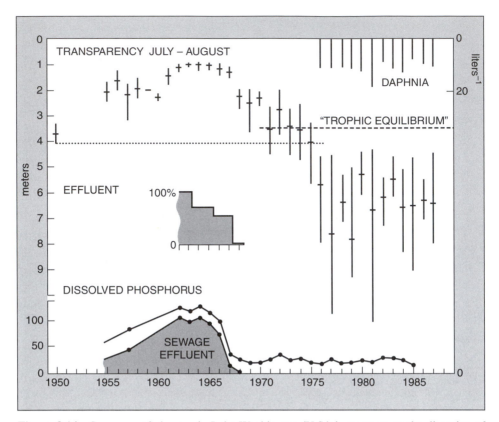

Figure 8.14 Summary of changes in Lake Washington (U.S.) in response to the diversion of sewage from the lake (after Edmondson 1991).

The fact that there is an accumulation of nutrients in the hypolimnion of stratified lakes can be used to accelerate the export of nutrients. Normally, the water flowing out of a lake comes from the relatively nutrient-poor epilimnion. Water can be released from the bottom of lakes by partly damming the outflow and inserting a tube to siphon the deep water (Olszewski tube), removing nutrient-rich water from the hypolimnion (Pechlaner 1979). This can also create problems, such as noxious odors produced by H_2S in the anoxic water. Also, when large quantities of water are released, the thermocline may be lowered, thereby raising the temperature of the hypolimnion, increasing the oxygen depletion, and releasing more phosphorus from the sediments. One can retain phosphorus in the lake bottom by improving the redox conditions in the sediments and hypolimnion. This can be accomplished either by introducing oxidizing substances into the lake (*aeration of the hypolimnion, injection of nitrate into the sediment*) or by pumping oxygenated water from the epilimnion into the hypolimnion. Forced mixing during the stagnation period is risky, however, since it adds nutrients to the epilimnion in addition to oxygenating the hypolimnetic water. *Intermittent mixing* is often successful, as the frequent changes in growth conditions prevent the gradual development of massive blooms of cyanobacteria and shift the species composition of the phytoplankton to rapidly growing and easily grazed nanoplankton (Reynolds et al. 1984).

The concept of *biomanipulation* (Shapiro and Wright 1984), introduced in Section 7.3.3, relies on the reduction or elimination of zooplanktivorous fish in order to relieve the predation pressure on the zooplankton, thereby intensifying the grazing pressure on the phytoplankton. Where this approach is successful, it not only reduces the biomass of the phytoplankton, but also decreases the sinking into the hypolimnion of algae and algal-generated detritus; this in turn reduces the oxygen depletion, improves the redox conditions in the hypolimnion and leads to greater immobilization of phosphorus. In shallow lakes, increased water transparency can allow for a recovery in the growth of macrophytes.

Complete removal of planktivorous fish usually has dramatic effects on water clarity. However, a community without planktivores is not stable (cf. Section 7.2). Fish must be continuously controlled in order to maintain the effect. Without additional nutrient removal the algal biomass will not be permanently suppressed by grazers, but will eventually shift to grazing-resistant forms. New biomanipulation strategies employ a compromise; planktivores are maintained at moderate abundance by enhancement of the stock of piscivores, requiring continual fisheries management (Benndorf et al. 1988, Shapiro 1990). Biomanipulation seems to be most effective in shallow lakes that have two stable states (cf. Section 7.5). Removal of fish can be the event that, within a year, can push the lake from a turbid system dominated by pelagic algae to a clear system dominated by macrophytes. Once the macrophytes have become established, fish can be reintroduced without shifting the system back (Reynolds 1994).

8.6.3 Acidification

Acidification is the second most widespread anthropogenic change in lakes and streams. Pure water with CO_2 in equilibrium with the atmosphere is slightly acidic, with a pH of 5.6. Today precipitation in parts of Europe and North America has a pH well below 4.7 ("acid rain"). The major sources of the elevated acidity in acid rain are SO_2 and nitrous oxides (NO_x) released from the burning of fossil fuels and transformed by photo-oxidation to sulfuric acid (ca. 70%) and nitric acid (ca. 30%). The impact of the acid loading depends on the *buffering capacity* of the water and the type of soil in the watershed. Lakes in calcium-rich regions are buffered by the calcium–carbon dioxide system (cf. Section 3.1.6) and show little or no effects of acid rain. Regions with granite bedrock (granite, gneiss) have much less buffering capacity (alkalinity < 0.1 mval/l). The incoming acid acts much like a chemical titration, first using up the alkalinity, then causing the pH to decrease rapidly. Many lakes and streams in granite regions (e.g., Scandinavia, Northeastern U.S., Southeastern Canada) have become strongly acidified (some to pH < 4.5).

The interactions between acidic emissions into the atmosphere and processes within the watershed can be demonstrated by the budgets of SO_4^{2-} and H^+ ions in a Norwegian lake (Fig. 8.15). Sulfur dioxide is one of the most important components of

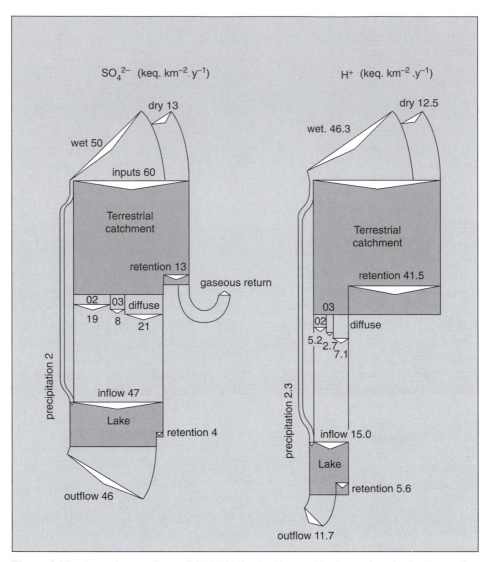

Figure 8.15 Annual mean fluxes (1974–1980) of sulfate and hydrogen ions in the Norwegian lake Langtjern and its catchment basin (after Wright 1983).

acidic emissions caused by the combustion of fossil fuels. In the atmosphere sulfur dioxide is oxidized to sulfur trioxide, which then combines with water to form sulfuric acid. Sulfuric acid dissociates into sulfate and hydrogen ions. After the ions are deposited in the watershed, their fates are quite different. Little of the sulfate is used by biological processes in the watershed, allowing about 80% of the deposited sulfate to reach the lake. Retention of sulfate within the lake is even less (<10%). In contrast, only 27.5% of the H^+ ions deposited on the water catchment area reach the lake, indicating a retention of over 70%. Retention of hydrogen ions within the lake is less important (about 37%), but still greater than for the sulfate ions. The major process contributing to the consumption of hydrogen ions is the weathering of minerals, as shown for muscovite in the following equation:

$$KAl_2(AlSi_3O_{10})(OH)_2 + 10\ H^+ \rightleftharpoons K^+ + 3\ Al^{3+} + 3\ H_4SiO_4$$

Hydrogen ion retention in the watershed of Langtjern is rather low because it lies in an area with crystalline rocks. In areas dominated by carbonate rocks it would be much higher and could approach 100%.

In addition to the direct physiological effects of increased H^+ ion concentrations, important biological effects are caused by altered solubility and speciation of many metals as the pH of the water decreases. For example, aluminum, iron, copper, zinc, nickel, lead, and cadmium become more soluble when acidified, but mercury and vanadium become less soluble in water. Many of the negative effects of acidification on organisms is attributed to the increased solubility of aluminum and the shift to the toxic Al^{3+} form. Increased mobilization of aluminum also causes precipitation of phosphorus and contributes to the tendency of acidified lakes to become oligotrophic. Acidified lakes also often become more transparent due to the precipitation of humic substances by aluminum.

The primary production of acidified lakes shifts to the benthos as a result of the oligotrophication and the increase in water transparency. The result is often the formation of a dense mat of filamentous algae such as *Mougeotia* on the bottom. In addition to the improved light conditions, the development of algal mats is also promoted by the loss of many benthic invertebrates that would otherwise graze on them. Angiosperm (flowering) macrophytes are often suppressed by *Sphagnum* moss. *Sphagnum* is an efficient ion exchanger, removing cations (nutrients) from the water and replacing them with protons (H^+), further intensifying the acidification process. The resulting acid conditions cause a general decrease in the number of species of aquatic organisms and the elimination of certain groups of phytoplankton such as the diatoms (Fig. 8.16).

There are clear pH preferences within the diatoms. Using this, one can reconstruct the pH history of lakes by examining the diatom frustules in the lake sediments. Each species is assigned an indicator value, based on its present day occurrence in relation to pH conditions. By knowing the age of the sediment layers examined, one can estimate the historic pH of the sediments from the diatom remains (Arzet et al. 1986).

Zooplankton can also be used as indicators. Most *Daphnia*, for example, disappear below a pH of 6, whereas *Bosmina longispina* still occurs at pH values of less than 4.1 (Brett 1989). Most fish are limited by pH values of 5.0–5.5, but populations of brook trout (*Salvelinus fontinalis*) may survive at pH 4.5. Some fish can live at low pH values as adults, but their eggs can not develop (Henriksen et al. 1989).

The most important aspects of acidification of lakes involve biotic changes. It is not clear whether the observed changes in communities in acidic lakes are due to direct pH effects (tolerance) or are caused by smaller changes in biotic interactions. For example, community structure could be altered by shifts in the competitive relationships

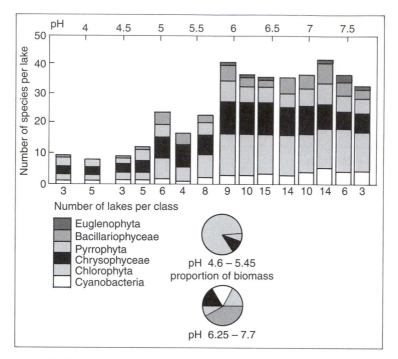

Figure 8.16 The number of species in various groups of phytoplankton in Swedish lakes with different pH conditions. Below: average proportions of biomass in the different taxa in acidic and circum-neutral lakes (from Almer et al. 1974).

of the algae or by the disappearance of keystone species (cf. Section 7.3.3) (Eriksson et al. 1980).

The success in restoring acidified lakes depends largely on the location of the lake. A short-term solution commonly used in Scandinavia is the periodic addition of lime or NaOH to lakes. In the long run, however, the problem of acidification can only be solved by reducing fossil fuel emissions from industry, automobiles, and domestic heating.

8.7 SUCCESSION

8.7.1 Long-term Succession and the Problem of the "Climax" Concept

Organisms alter their own environment as a result of their activities. This also changes the natural selection pressures to which they are subjected. Species that were originally well adapted may lose their fitness, while other species may be better adapted to the new conditions. New species will exclude the original species, as they themselves will in turn be excluded in a similar fashion. This process is referred to as *succession*. It includes the temporal change in the community as well as the temporal change in the ecosystem.

Succession that is caused exclusively by the organisms themselves is called *auto-genic* succession, whereas long-term changes resulting from altered external factors (e.g., climate) are called *allogenic* succession. A sudden disturbance can revert a system to its earlier condition (*reversion*), from which autogenic succession begins again in the original direction. Assume that a succession consists of the stages A → B → C → D → E. A disturbance could cause A → B → C → D → disturbance → C → D → E, but not A → B → C → D → disturbance → C → B → A (Reynolds 1984).

The monoclimax theory of Clements (1936) maintains that within a major climatic zone all succession will proceed to the same final climax stage, whose species cannot be excluded by others. The climax stage can be viewed as a homeostatic system in which the losses and gains of all populations and all chemicals are in a nearly perfect steady-state equilibrium. According to the monoclimax theory, the floristic and faunistic composition of the climax stage does not depend on the starting conditions of the succession or the geological conditions. Even successions starting from bare rocks ("xerosere") or from a lake ("hydrosere") would finally arrive at the same end point, which at low and intermediate altitudes in middle Europe would be a red beech forest or in North America beech–maple.

The geological basis of the hydrosere is the filling in of lakes. As the lake basin fills in with sediments, its volume is decreased and, with a constant total loading of nutrients, it becomes increasingly eutrophic (cf. Section 8.6.1). Eutrophication likewise increases the rate of sedimentation in the basin. Lakes also become overgrown from the shoreline. In shallow shore regions emergent macrophytes (reeds, rushes) accumulate nutrients that contribute to eutrophication. These lakes gradually transform into *marshlands*, into which moisture-tolerant trees (willows, alder) invade and eventually form a forest. Trees with intermediate moisture tolerance follow as the soils develop. A *floating mat* made of peat moss (*Sphagnum*) and sedges first develops out from the edges, where there are steep banks. Peat forms under the floating mat from the accumulation of organic matter that is not easily decomposed. Once the floating mat becomes secured, shrubs (mostly Ericaceae, e.g., blueberry) and later trees become established, eventually transforming the landed lake into a forest.

Such succession has not been directly observed, since it may take centuries, millennia, or even millions of years (for very large lakes). It is generally assumed that ecosystems and communities observed today correspond to specific temporal stages of succession *(space-for-time substitution)*.

There are a number of uncertainties and assumptions in the space-for-time substitution, since it is almost impossible to prove whether all successions within a major climatic zone converge on the same climax stage (*monoclimax theory*), or whether there are different climax stages within a zone, depending on the starting conditions or on the geology of the region (*polyclimax theory*). Furthermore, succession may take so long that long-term climatic conditions may change. If the climate change alters the course of the succession, it would be impossible to determine whether the climate caused the change to a new monoclimax or whether the succession would have changed regardless of the climate change. The monoclimax–polyclimax controversy must be regarded as an insoluble question (nonfalsifiable).

8.7.2 General Trends

Regardless of whether there is a single climax stage, distinctions can be made between early and late (mature) succession stages. This is especially true for "primary" succession, involving the first settlement of a habitat not previously colonized. The trends described (Odum 1959) are largely deductive reasoning regarding the abilities organisms would need to be successful as a habitat becomes more and more populated. Empirical tests of this concept have been few and incomplete, since, again, one runs into the problem of long-term succession and space-for-time substitution.

At first, colonization of an empty habitat is largely a matter of *accumulation of biomass*. The accumulation of biomass corresponds to an increasing consumption of abiotic resources. Generally, during early succession in aquatic habitats most of the nutrients are dissolved in the water and are thus available. In later stages an increasing percentage of the nutrients is tied up in the biomass.

If a succession in a habitat begins with an insignificant addition of organic matter, the first organisms to settle must be autotrophic ("*autotrophic succession*"). At first there is an excess of primary production compared to respiration. Once the primary producer biomass is present, consumers and detritivores can follow. This increases the respiration and eventually it will move toward an equilibrium between photosynthesis and respiration.

On the other hand, the first settlement is by heterotrophs ("*heterotrophic succession*") in habitats with heavy organic pollution (e.g., in waste water). The primary producers gradually become important as the organic substrates are mineralized. In a heterotrophic succession there is an excess of respiration over primary production in the beginning. Both heterotrophic and autotrophic succession *tend to converge toward a relationship of primary production: respiration* of 1:1.

Competition increases and there is limitation as the abiotic resources are consumed. Thus the successful species in succession are those that can rapidly take up available resources and hold on to them. When resources are in short supply, populations cannot achieve adequate reproductive rates to compensate for high mortality. Therefore, successful species are also those that have some resistance to mortality, especially since as higher trophic levels are added it also increases the mortality. This shift toward species that are strong competitors and can resist mortality leads to *increasingly closed nutrient cycles.*

As primary producers become more resistant to grazing, the *herbivores become less important* and *detritivores take on greater importance*. There is also an *accumulation of refractory substances,* since some of the grazing resistance comes from the formation of grazing resistant structures that are not easily decomposed (cellulose, lignin, chitin).

Relative metabolic rates (e.g., the P:B ratio) decline as the resources become more limiting and more materials are invested in protective structures and larger bodies. These changes during succession represent a *transition from r to k selection* (cf. Section 5.6).

Margalef (1968) referred to this concept as the "*maturation*" of ecosystems. This idea has been proposed for terrestrial systems in which long-lived primary producers represent an important element. We have already mentioned that there is no such climax stage in aquatic systems, but rather that they ultimately become terrestrial systems. They are only the beginning stages of a succession. One should not expect to find features identifying lakes as "mature" or "less mature." The pelagic region of a lake cannot achieve a "mature" condition, since it is constantly being reverted to an earlier succession niveau (cf. Section 8.7.1). At temperate latitudes, these disturbances are caused by the seasons and the circulation of the water column. In a sense, there is certain degree of "maturing" of the system that occurs every year as the lake succession repeats itself in a yearly cycle.

8.7.3 Seasonal Succession in Plankton

The pattern of succession in a lake can be observed in the seasonal changes in the biomass and species composition of the plankton. An advantage of the short time course is that the problem of space-for-time substitution is avoided. Although the time scale is small, the plankton succession can be compared to terrestrial succession, for the level of primary producers involves ca. 30 to 100 generations of organisms and a series of distinctive floristic stages.

Because plankton are well suited for experimental studies, the seasonal succession of plankton is one of the best-studied and best-understood examples of succession. It is difficult, however, to separate the autogenic and allogenic components, due to the overlay of autogenic succession with the physical changes of the season. Originally, the seasonal periodicity of the phytoplankton was attributed almost solely to allogenic effects (temperature, stratification, light). Meanwhile, it has been possible to develop a relatively clear picture of the causal mechanisms of plankton succession from numerous studies that have manipulated the physical conditions (e.g., artificial circulation; Reynolds et al. 1984), categorized the various components of population dynamics for the important species (Reynolds 1984, Sommer 1987), and compared growth and losses with seasonal changes in the species composition (Sommer 1989).

Thermal stratification of a lake is the single most important factor controlling the seasonal succession of the plankton. Succession begins with the breakup of the ice or the time when the ice becomes transparent. If the previous winter period was long enough to reduce the standing crop of the plankton to low levels, the succession has characteristics of a primary succession that begins with "primary colonization." Constant or increasing stability of the stratification allows for the development of an autogenic succession. Temporary short-term expansion of the depth of mixing (e.g., by storms) led to reversions. Finally, erosion of the thermocline in the fall causes allogenic shifts. The PEG model of seasonal succession of phytoplankton and zooplankton is a general diagram that was developed by in a comparative study of the international Plankton Ecology Group (PEG). It summarizes the course of seasonal succession in 24 sequential events (Sommer et al. 1986). Lake Constance and a number of comparable eutrophic lakes served as model systems. Although the sequence of events may not apply to all lakes with different specific conditions, the causal explanations should nevertheless apply. The applicability of the model to soft-water lakes, shallow lakes, and tropical lakes is questionable, since this has not been adequately examined.

Figure 8.17 summarizes the most important predictions of the PEG model. Succession in eutrophic lakes typically has two phytoplankton maxima, a spring maximum of small algae, and a summer maximum of large, grazing-resistant forms. These maxima are separated by a clear-water phase. The clear-water phase is caused by a maximum of large zooplankton that is later replaced by smaller zooplankton species. The seasonal pattern is different in oligotrophic lakes. It appears that the entire process proceeds more slowly, so that only the first part of the succession is completed and there is no clear-water phase.

The intensity of controlling factors is indicated by the thickness of the black horizontal bar. Their effect on the accumulation of biomass and the selection of certain species was described in earlier chapters (cf. Chapter 6). It is clear that the physical factors create a "frame," within which the biological factors form the specific "picture." The autogenic succession takes place within this temporal "window." The sea-

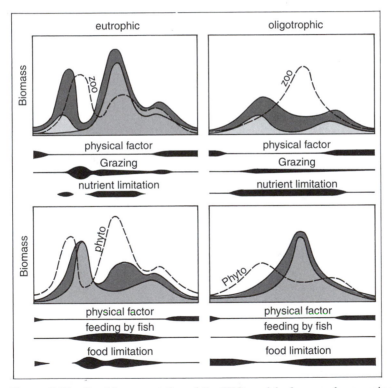

Figure 8.17 Graphic presentation of the PEG model of seasonal succession. Seasonal development of phytoplankton (above) and zooplankton (below) in eutrophic (left) and oligotrophic (right) lakes. Phytoplankton: dark shading, small species; medium shading, large nonsiliceous species; light shading, large diatoms. Zooplankton: dark shading, small species; medium shading, large species. The black horizontal symbols indicate the relative importance of the selection factors (from Sommer et al. 1986).

sonal succession of the phytoplankton in Lake Constance is presented in Fig. 8.18 for comparison with the predictions of the PEG model.

Climatic conditions can affect the seasonal succession, but the basic pattern remains unchanged. Definite trends during the "maturation" of an ecosystem, as described in Section 8.7.2, can be seen in the plankton during the course of a season:

Accumulation of biomass: Eutrophic lakes have their seasonal biomass maximum during the summer, but this is not true for oligotrophic lakes (Fig. 8.17). It should be noted, however, that nutrients and matter continually settle out of the epilimnion.

Decrease in the relative metabolic rates: The ratio of production to biomass of the phytoplankton decreases during the course of the autogenic succession. This is interrupted by the clear-water phase, during which the yearly maximum P:B ratio is reached (Fig. 8.19).

Increase in resource limitation and competition: Nutrient limitation increases after the clear-water phase, resulting in increasing competition within the phytoplankton.

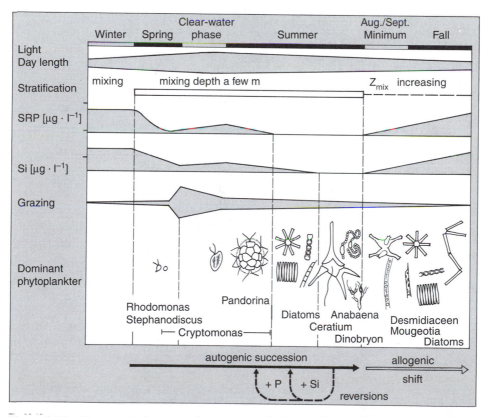

Figure 8.18 Diagram of the seasonal succession of phytoplankton and the relevant environmental conditions in Lake Constance (from Sommer 1987).

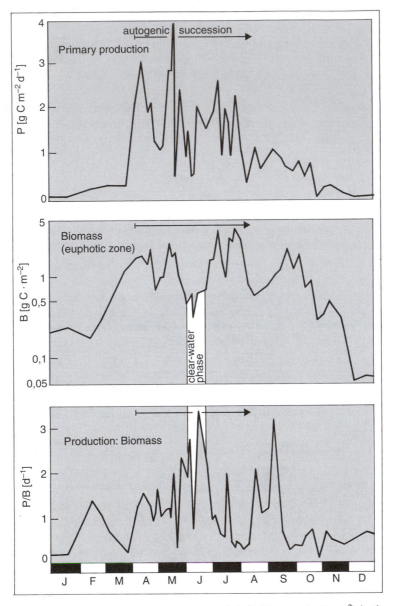

Figure 8.19 Primary production (g C m^{-2} d^{-1}), biomass (g C m^{-2}) in the euphotic zone and the production–biomass ratio of the phytoplankton in Lake Constance in 1981 (from Tilzer 1984b).

The water transparency decreases with increasing biomass, creating competition for light. Food limitation for the zooplankton is most intense during the clear-water phase. Zooplankton also tend to be food limited during the summer, when there are fewer edible algae than during the spring algal maximum.

Closing of the nutrient cycles: During the spring maximum and following the clear-water phase, nutrients are increasingly shifted from the dissolved to the incorporated pool. After the diatoms disappear, it is mainly the motile species that remain (flagellates, cyanobacteria with gas vacuoles). This results in a drastic reduction in the sedimentation and an increase in the retention of nutrients in the epilimnion.

Shift from herbivory to detritivory: After the clear-water phase, the herbivores utilize a smaller fraction of the primary production due to the shift to inedible algal species. During this time the large and poorly edible algae are utilized mainly by parasites (fungi) and detritivores.

8.7.4 Self-Purification as Heterotrophic Succession

If a pulse of allochthonous organic matter is introduced into a body of water, it is decomposed and eventually remineralized. Since the organic matter changes and the organisms also change their environment as they decompose, the organic substrate a type of succession takes place. Such *heterotrophic succession* occurs in lakes, but the individual stages can be more easily analyzed in streams, where various horizontal zones along a "*self-purification stretch*" correspond to the various stages of heterotrophic succession. The distance downstream from the source of organic matter corresponds to the increasing age of a lake (*flowing wave*), so that the temporal course of lake succession is translated to a spatial course in streams.

Over long periods the remineralization of organic matter reflects the amount of production, unless organic matter is added to the lake. In undisturbed water bodies, the sum of all decomposition processes should approach equilibrium with the sum of all production processes. In late succession the primary production–respiration quotient approaches 1. If decomposable organic matter is added to a water body, the quotient will drop well below 1. The return to equilibrium by decomposition of these substances is referred to as *self-purification*. In contrast to trophic condition, which refers to the processes of production (cf. Section 8.5.4), the term *saproby* is used to describe the intensity of the decomposition processes.

Allochthonous loading of organic substances is the rule in undisturbed streams (cf. Section 8.4). However, the concept of "self-purification" is normally only used to describe the breakdown of anthropogenic disturbances (e.g., waste water). In applied limnology "*water quality*" is sometimes based on the succession stages used to classify streams. Sections of the stream are divided according to chemical criteria or indicator organisms. A system widely used in Central Europe is the "*saprobic system*," in which four saprobic steps are distinguished based on the chemical conditions and the types of organisms present. (Figs. 8.18 and 8.19). The categories of organisms are based on descriptive studies relating the abundance of species in particular stream types. The saprobic system requires extensive identification at the species level for organisms ranging from bacteria to fish. Despite the lack of experimental evidence for placing each species in a particular category, the saprobic system is a useful means for characterizing streams, for there is a distinct pattern in the occurrence of various groups of organisms in the self-purification stretch (Fig. 8.20). In North America a similar, but

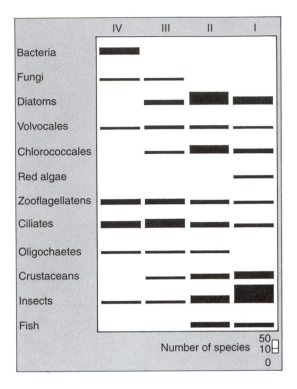

Figure 8.20 The distribution of indicator species of the saprobic system in the various water quality classes (from Uhlmann 1975).

much simplified, biotic index using only arthropods has proved to be a useful tool for monitoring the degree of organic and nutrient pollution in streams (Hilsenhoff 1977, cf. Box 8.3).

Several critical points must be considered, however. There is disagreement concerning the classification of cyanobacteria, algae, and higher plants in the saprobic system, since as photolithoautotrophic organisms they should respond to the trophic condition rather than the saprobic condition of a water body. Also, it is essential to make exact identifications of animals that are used as indicator organisms, as their specific environmental requirements are what make them useful as indicators. For this reason, groupings such as "caddisfly larvae" have little predictive power. It should also be noted that the limiting environmental factors are sometimes only indirectly related to the pollutant. This is the case with oxygen, for example. A mountain stream with a high physical mixing of air would have higher dissolved oxygen than a lowland stream with the same amount of pollution. Organisms that are limited by oxygen would have a very different indicator value in the two types of streams.

The changes in the environmental conditions in the course of a self-purification stretch are presented in Fig. 8.21. The most strongly polluted zone is called the "polysaprobic zone" (water class IV). Below the input of organic substances there is first a dilution and physical purification processes (e.g., flocculation of organic matter). There is a rapid oxygen consumption by oxic decomposition of organic substances. The oxygen depletion leads to the development of an anoxic zone, not far from the input of

BOX 8.3   **BIOASSESSMENT OF STREAM WATER QUALITY**

It is not easy to characterize the water quality of a stream. Chemical analysis can provide precise measurements, but pollution levels usually vary, for example, with the time of year and even with the time of day. If the sample for the chemical analysis happens to be taken at the wrong time, it may be entirely misleading. Water samples can be collected over a longer period and combined, but such averages underestimate short-term, catastrophic events, such as pulses of toxins. Macroinvertebrates in a streams generally have life cycles of a year or more, which exposes them to pollutants over long periods and integrates the effects of short-term episodes. Thus there has been much emphasis on finding organisms that can be used to indicate the average water quality in a stream.

A *biotic index* based on communities of insects, amphipods, and isopods was developed by Hilsenhoff (1977, 1987) to assess the degree of organic and nutrient pollution in Wisconsin streams. Streams with high levels of organic matter develop low oxygen concentrations during the summer and especially at night. The range of tolerance of a species to dissolved oxygen in water affects its ability to survive or compete with other species. Species tolerant of oxygen deficiencies, such as chironomid midge larvae, tend to dominate the benthic community in organically polluted streams, whereas intolerant species, such as stoneflies and mayflies, tend to be rare or absent.

To calculate the Hilsenhoff biotic index (HBI), 100 or more arthropods must be first collected from a rocky or sand and gravel riffle area of a stream. The shallow, turbulent water in this habitat provides the highest possible oxygen conditions in the stream. Next, organisms must be identified to the lowest taxonomic level possible. Tables of species list their "tolerance" to organic pollution (low oxygen), on a scale of 0 (least tolerant) to 10 (most tolerant). Tolerance values were derived from a comparison of water quality conditions and the occurrence of arthropod species in more than 2000 stream samples. The biotic index represents an average tolerance for the arthropod community in a particular stream or location within a stream:

$$HBI = \frac{\sum tn}{\sum n}$$

Example:

Species	n, Number counted	t, Tolerance value	t n
Clioperla clio	27	1	27
Heptagenia diabasia	58	3	174
Hydropsyche arinale	23	5	115
Σ	108		316

HBI = 316/108 = 2.9 (excellent water quality with no detectable organic pollution; see evaluation table below). Categories used to evaluate water quality using the Hilsenhoff biotic index (Hilsenhoff 1987):

Biotic index	Water quality	Degree of organic pollution
0.00–3.50	Excellent	No apparent organic pollution

BOX 8.3	(Continued)	
3.51–4.50	Very good	Possible slight organic pollution
4.51–5.50	Good	Some organic pollution
5.51–6.50	Fair	Fairly significant organic pollution
6.51–7.50	Fairly poor	Significant organic pollution
7.51–8.50	Poor	Very significant organic pollution
8.51–10.00	Very poor	Severe organic pollution

The Hilsenhoff biotic index focuses on a narrower range of organisms than the saprobic system, but species identifications require expertise and are time consuming. Ideally all organisms should be identified to genus and species. This is not always practical, however, and useful, but less sensitive measures of water quality can be gained from estimates of the biotic index using only family-level identification (Hilsenhoff 1988). Such *bioassessment* of water quality at the family level requires little taxonomic training and can be completed in the field without identification keys in as little as 23 min (Hilsenhoff 1988). However, because of the broad range of species requirements within a family of arthropods, use of average family-level tolerance values may lead to inaccurate estimates of water quality.

the organic matter. The anoxic decomposition of organic matter results in CO_2 as the oxidized end product, as well as alcohols, organic acids, H_2, CH_4, NH_4^+, and H_2S as reduced end products (cf. Section 4.3.10). The anoxic respiration leads to the consumption of nitrates (nitrate ammonification and dentrification) as well as sulfate (desulfurication). The availability of H_2S and adequate light promote anoxic photosynthesis by sulfur bacteria. At the aerobic–anoxic boundary chemolithotrophic microorganisms appear because of the supply of reduced substrates (cf. Section 4.3.8). When the supply of atmospheric oxygen is sufficient to maintain a dissolved oxygen concentration of ca. 3 mg/l, there are often blooms of the bacterium *Sphaerotilus natans*, referred to as "sewage fungi." Bacteria-feeding animals follow, with a spatial lag. The range of species is very limited because of the low or lacking dissolved oxygen. Bacteria-feeding protozoans dominate, since relatively few metazoans are adapted to living at low oxygen concentrations (e.g., *Tubifex, Chironomus*).

The release of mineral nutrients by bacteria feeders allows for an increase in photosynthesis downstream. There is usually also an increase in oxygen due to atmospheric inputs and lower oxygen consumption in this region, as some of the organic matter has already been utilized. The improved supply of oxygen permits chemosynthetic oxidation of reduced substances, especially the nitrification of ammonium. The number of animal species increases in response to the improved oxygen conditions and the additional food available due to the cyanobacteria and algae. This large supply of food can sometimes lead to massive developments of certain animals. This zone is called the "α mesosaprobic zone" (class III). Fish reappear in large numbers in the " mesosaprobic zone" (class II). Rarely does a stream reestablish itself to an "oligosaprobic" condition

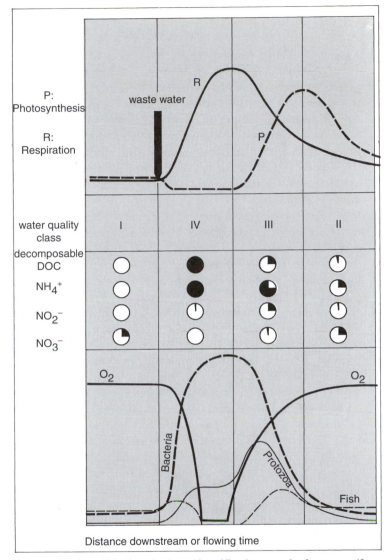

Figure 8.21 A diagrammatic view of the length of a self-purification stretch of a stream (from Uhlmann 1975).

(class I), for the nutrients released by the remineralization of the organic pollution stimulate increased primary production, which in turn produces a secondary loading of organic matter in the stream. Nutrients are removed from the nutrient "spirals" (cf. Section 8.4) only after moving over long stretches of a stream. Often, however, before that occurs in our anthropogenically affected landscape, a new source of organic pollution entering the stream once again restarts the heterotrophic succession.

[335]

REVIEW QUESTIONS FOR CHAPTER 8

1. Assume a simple trophic chain consisting of nanophytoplankton–*Daphnia* –whitefish and a microbial chain of picophytoplankton–heterotrophic nanofla- gellates–ciliates–*Daphnia*–whitefish. Assume the ecological efficiency within the short chain is 10% and the long chain can support the same amount of fish production as the short chain. (a) If the ecological efficiency within the long chain is equal to the short chain: How much higher must the picoplank- ton production be than the nanophytoplankton production? (b) If the pi- coplankton and nanoplankton production are equal: What must the ecological efficiency be for the following links: picoplankton–HNF, HNF–ciliates, and ciliates–*Daphnia*?

2. What common feature distinguishes the carbon and nitrogen cycle in lakes from the phosphorus and silica cycle? How does this feature affect the cycling of car- bon and nitrogen?

3. Are the concentrations of dissolved phosphate, nitrate, and ammonium in the epil- imnion during the summer a good measure of the nutrient richness of a lake? Ex- plain.

4. Nutrient ratios are important for competitive interactions in phytoplankton com- munities. Which ratios do you expect to decrease or to increase during the strati- fied period: Si:N; Si:P; NH_4:NO_3.

5. Use the IBP- and OECD-type equations presented in Sections 8.5.2, 8.5.3, and 8.5.4 to calculate for (a) an oligotrophic (10 μg P_{tot} l^{-1}), (b) a mesotrophic (30 μg P_{tot} l^{-1}), and (c) a eutrophic (100 μg P_{tot} l^{-1}) lake characteristic values for: (i) the mean annual chlorophyll concentration; (ii) the annual primary production; (iii) the annual secondary production; (iv) the fish yield.

6. For each of the lakes in question 5, make some calculations concerning the light climate in each lake. Assume a background vertical extinction coefficient (ex- tinction coefficient without chlorophyll) of 0.2 m^{-1} and assume that for each mg Chl l^{-1} the extinction coefficient increases by 0.015 units. Calculate: (a) the an- nual mean extinction coefficient; (b) the annual mean thickness of the euphotic zone (you can find Lambert–Beer's equation in Section 3.2.1).

7. Half a century ago, fisheries scientists complained about the infertility of olig- otrophic lakes and suggested they be fertilized. Now, after the widespread eu- trophication of lakes in recent decades, lake managers have not been happy about the increase in lake fertility. Why?

8. It has been suggested that denitrification in sewage treatment plants could reduce the nitrogen loading to marine coastal waters, especially if the water from the treat- ment plant must travel a long distance as it is discharged into the ocean. Which process could severely limit the success of denitrification?

9. Why has the reduction of phosphates in detergents been an effective measure in the control of eutrophication?

10. Under which conditions and at what time of the year do you expect diatoms, dinoflagellates, and cyanobacteria to be dominant phytoplankton?
11. Describe and explain the seasonal size trends in phytoplankton and zooplankton in oligotrophic and eutrophic lakes.
12. The Hilsenhoff Index is based on the composition of benthic organisms in riffle habitats. Why?

CHAPTER 9

Final Remarks

9.1 ECOLOGY AND THE BIOLOGICAL SCIENCES

In the beginning of the book we defined ecology according to Krebs (1985): Ecology is the scientific study of the interactions that determine the distribution and abundance of organisms. This definition is somewhat too narrow; it includes the analysis of the environmental requirements of individuals, of growth and distribution of populations, of the interactions between populations, and of the function of communities. It does not include, however, the analysis of the function of ecosystems, which has become an increasingly important part of ecology.

We have not attempted to redefine the term *ecology*, but examples described in our text illustrate the breadth and limits of ecological research. We have restricted the subject in this book to the ecology of freshwater, although the same questions and problems are found in terrestrial and marine ecology. One exception is the analysis of biogeochemical cycles that occur on spatial scales beyond the individual ecosystem, such as landscapes, regions, and the entire biosphere. A textbook about terrestrial or marine ecology might be expected to have a section devoted to global transport. However, the percentage of the earth's surface that contains limnetic ecosystems is negligible in terms of global flow of matter. It is thus reasonable to conclude a textbook in limnoecology with a section devoted to ecosystems.

We have limited our approach to ecosystems and smaller units, and also focused on short-term events. We assume that interactions such as competition and predation take place between existing species or genotypes; that is, no new genotypes are created during the ecological process. It would be otherwise impossible to predict the outcome of the interaction on the basis of the physiological condition of the organisms. The creation of new genotypes lies in the realm of genetics, not ecology. Evolution also requires selection, and the study of natural selection factors is an objective of ecology (cf. Section 1.1).

In his famous essay "The Ecological Theater and the Evolutionary Play," Hutchinson (1965) characterized evolution as a play that is performed within the theater of "ecology." This picture can be modified to clarify the distinction between ecology and related sciences. If one views the interactions between organisms and their environment as a play, then there are three different ways to explain it: One can analyze the actors, the stage, or the play itself. Numerous biological disciplines analyze the actors, for example, taxonomy, functional morphology, physiology, and genetics. The analysis of the stage is "geography" in the broadest sense of the word and includes many nonbiological sciences, such as geology, hydrology, lake physics, water chemistry, etc. The analysis of the play itself is ecology, and it must be qualified that the methods used in ecology cannot be applied to evolutionary and geological time scales.

It should be obvious that we view ecology as a natural science. At several points we have emphasized that some ecological concepts do not conform to the rigorous principles of the scientific method, especially the principle of falsifiability. Some hypotheses such as the diversity–stability hypothesis have created much controversy among academics as well as in the popularized ecology. We view this as an indication of immaturity of the science and by no means as evidence that ecology is unscientific.

9.2 ECOLOGY AND SOCIETY

The emphasis on the natural sciences character of ecology is not entirely consistent with the popular view in society. Many people would like to use ecology as the basis for a new political doctrine. These ideas can be seen in expressions such as "reconciliation of ecology and economy," "ecological world market," and "ecosocialism." Such uses of a natural science are contrary to the meaning of science as well as the meaning of a democratic society. Natural sciences provide theories describing how nature functions. They cannot, however, show which aspects of nature are more worthy of conserving or which goals humans should strive towards. Determining values and goals are tasks for a democratic process and not for science. Ecologists and other experts have no more rights in this decision making than other people. The role of the expert should be to indicate possible ways to achieve goals and to point out possible conflicts between goals that have been selected.

Even if a society were to agree that the health and preservation of ecosystems is valuable and should be given the highest priority, ecology could not help further define this value. Having seen the prescientific character of the "balance of nature" and the superorganism concept, there is no scientific criterion that can be used to place a value on the health of ecosystems. There is no scientific basis for selecting a particular historic condition of an ecosystem as an ideal or "target condition." Even the most recognized and most agreed-upon goals in environmental politics are not closely tied to ecological knowledge. In addition to the reduction of emissions that are harmful to human health, most other attempts toward protection of the environment and nature fall into two major goals: The maintenance of diversity and the re-establishment or retention of the largest possible closure of biogeochemical cycles. Both are goals that have recently become accepted by many people; they are human value judgments, however, and are thus not scientific principles. Ecology can show that anthropogenic opening of cycles of matter has undesirable effects (e.g., more nutrient losses from soils → greater need for fertilizing → more eutrophication of lakes); it is not capable of showing how undesirable these effects are.

9.3 ECOLOGY AND THE ENVIRONMENT

This book has included little of the so-called "applied ecology." Lacking a clear distinction between "ecology" and "environmental science," the word *ecology* has become a hollow word that can be filled with any meaning to fit the need. There is no doubt that the greatest challenge of our time is the protection of our habitat and the resources as the world population increases. By training us to think in terms of interconnected systems, ecology has contributed significantly to a better appreciation of environmental problems. Demands will also be made on ecology to make predictions about the consequences of human activities. Such predictions can be made only if there is a strong natural science theory.

Environmental problems have become widely recognized. This is a very positive sign, but it has led to a strong shift toward applied ecology. The objective of this applied ecology is not the development of generalized theories, but the optimization of human living conditions, and this cannot always be based on scientific criteria. According to Fenchel (1987), who has critically evaluated this problem, the relationship between ecology and environmental sciences can be compared to the relationship between physics and engineering science. Physics provides the fundamental materials from which the engineers make products. Physics and engineering science are relatively clearly distinguished from one another, but this is not the case for ecology and environmental sciences. Only a small fraction of the support that is intended for "ecological research" goes into basic research in ecology. The largest percentage is used to deal with serious environmental problems and for "monitoring." Such programs are badly needed and should be strengthened, but they result in little scientific progress, since as a rule they apply known concepts and methods, rather than develop them. For this reason, increases in research funding have unfortunately not resulted in corresponding increases in ecological theories.

Hopefully it has become clear that the purpose of this book is to highlight the possibilities and limitations of ecological theories in the field of aquatics. We would like to see the concept of "ecology" once again used as it was earlier, as a subdiscipline of biology with its own methods and its own theoretical concepts, equal to other subdisciplines. It is neither an "integrated science" nor a "doctrine of salvation."

Glossary

abundance: the number of individuals per unit surface area or volume

aerobic: in the presence of oxygen

akinetes: the resting stage (cysts) of cyanobacteria

albedo: reflection of light from surfaces

alkalinity: the capacity of water to buffer against acidity

allele: one of the various forms of a gene at the same locus

allelopathy: inhibition of competing organisms through excreted chemicals

allocation: distribution of available material for mutually exclusive purposes, e.g., growth and reproduction

allochthonous: imported from outside the system

allozyme: one of several alleles of an enzyme that have the same function, but differ in their structure so that they migrate different distances in an electrophoretic assay. They are used as genetic markers to identify a genotype

amictic: lakes without complete circulation

ammonification: formation of ammonium from nitrate during nitrate respiration

anabolism: constructive metabolism in which energy is used to transform simple to more complex molecules

anabolism: constructive metabolism, formation of body matter from simple molecules

anaerobic: metabolic activity in the absence of oxygen

anoxibiosis: life without oxygen, e.g., use of fermentation

anoxic: without oxygen

antibiosis: allelopathy among bacteria and fungi

aphotic zone: dark, deep zone where light is insufficient for photosynthesis

assimilation: formation of body matter from food molecules

assimilation: incorporation of matter into an organism's biomass

attenuation: reduction of light intensity penetrating a water column

autochthonous: originating from within a system

autotrophy: use of inorganic carbon sources (CO_2, HCO_3) for the formation of an organism's own body matter

bacterioplankton: plankton consisting of bacteria

bacterivory: feeding on bacteria

batch culture: culture of algae or bacteria in a closed volume. Population growth is typically along a sigmoid (logistic) growth curve

benthos: community of organisms living on the bottom

biocoenosis: community

biogeochemical cycles: material cycles consisting of biological and abiotic transformations

biomanipulation: food chain manipulation. A method of restoration by "top-down" control of the food chain. Removal of fish causes increased grazing and lower algal density

biomass: mass of living organisms in a given volume or within a given surface area. Can be used for a population, a community or a functional subgroup

bioturbation: destratification of the uppermost layers of sediment by organisms

birth rate: the number of births per unit population density and time

Blackman model: saturation model showing the biological response to resource availability, having an abrupt transition from limitation to saturation

bloom: massive development of phytoplankton

bottom-up control: regulation of structure and dynamics of a community by the amount of resources

C-14 method: measurement of photosynthesis by incorporation of radioactively labeled CO_2

carnivory: nutrition by feeding on animals

carrying capacity: maximum sustainable population density (biomass) in a given ecosystem

carrying capacity: the upper limit of population density (biomass) for a given ecosystem

catabolism: breakdown metabolism in which energy is gained by transforming large complex molecules into simpler molecules (final product: CO_2)

cell quota: intracellular concentration of a biogenic element, important for the Droop model

chemocline: zone of rapid vertical change in the chemistry of a water body, usually associated with the transition from aerobic to anaerobic conditions

chemostat: a culture apparatus (algae, bacteria) with a continuous exchange of medium. In a steady state, the growth rate equals the flow through rate

chemotaxis: orientation of movements relative to chemical gradients

chemotrophy: the use of exergonic, chemical reactions as an energy source for the synthesis of body matter

chlorophyll: main pigment of photosynthesis

clear-water stage: phytoplankton biomass minimum in the middle of the growth period caused by grazing

climax: the final stage of a succession

coprecipitation: settling out as a complex with an insoluble salt (e.g., phosphorus with ferric hydroxide)

coevolution: mutually dependent evolution of several species

cohort: a group of individuals of the same age that can be identified within a population

community: assemblage of interacting populations

compensation depth: depth where photosynthesis and respiration are in balance

competition: mutually negative interaction between species that have common resources

competitive exclusion principle: ecological theory that predicts the eventual exclusion of inferior competitors

consumption rate: amount of resources consumed per unit time

convection: vertical mixing caused by density differences

copepodid: late juvenile stages of copepods

cyclomorphosis: seasonal changes in the morphology of organisms within a population

cyst: overwintering cells of algae and protozoa

decalcification: precipitation of calcite caused by photosynthesis

decomposition: breakdown of organic matter

denitrification: transformation of nitrates to N_2, oxidizing nitrates in nitrate respiration

density dependence: decline of growth rates with increasing abundance

desulfuration: microbial reduction of sulfate to hydrogen sulfide

detrital food chain: food chain starting with detritivores

detritivory: nutrition by feeding on detritus

detritus: dead organismal matter

diapause: interruption of the ontogenetic development by a resting period

DIC: dissolved inorganic carbon

dimictic lake: a lake with two complete circulation periods per year

DIN: dissolved inorganic nitrogen

diversity: a measure of species variety

DOC: dissolved organic carbon

DOM: dissolved organic matter

Droop model: an equation that relates algal reproductive rates to internal concentrations (cell quota) of the limiting nutrient

ecological efficiency: the ratio of production rates for adjacent trophic levels

electrophoresis: a method for separating molecules with different electrical charges. It involves differences in the speed of migration of such molecules in a gel after an electrical charge has been applied

encystment: formation of cysts

endemism: restriction of a species to a single place

energy flow: transport of energy through trophic chains and food webs

epilimnion: warm surface layer in a stratified lake

epilithic: growing on rocks

epipelic: growing on mud

epiphytic: growing on plants

equitability: a measure of the evenness of abundance of different species in a community. Maximum *e* indicates that all species are represented in approximately the same proportions. Minimum *e* means one species is dominant.

euphotic zone: water layer with sufficient light for photosynthesis

eurythermic: tolerates a wide range of temperatures

eutrophic: nutrient-rich and with a high productivity

fitness: a measure of the contribution of a genotype to the next generation, relative to other genotypes

excess density: difference between the density of a suspended particle and the density of the medium

excretion: release of organic product of metabolism

exoenzyme: enzyme released into the medium

exploitative competition: competition where competitors are disadvantaged by removal of shared resources

exponential growth: population growth at a constant rate

extinction: see attenuation

feces: particulate matter released (egested) by animals

femtoplankton: plankton < 0.2 μm (viruses and phages)

fermentation: anaerobic respiration where energy is gained by splitting an organic compound into a reduced and an oxidized fragment

filter feeding: uptake of food particles by sievelike structures

filtering rate: volume of water filtered per unit time

fluctuation: irregular changes of a state variable

food web: network of connected trophic chains

form resistance: deviation of a body from the shape of a sphere, causing it to sink more slowly

functional response: the change in rate of consumption of a resource as a function of the resource density

gas vacuoles: gas-filled vesicles in cyanobacteria cells

grazing: removal of entire plants through feeding. In freshwater used to describe the removal of phytoplankton by herbivorous zooplankton or the feeding of benthic animals on periphyton.

gross production: potential production excluding catabolic losses

growth rate: rate of change of a biotic variable; important distinctions are somatic vs. population growth rate, absolute vs. relative growth rate, gross vs. net growth rate

habitat: characteristic residence of a species

hard water: fresh water with high concentrations of calcium and magnesium

Henry's law: dependence of the solubility of gases on partial pressure

herbivory: in the most strict sense, feeding on plants. Used in freshwater studies to describe all types of feeding on living plants and cyanobacteria (e.g., algae)

heterocysts: specialized N_2-fixing cells of cyanobacteria

heterotrophy: use of organic carbon sources for the synthesis of body substances

HNF: heterotrophic nanoflagellates: unpigmented flagellates between 2 and 20 μm

holomictic lake: a lake that circulates to the bottom at least once a year

humus: refractory dissolved organic substances

hypertonic: a higher osmotic pressure in the cell, relative to the environment

hypolimnion: the cold, deep layer in stratified lakes

hyporheal (hyporheic interstitial): the habitat in the spaces of loosely packed material beneath the bottom of a stream and through which some water flows

hypotonic: a lower osmotic pressure within a cell, relative to the environment

IBP: International Biological Program of UNESCO

ingestion rate: food uptake per animal per unit time

interference competition: competition involving direct impairment of competitors

intermediate disturbance hypothesis: hypothesis predicting peak diversity of species at intermediate frequencies and intensities of disturbances

interspecific: between species

intraspecific: within a species

isotonic: having the same osmotic value as the surrounding medium

kairomone: a substance released by a predator that induces an antipredator response (behavioral or morphological change) by the prey

k_2 value: the concentration of a resource at which one-half the maximum growth rate is attained

laminar flow: flow with parallel stream lines

Liebig's law: states that the biotic response variables such as growth rates or growth yield are limited by only the single growth factor in shortest supply (e.g., phosphorus or nitrogen)

light adaptation: physiological adaptation to prevailing light conditions altering the photosynthetic response to light intensities

light inhibition: decreased photosynthesis caused by excessively high light intensity

light limitation: limitation of photosynthesis by low light intensity

light saturation: light conditions allowing for maximum photosynthesis, i.e., neither light limitation nor light inhibition

limiting factor: any factor that limits a biological response (see Liebig's law)

lithotrophy: use of inorganic substances as electron donors

littoral: the zone near the lake shore

locus: the position of a gene on the chromosome (DNA strand)

logistic growth: a population growth curve that approaches the carrying capacity asymptotically

macrophytes: higher aquatic plants, including the stoneworts (characeae)

meromictic lakes: lakes with a deep layer (monimolimnion) that never mixes with the upper water

meroplankton: organisms that spend only a part of their life cycle in the plankton community

mesocosm: an artificially enclosed sector of an ecosystem used for experiments

metalimnion: the zone of greatest temperature change in stratified lakes

methanogenesis: anaerobic decomposition of organic matter with methane as the final product

methylotrophy: heterotrophy in which C_1 molecules (methane, methanol, formaldehyde, etc.) serve as carbon sources

Michaelis–Menten equation: the relationship between consumption rate and resource availability described as a curvilinear function with a smooth transition from limitation to saturation

microbial loop: a pathway in the pelagic cycling of matter in which DOC released by phytoplankton is taken up by bacteria, which in turn is eaten by bactivorous protozoans and then may be passed on to metazoan plankton

mixolimnion: the region of a meromictic lake that mixes

mixotrophy: nutrition by both autotrophic and heterotrophic processes

monimolimnion: the deep layer of meromictic lakes that is excluded from mixing

Monod equation: the relationship between reproductive rates of bacteria or algae and the resource concentration described as a curvilinear function with a gradual transition from limitation to saturation

monomictic lakes: lakes with one complete circulation (holomixis) per year

nanoplankton: plankton with a body size of 2 (3) to 20 (30) μm

nauplius: larval stage of certain crustaceans, e.g., copepods

nekton: the community of organisms capable of active swimming

net production: production after removal of metabolic losses

neuston: the community living in the surface film

nitrate respiration: use of the oxygen in nitrate by bacteria under anaerobic conditions. See denitrification

nitrification: chemical oxidation of ammonium to nitrite and nitrate

nitrogen fixation: use of N_2 as a nitrogen source by cyanobacteria

numerical response: dependence of growth rate of a population of consumers on the resource density

nutrient limitation: limitation of growth rates or biomass yield by a shortage of mineral nutrients

OECD model: empirical eutrophication model for lakes developed by an OECD- sponsored comparative eutrophication study

Oksanen model: a model describing alternating bottom-up and top-down control in communities

oligomictic lakes: lakes that do not mix completely every year

oligotrophic: nutrient poor, with low production

optimal foraging: the searching for and selection of prey in a way that optimizes the net energy gain

organotrophy: the use of organic substances as electron donors

oscillation: regular, periodic changes in a state variable (e.g., abundance)

osmoregulation: the organism's control of its internal osmotic condition

P–I curve: saturation curve describing the dependence of photosynthesis on light intensity

P/B ratio: the relationship between production and biomass

PAR: (photosynthetically active radiation) radiation that can be used for photosynthesis

paradox of the plankton: the apparent contradiction between the richness of phytoplankton species and the competitive exclusion principle

pelagic region: open water zone

periphyton: the community of microalgae growing on submersed surfaces (aufwuchs)

phagotrophy: ingestion of particles by unicellular organisms through their cell surface

phototaxis: movement towards higher (positive p.) or lower (negative p.) light intensities

phototrophic: the use of light as an energy source for the synthesis of body substances

phytoplankton: plant plankton

picoplankton: plankton with a body size of < 2 (3) μm

planktivorous fish: fish that feed on zooplankton

plankton: the community of organisms suspended in the water

plastron: an incompressible layer of air between the fine hairs of aquatic insects that can serve as a permanent "physical gill"

POC: particulate organic carbon

polymictic lakes: lakes that mix more than twice a year; sometimes even daily in the tropics

polymorphism: the appearance in a population of different forms, which are genetically determined, but are not only the result of mutations. It can refer to morphological features, behavior, or allozymes

POM: particulate organic matter

PON: particulate organic nitrogen

predator–prey relationship: an interaction an organism that serves as a food source (prey) for another organism (predator), used in this book in the widest sense to include herbivory and parasitism

primary production: the synthesis of organic matter from inorganic components (through photosynthesis or chemosynthesis)

production: newly formed biomass (B) of a population or trophic level, including the organic matter eliminated (e) during the period of observation: $p = \Delta b + e$

proximate factor: an environmental factor causing a behavioral or morphological response

q_{10}: the factor by which a chemical or biological reaction rate increases with a temperature increase of 10°C

R^* value: the concentration of a resource that results in an equilibrium between birth and death rates

r: mathematical symbol for the growth rate of a population

Redfield ratio: stoichiometric C:N:P ratio of phytoplankton not limited by N or P (106:16:1)

remineralization: transformation of elements from the organic to inorganic phase

resource limitation: limitation of physiological rates or growth rates by resources

resource ratio: ratio of the concentration of two or more potentially limiting resources; according to Tilman's theory, decisive in predicting the outcome of exploitative competition

resources: consumable factors (energy, matter, prey, space) needed for the growth and reproduction of organisms

Reynolds number (Re): a dimensionless number that describes the forces affecting a body moving relative to a fluid (e.g., sinking algae). Re describes the relationship between the forces of inertia and viscosity

rheotaxis: the response of an organisms to flow

Secchi depth: depth to which a white (black and white) standard disk (Secchi disk) remains visible when lowered into the water

secondary production: production by heterotrophic organisms

sedimentation: sinking of particles heavier than water

selection: (a) in terms of evolutionary biology: "natural selection" of better-adapted genotypes; (b) physiologically defined: selection between different food types (by active selection or passively through differential retention, e.g., in filter feeding)

selectivity coefficient: a measure for the relative preference or rejection of a food type

size-efficiency hypothesis: predicts that large herbivorous zooplankton are better resource competitors, while smaller zooplankton are better protected from fish predation

soft water: fresh water with a low concentration of calcium and magnesium

spring bloom: explosive growth of plankton during the spring

steady state: dynamic equilibrium where production and removal processes are in balance; equilibrium in a chemostat

stenotherm: an organism limited in its tolerance to a narrow temperature range

succession: the long-term changes in species through time within a habitat (e.g., lake aging) or the regular species changes following disturbances (e.g., seasonal succession of plankton). one must distinguish between *allogenic* succession, driven by external factors (e.g., climatic changes), and autogenic succession, in which the appearance of a species is dependent on the activities of preceding species (e.g., nutrient consumption)

sulfate respiration: anaerobic respiration using sulfate as an oxidizing agent

surface bloom: visible surface aggregation of buoyant cyanobacteria

symbiosis: mutually beneficial interaction between organisms

thermocline: the region of the greatest vertical temperature change in a stratified water body

thymidine method: measurement of bacterial production by incorporation of radioactively labeled thymidine

Tilman's theory: theory of exploitative competition based on the numerical response of competitors to limiting resources

top-down control: control of the structure and dynamics of a community by predators, i.e., control from the highest to the lowest trophic levels

trophic level: position in the food chain, defined by the number of energy transfer steps to that position

trophy: nutrient richness of a water body

turbulent flow: flow with spiral-shaped stream lines

turnover time: the theoretical time needed to replace a particular pool size at a given input flux rate

ultimate factor: a natural selection factor that has been responsible for the evolution of a trait

uptake rate: the rate at which microorganisms take up dissolved substances

ZNGI line: (zero net growth isocline) in a graphic model of competition it is the line that connects all points on the graph where there is zero growth

zooplankton: animal plankton

Literature Cited

Adrian, R., T.M. Frost: Omnivory in cyclopoid copepods: Comparisons of algae and inverte-brates as food for three, differently sized species. *J. Plankton Res.* **15** (1993) 643–658.

Ahlgren, G.: Growth of *Oscillatoria agardhii* in chemostat culture. II. Dependence of growth constants on temperature. *Mitt. Int. Ver. Limnol.* **21** (1978) 88–102.

Alcaraz, M., J. R. Strickler: Locomotion in copepods: pattern of movements and energetics of *Cyclops*. *Hydrobiologia* **167/168** (1988) 409–14.

Aliotta, G., A. Molinaro, P. Monaco, G. Pinto, L. Previtera: Studies on aquatic plants. 20. Three biologically active phenylpropanoid glucosides from *Myriophyllum verticillatum*. *Phyto-chemistry* **31** (1992) 109–11.

Allan, J. D., A. S. Flecker, N. L. McClintock: Prey preference of stoneflies: sedentary vs mo-bile prey. *Oikos* **49** (1987) 323–31.

Almer, B., W. Dickson, C. Eckström, E. Hornström, U. Miller: Effects of acidification on Swedish lakes. *Ambio* **3** (1974) 30–36.

Ambühl, H.: Die Bedeutung der Strömung als ökologischer Faktor. *Schweiz. Z. Hydrol.* **21** (1959) 133–264.

Anderson, R. S.: Effects of rotenone on zooplankton communities and a study of their recovery patterns in two mountain lakes in Alberta. *J. Fish. Res. Bd. Canada* **27** (1970) 1335–65.

Apperson, C. S., D. Yows, L. Madison.: Resistance to methyl parathion in *Chaoborus asticto-pus* from Clear Lake California. *J. Econ. Entomol.* **71** (1978) 772–73.

Arens, W.: Comparative functional morphology of the mouth parts of stream animals feeding on epilithic algae. *Arch. Hydrobiol. Suppl.* **83** (1989) 253–354

Arzet, K., C. Steinberg, R. Psenner, N. Schulz: Diatom distribution and diatom inferred pH in the sediment of four alpine lakes. *Hydrobiologia* **143** (1986) 247–54.

Azam, F., T. Fenchel, J. G. Field, J. S. Ray, L. A. Meyer-Reil, F. Thingstad: The ecological role of water-column microbes in the sea. *Mar. Ecol. Progr. Ser.* **10** (1983) 257–63.

Bailey, P. C. E.: The feeding behaviour of a sit-and-wait predator, *Ranatra dispar* (Heteroptera: Nepidae): optimal foraging and feeding dynamics. *Oecologia* **68** (1986) 291–97.

Bannister, T. T.: Production equations in terms of chlorophyll concentration, quantum yield, and upper limit to production. *Limnol. Oceanogr.* **19** (1974) 1–12

Banse, K.: Cell volumes, maximal growth rates of unicellular algae and ciliates, and the role of ciliates in the marine pelagial. *Limnol. Oceanogr.* **27** (1982) 1059–71.

Barnhisel, D. R.: The caudal appendage of the cladoceran *Bythotrephes cederstroemi* as defense against young fish. *J. Plankton. Res.* **13** (1991) 529–37.

Barry, M. J.: The costs of crest induction for *Daphnia carinata*. *Oecologia* **97** (1994) 278–88.

Bauer, G.: Reproductive strategy of the freshwater pearl mussel *Margaritifera margaritifera*. *J. Animal Ecol.* **56** (1987) 691–704.

Beadle, L. C.: Osmotic regulation and the faunas of inland waters. *Biol. Rev.* **18** (1943) 172–83.

Bechara, J. A., G. Moreau, L. Haré: The impact of brook trout (*Salvelinus fontinalis*) on an experimental stream benthic community: the role of spatial and size refugia. *J. Animal Ecol.* **62** (1993) 451–64.

Bechara, J. A., G. Moreau, D. Planas: Top-down effects of brook trout (*Salvelinus fontinalis*) in a boreal forest stream. *Can. J. Fish. Aquat. Sci.* **49** (1992) 2093–103.

Benndorf, J., H. Schultz, A. Benndorf, R. Unger, E. Penz, H. Kneschke, K. Kossatz, R. Dumke, V. Hornig, R. Kruspe, S. Reichel: Food-web manipulation by enhancement of piscivorous fish stocks: Long-term effects in the hypertrophic Bautzen Reservoir. *Limnologica (Berlin)* **19** (1988) 97–110.

Bergh, O, K. Y. Borsheim, G. Bratbak, M. Heldal: High abundance of viruses found in aquatic environments. *Nature* **340** (1989) 467–68.

Bermingham, E., J.C. Avise: Molecular zoogeography of freshwater fishes in the Southeastern United States. *Genetics* **113** (1986) 939–65.

Bird, D. F., J. Kalff: Algal phagotrophy: Regulating factors and importance relative to photosynthesis in *Dinobryon* (Chrysophyceae). *Limnol. Oceanogr.* **32** (1987) 277–84.

Bird, D. F., J. Kalff: Protozoan grazing and the size-activity structure of limnetic bacterial communities. *Can. J. Fish. Aquat. Sci.* **50** (1993) 370–80.

Birge, E. A.: Plankton studies on Lake Mendota: II. The Crustacea from the plankton from July 1894, to December 1896. *Trans. Wisconsin Acad. Sci. Arts Lett.* **11** (1897) 274–448.

Black, R. W.: The genetic component of cyclomorphosis in *Bosmina*. In: W. C. Kerfoot (ed.): *Evolution and ecology of zooplankton communities.* University Press of New England, Hanover, N.H. (1980) 456–69.

Black, R. W. II, L. B. Slobodkin: What is cyclomorphosis? *Freshw. Biol.* **18** (1987) 373–78.

Bohl, E.: Diel pattern of pelagic distribution and feeding in planktivorous fish. *Oecologia* **44** (1980) 368–75.

Bollens, S. M., B. W. Frost: Predator-induced diel vertical migration in a planktonic copepod. *J. Plankton Res.* **11** (1989) 1047–65.

Borgström, R., G. R. Hendrey: pH tolerance of the first fish larval stages of *Lepidurus arcticus* and of adult *Gammarus lacustris*. SNSF-Project, IR 22/76 (1976).

Bottrell, H. H.: The relationship between temperature and duration of egg development in some epiphytic Cladocera and Copepoda from the river Thames, Reading, with a discussion of temperature functions. *Oecologia* **18** (1975) 63–84.

Bottrell, H. H., A. Duncan, Z. M. Gliwicz, E. Grygierek, A. Herzig, A. Hillbricht-Ilkowska, H. Kurasawa, P. Larsson, T. Weglenska: A review of some problems in zooplankton production studies. *Norw. J. Zool.* **24** (1976) 419–56.

Bowers, G.: Aquatic plant photosynthesis: strategies that enhance carbon gain. In: R. M. Crawford (ed.): *Plant life in aquatic and amphibian habitats.* Blackwell, Oxford (1987) 79–98.

Bråband, A., B. Fåafeng: Habitat shift in roach (*Rutilus rutilus*) induced by pikeperch (*Stizostedion lucioperca*) introduction: Predation risk versus pelagic behavior. *Oecologia* **95** (1993) 38–46.

Braunwarth, C.: Populationsdynamik natürlicher Phytoplankton- populationen: Analyse der in-situ Wachstums- und Verlustraten. Ph.D. Thesis, Univ. Konstanz, Germany (1988).

Braunwarth, C., U. Sommer: Analyses of the in situ growth rates of Cryptophyceae by use of the mitotic index technique. *Limnol. Oceanogr.* **30** (1985) 893–97.

Brendelberger, H.: Filter mesh size of cladocerans predicts retention efficiency for bacteria. *Limnol. Oceanogr.* **36** (1991) 884–94.

Brett, J. R.: The relation of size to rate of oxygen consumption and sustained swimming speed of Sockeye Salmon (Oncorhynchus nerka). *J. Fish. Res. Bd. Canada* **22** (1965) 1491–501.

Brett, M. T.: Zooplankton communities and acidification processes (a review). *Water Air Soil Poll.* **44** (1989) 387–414.

Brönmark, C.: Effects of tench and perch on interactions in a freshwater, benthic food chain. *Ecology* **75** (1994) 1818–28.

Brönmark, C., B. Malmqvist: Interactions between the leech *Glossiphonia complanata* and its gastropod prey. *Oecologia* **69** (1986) 268–76.

Brönmark, C., L. B. Pettersson: Chemical cues from piscivores induce a change in morphology in crucian carp. *Oikos* **70** (1994) 396–404.

Brooks, J. L.: The effects of prey size selection by lake planktivores. *Syst. Zool.* **17** (1968) 272–91.

Brooks, J. L., S. I. Dodson: Predation, body-size and composition of plankton. *Science* **150** (1965) 28–35.

Bruning, K., J. Ringelberg: The influence of phosphorus limitation of the diatom *Asterionella formosa* on the zoospore production of its fungal parasite *Rhizophydium planktonicum*. *Hydrobiol. Bull.* **21** (1987) 49–54.

Brylinsky, M.: Estimating the productivity of lakes and reservoirs. In: E. D. LeCren, R. H. Lowe-McConnell (eds.): *The functioning of freshwater ecosystems.* Cambridge Univ. Press (1980) 411–53.

Burns, C. W.: The relationship between body size of filter-feeding cladocera and the maximum size of particles ingested. *Limnol. Oceanogr.* **13** (1968) 675–78.

Burton, R. S., S. G. Swisher: Population structures of the intertidal copepod *Tigriopus californicus* as revealed by field manipulation of allele frequencies. *Oecologia* **65** (1985) 108–11.

Byron, E. R.: The adaptive significance of calanoid copepod pigmentation: A comparative and experimental analysis. *Ecology* **63** (1982) 1871–86.

Calow, P.: Can ecosystems be healthy? Critical consideration of concepts. *J. Aquat. Ecosyst. Health* **1** (1992) 1–15.

Cannell, R. J. P., P. Farmer, J. M. Walker: Purification and characterization of pentagalloylglucose, a a-glucosidase inhibitor/antibiotic from a freshwater green alga *Spirogyra varians*. *Biochem. J.* **255** (1988) 937–41.

Carmichael, W. W.: Cyanobacteria secondary metabolites-the cyanotoxins. *J. Appl. Bact.* **72** (1992) 445–59.

Carpenter, S. R., J. F. Kitchell, J. R. Hodgson: Cascading trophic interactions and lake productivity. *BioScience* **35** (1985) 634–39.

Carpenter, S. R., R. C. Lanthrop, A. Munoz del Rio: Comparison of dynamic models for edible phytoplankton. *Can. J. Fish. Aquat. Sci.* **50** (1993) 1757–67.

Cattaneo, A.: Grazing on epiphytes. *Limnol. Oceanogr.* **28** (1983) 124–32.

Chesson, J.: Measuring preference in selective predation. *Ecology* **9** (1978) 923–47.

Chesson, J.: The estimation and analysis of preference and its relationship to foraging models. *Ecology* **64** (1983) 1297–1304.

Clements, F. E.: Structure and nature of the climax. *J. Ecol.* **24** (1936) 252–84.

Coleman, J. R., B. Coleman: Inorganic carbon accumulation and photosynthesis in a blue-green alga as a function of external pH. *Plant Physiol.* **67** (1981) 917–21.

Connell, J. H.: Diversity in tropical rainforests and coral reefs. *Science* **109** (1978) 1304–10.

Connell, J. H.: Diversity and coevolution of competitors, or the ghost of competition past. *Oikos* **35** (1980) 131–38.

Cowan, C. A., Peckarsky, B. L.: Diel feeding and positioning periodicity of a grazing mayfly in a trout stream and fishless stream. *Can. J. Fish. Aquat. Sci.* **51** (1994) 450–59.

Cowgill, U. M., D. P. Milazzo: The sensitivity of two cladocerans to water-quality variables: salinity and hardness. *Arch. Hydrobiol.* **120** (1990) 185–96.

Crawford, D. W.: Metabolic cost of motility in planktonic protists: Theoretical considerations on size scaling and swimming speed. *Microb. Ecol.* **24** (1992) 1–10.

Crowl, T. A., A. P. Covich: Predator-induced life history shifts in a freshwater snail. *Science* **247** (1990) 949–51.

Darwin, C. R.: On the origin of species by means of natural selection, or the preservation of favoured races in the struggle for life. John Murray, London 1859. *Contemporary Edition.*

Dawidowicz, P., Loose, C. J.: Metabolic costs during predator-induced diel vertical migration of *Daphnia. Limnol. Oceanogr.* **37** (1992) 1589–95.

Dawidowicz, P., J. Pijanowska, K. Ciechomski: Vertical migration of *Chaoborus* larvae is induced by the presence of fish. *Limnol. Oceanogr.* **35** (1990) 1631–37.

Dejours, P.: Principles of comparative respiratory physiology. North-Holland Amsterdam (1975).

DeMeester, L., H. J. Dumont: The genetics of phototaxis in *Daphnia magna*: Existence of three phenotypes for vertical migration among parthenogenetic females. *Hydrobiologia* **162** (1988) 47–55.

DeMott, W. R.: The role of taste in food selection by freshwater zooplankton. *Oecologia* **69** (1986) 334–40.

DeMott, W. R.: Discrimination between algae and artificial particles by freshwater and marine copepodes. *Limnol. Oceanogr.* **33** (1988) 397–408.

DeMott, W. R.: The role of competition in zooplankton succession. In. U. Sommer (ed.): *Plankton Ecology. Succession in plankton communities.* Springer-Verlag, Heidelberg (1989) 195–252.

DeMott, W. R., F. Moxter: Foraging on cyanobacteria by copepods: responses to chemical defenses and resource abundance. *Ecology* **72** (1991) 1820–34.

Dini, M. L., S. R. Carpenter: Variability in *Daphnia* behavior following fish community manipulations. *J. Plankton Res.* **10** (1988) 621–635.

Dodds, W. K.: Community interactions between the filamentous alga *Cladophora glomerata* (L) Kuetzing, its epiphytes, and epiphyte grazers. *Oecologia* **85** (1991) 572–80.

Dodson, S. I.: Adaptive change in plankton morphology in response to size-selective predation: A new hypothesis of cyclomorphosis. *Limnol. Oceanogr.* **19** (1974a) 721–29.

Dodson, S. I.: Zooplankton competition and predation: an experimental test of the size-efficiency hypothesis. *Ecology* **55** (1974b) 605–13.

Dodson, S. I.: The ecological role of chemical stimuli for the zooplankton: predator-induced morphology in *Daphnia. Oecologia* **78** (1989) 361–67.

Douglas, M., P. S. Lake: Species richness of stream stones: an investigation generating the species-area relationship. *Oikos* **69** (1994) 387–96.

Downing, J. A., F. H. Rigler (eds.): *A manual on methods for the assessment of secondary production in fresh waters.* IBP Handbook **17,** 2nd edn.. Blackwell, Oxford (1984).

Drenner, R. W., F. De Noyelles, Jr., D. Kettle: Selective impact of filter-feeding gizzard shad on zooplankton community structure. *Limnol. Oceanogr.* **27** (1982) 965–68.

Drenner, R. W., J. R. Strickler, W. J. O'Brien: Capture probability: the role of zooplankter escape in the selective feeding of planktivorous fish. *J. Fish. Res. Board. Can.* **35** (1978) 1370–73.

Droop, M. R.: 25 years of algal growth kinetics. *Bot. Mar.* **26** (1983) 99–112.

Dudgeon, D.: Phoretic Diptera (Nematocera) on *Zygonyx iris* (Odonata: Anisoptera) from a Hong Kong river: incidence, composition and attachment sites. *Arch. Hydrobiol.* **115** (1989) 433–39.

Edmondson, W. T.: A graphical method for evaluating the use of the egg ratio for measuring birth and death rates. *Oecologia* **1** (1968) 1–37.

Edmondson, W. T.: Instantaneous birth rates of zooplankton. *Limnol. Oceanogr.* **17** (1972) 792–95.

Edmondson, W. T.: *The uses of ecology. Lake Washington and beyond.* Univ. Washington Press, Seattle (1991).

Edmondson, W. T.: Experiments and quasi-experiments in limnology. *Bull. Mar. Science* **53** (1993) 65–83.

Edmondson, W. T., G. G. Winberg (eds.): *A manual on methods for the assessment of secondary production in fresh waters.* IBP Handbook **17,** Blackwell, Oxford (1971).

Einsle, U.: The long-term dynamics of crustacean communities in Lake Constance (Obersee, 1962–1986). *Schweiz. Z. Hydrol.* **50** (1988) 136–165.

Elser, J. J., N. C. Goff, N. A. MacKay, A. L. St. Amand, M. M. Elser, S. R. Carpenter: Species-specific algal responses to zooplankton: experimental and field observations in three nutrient limited lakes. *J. Plankton Res.* **9** (1987) 699–718.

Elster, H.-J.: History of limnology. *Mitt. Int. Verein. Limnol.* **20** (1974) 7–30.

Eriksson, M. O. G., L. Henrikson, B. I. Nilsson, G. Nyman, H. G. Oscarson, A. E. Stenson, K. Larsson: Predator-prey relations important for the biotic changes in acidified lakes. *Ambio* **9** (1980) 248–49.

Feminella, J. W., M. E. Power, V. H. Resh: Periphyton responses to invertebrate grazing and riparian canopy in three northern California coastal streams. *Freshw. Biol.* **22** (1989) 445–57.

Fenchel, T.: *Ecology—Potentials and limitations.* Ecology Institute, Oldendorf/Luhe (1987).

Findenegg, I.: Untersuchungen über die Ökologie und die Produktionsverhältnisse des Planktons im Kärntner Seengebiet. *Int. Rev. ges. Hydrobiol.* **43** (1943) 368–429.

Flecker, A. S.: Fish predation and the evolution of invertebrate drift periodicity: evidence from neotropical streams. *Ecology* **73** (1992) 438–48.

Flores, E., C. P. Wolk: Production by filamentous, nitrogen-fixing cyanobacteria, of a bacteriocin and of other antibiotics that kill related strains. *Arch. Microbiol.* **145** (1986) 215–19.

Folt, C., Goldman, C. R.: Allelopathy between zooplankton: A mechanism for interference competition. *Science* **213** (1981) 1133–35.

Frisch, K. von: Über einen Schreckstoff der Fischhaut und seine biologische Bedeutung. *Z. vergl. Physiol.* **29** (1941) 46–145.

Frost, B. W.: Variability and possible adaptive significance of diel vertical migration in *Calanus pacificus*, a planktonic marine copepod. *Bull. Mar. Sci.* **43** (1988) 675–94.

Fuhrman, J. A., F. Azam: Thymidine incorporation as a measure of heterotrophic bacterioplankton production in marine surface waters: evaluation and field results. *Mar. Biol.* **66** (1982) 109–20.

Gabriel, W.: Modelling reproductive strategies of *Daphnia. Arch. Hydrobiol.* **95** (1982) 69–80.

Gabriel, W., B. E. Taylor, S. Kirsch-Prokosch: Cladoceran birth and death rates estimates: experimental comparisons of egg-ratio methods. *Freshw. Biol.* **18** (1987) 361–72.

Gaedeke, A., U. Sommer: The influence of the frequency of periodic disturbances on the maintenance of phytoplankton diversity. *Oecologia* **71** (1986) 25–28.

Gaedke, U.: The size distribution of plankton biomass in a large lake and its seasonal variability. *Limnol. Oceanogr.* **37** (1992) 1202–20.

Galbraith, M. G., Jr.: Size-selective predation on *Daphnia* by rainbow trout and yellow perch. *Trans. Amer. Fish. Soc.* **96** (1967) 1–10.

Gaudet, C. L., P. A. Keddy.: Predicting competitive ability from plant traits: a comparative approach. *Nature* **334** (1988) 242–243.

Gause, G. J.: *The struggle for existence.* Williams and Wilkins, Baltimore (1934).

Geller, A.: Degradation and formation of refractory DOM by bacteria during simultaneous growth on labile substrates and persistent lake water constituents. *Schweiz. Z. Hydrol.* **47** (1985) 27–44.

Geller, W.: Die Nahrungsaufnahme von *Daphnia pulex* in Abhängigkeit von der Futterkonzentration, der Temperatur, der Körpergröße und dem Hungerzustand der Tiere. *Arch. Hydrobiol. Suppl.* **48** (1975) 47–107.

Geller, W., H. Müller: The filtration apparatus of cladocera: Filter mesh sizes and their implications on food selectivity. *Oecologia* **49** (1981) 316–21.

Gerritsen, J., R. Strickler: Encounter probabilities and community structure in zooplankton: a mathematical model. *J. Fish. Res. Bd. Can.* **34** (1977) 73–82.

Gilbert, J. J.: Rotifer ecology and embryological induction. *Science* **151** (1966) 1234–37.

Gilbert, J. J.: Asplanchna and postero-lateral spine production in Brachionus calyciflorus. *Arch. Hydrobiol.* **64** (1967) 1–62.

Gilbert, J. J.: Suppression of rotifer populations by *Daphnia*: A review of the evidence, the mechanisms, and the effects on zooplankton community structure. *Limnol. Oceanogr.* **33** (1988) 1286–313.

Giller, P. S.: *Community structure and the niche.* Chapman and Hall, London (1984).

Gleason, H. A.: The individualistic concept of the plant association. *Torrey Bot. Club Bull.* **53** (1926) 1–10.

Gliwicz, Z. M.: Filtering rates, food size selection, and feeding rates in cladocerans—another aspect of interspecific competition in filter-feeding zooplankton. In: W.C. Keerfoot (ed.): *Evolution and ecology of zooplankton communities.* University Press of New England, Hanover, N.H. (1980) 282–91.

Gliwicz, Z. M.: Food thresholds and body size in cladocerans. *Nature* **343** (1990a) 638–40.

Gliwicz, Z. M.: Why do cladocerans fail to control algal blooms? *Hydrobiologia* **200/201** (1990b) 83–97.

Gliwicz, Z. M., J. Pijanowska: Effect of predation and resource depth distribution on vertical migration of zooplankton. *Bull. Mar. Sci.* **43** (1988) 695–709.

Gliwicz, Z. M., A. Rykowska: Shore avoidance in zooplankton: A predator-induced behavior or predator-induced mortality. *J. Plankton. Res.* **14** (1992) 1331–42.

Gliwicz, Z. M., E. Siedlar: Food size limitation and algae interfering with the food collection in *Daphnia. Arch. Hydrobiol.* **88** (1980) 155–77.

Goldman, J. C., J. J. Mc Carthy, D. G. Peavey: Growth rate influence on the chemical composition of phytoplankton in oceanic waters. *Nature* **279** (1979) 210–15.

Goodman, D.: The theory of diversity-stability relationships in ecology. *Quart. Rev. Biol.* **50** (1976) 237–66.

Grant, J. W. G., I. A. E. Bayly: Predator induction of crests in morphs of the *Daphnia carinata* King complex. *Limnol. Oceanogr.* **26** (1981) 201–18.

Grinnell, J.: The niche-relationships of the California trasher. *Auk* **34** (1917) 427–33.

Grodzinski, W., R. Z. Klekowski, A. Duncan (eds.). *Methods for Ecological Bioenergetics.* IBP Handbook **24**, Blackwell, Oxford (1975).

Gross, E. M., C. P. Wolk, F. Jüttner: Fischerellin, a new allelochemical from the freshwater cyanobacterium *Fischerella muscicola. J. Phycol.* **27** (1991) 686–92.

Güde, H.: Loss processes influencing growth of bacterial populations in Lake Constance. *J. Plankton Res.* **8** (1986) 795–810.

Haag, W. R., D. J. Berg, D. W. Carton, J. L. Farris: Reduced survival and fitness in native bivalves in response to fouling by the introduced zebra mussel (*Dreissena polymorpha*) in western Lake Erie. *Can. J. Fish. Aquat. Sci.* **50** (1993) 13–19.

Hairston, N.G., Jr: The interaction of salinity, predators, light and copepod color. *Hydrobiologia* **81** (1981) 151–58.

Hairston, N. G. Jr., B. T. DeStasio, Jr.: Rate of evolution slowed by a dormant propagule pool. *Nature* **336** (1988) 239–42.

Hairston, N. G., Jr., E. J. Olds: Population differences in the timing of diapause: a test of hypotheses. *Oecologia* **71** (1987) 339–44.

Hairston, N. G., Jr., R. A. Van Brunt, C. M. Kearns, D. R. Engstrom: Age and survivorship of diapausing eggs in a sediment egg bank. Ecology 76 (1995) 1706–1711.

Halbach, U.: Das Zusammenwirken von Konkurrenz und Räuber-Beute- Beziehungen bei Rädertieren. *Zool. Anz. Suppl.* **33** (1969) 72–79.

Hall, D. J., S. T. Threlkeld, C. W. Burns, P. H. Crowley: The size-efficiency hypothesis and the size structure of zooplankton communities. *Ann. Rev. Ecol. Syst.* **7** (1976) 177–208.

Hamilton, D. J., C. D. Ankney, R. C. Bailey: Predation of zebra mussels by diving ducks: an exclosure study. *Ecology* **75** (1994) 521–31.

Hamrin, S. F., L. Persson: Asymmetrical competition between age classes as a factor causing population oscillations in an obligate planktivorous fish species. *Oikos* **47** (1986) 223–32.

Haney, J. F.: An in situ method for the measurement of zooplankton grazing rates. *Limnol. Oceanogr.* **16** (1971) 970–77.

Haney, J. F.: An in situ examination of the grazing activities of natural zooplankton communities. *Arch. Hydrobiol.* **72** (1973) 87–132.

Haney, J. F.: Environmental control of diel vertical migration behaviour. *Arch. Hydrobiol. Beih.* **39** (1993) 1–17.

Haney, J. F., T. R. Beaulieu, R. P. Berry, D. P. Mason, C. R. Miner, E. S. McLean, K. L. Price, M. A. Trout, R. A. Vinton, S. J. Weiss: Light intensity and relative light change as factors regulating stream drift. *Arch. Hydrobiol.* **97** (1983) 73–88.

Haney, J. F., D. J. Hall: Diel vertical migration and filter-feeding activities of *Daphnia*. *Arch. Hydrobiol.* **75** (1975) 413–41.

Hargrave, B. T.: An energy budget for a deposit-feeding amphipod. *Limnol. Oceanogr.* **16** (1971) 99–103.

Harper, J. L.: A Darwininian approach to plant ecology. *J. Ecol.* **55** (1967) 247–70.

Harris, G. P.: Photosynthesis, productivity and growth: the physiological ecology of phytoplankton. *Arch. Hydrobiol. Beih. Ergebn. Limnol.* **10** (1978) 1–177.

Harris, G. P.: *Phytoplankton Ecology*. Chapman & Hall, London (1986).

Hart, D. D.: The importance of competitive interactions within stream populations and communities. In: J. R. Barnes, G. W. Minshall (eds.): *Stream Ecology*. Plenum Press, New York (1983) 99–136.

Hart, D. D.: Causes and consequences of territoriality in a grazing stream insect. *Ecology* **66** (1985a) 404–14.

Hart, D. D.: Grazing insects mediate algal interactions in a stream benthic community. *Oikos* **44** (1985b) 40–46.

Hart, D. D.: The adaptive significance of territoriality in filter-feeding larval blackflies (Diptera: Simuliidae). *Oikos* **46** (1986) 88–92.

Hartmann, O.: Studien über den Polymorphismus der Rotatorien mit besonderer Berücksichtigung von *Anuraea aculeata*. *Arch. Hydrobiol.* **12** (1920) 209–310.

Havel, J. E., S. I. Dodson: *Chaoborus* predation on typical and spined morphs of *Daphnia pulex*: Behavioral observations. *Limnol. Oceanogr.* **29** (1984) 487–94.

Havel, J. E., P. D. N. Hebert: *Daphnia lumholtzi* in North America: another exotic zooplankter. *Limnol. Oceanogr.* **38** (1993) 1823–27.

Healey, F. P.: Characteristics of phosphorus deficiency in *Anabaena*. *J. Phycol.* **9** (1973) 383–94.

Hebert, P. D. N., B. W. Muncaster, G. L. Mackie: Ecological and genetic studies on *Dreissena polymorpha* (Pallas): a new mollusk in the Great Lakes. *Can. J. Fish. Aquat. Sci.* **46** (1989) 1587–91.

Henriksen, A., L. Lein, B. O. Rosseland, T. S. Traaen, I. S. Sevaldrup: Lake acidification in Norway: present and predicted fish status. *Ambio* **18** (1989) 314–21.

Hessen, D. O., A. Lyche: Interspecific variations in zooplankton element composition. *Arch. Hydrobiol.* **121** (1991) 343–53.

Hessen, D. O., E. van Donk: Morphological changes in *Scenedesmus* induced by substances released from *Daphnia. Arch. Hydrobiol.* **127** (1993) 129–40.

Hildrew, A. G., C. R. Townsend: Predators and prey in a patchy environment. A freshwater study. *J. Anim. Ecol.* **51** (1982) 797–815.

Hilsenhoff, W. L.: Use of arthropods to evaluate water quality of streams. *Tech. Bull. Wisconsin Dept. Nat Resour.* **100** (1977) 15 pp.

Hilsenhoff, W. L.: An improved biotic index of organic stream pollution. *The Great Lakes Entomologist* **20** (1987) 31–39.

Hilsenhoff, W. L.: Rapid field assessment of organic pollution with a family-level biotic index. *J. N. Am. Benthol. Soc.* **7** (1988) 65–68.

Hochachka, P. W., G. N. Somero: Biochemical adaptation. Princeton University Press, Princeton 1984, 538 pp.

Höfle, M.: Bacterioplankton community structure and dynamics after a large-scale release of nonindigenous bacteria as revealed by low-molecular-weight-RNA analysis. *Appl. Environ. Microbiol.* **58** (1992) 3387–94.

Höfle, M.: Community structure and dynamics of bacterioplankton as revealed by low-molecular weight RNA analysis. In R. Guerrero, C. Pedros-Alio (eds.): Trends in microbial ecology. *Span. Soc. Microbiol.* 1993, pp. 391–395.

Holling, C. S.: The components of predation as revealed by a study of small-mammal predation of the European pine sawfly. *Can. Entomol.* **91** (1959) 293–320.

Hrbacek, J.: Species composition and the amount of zooplankton in relation to the fish stock. *Rozpr. Cesk. Akad. Ved. Rada. Mat. Prir.* **72** (1962) 1–116.

Huisman, J., F. J. Weissing: Light-limited growth and competition for light in well-mixed aquatic environments: an elementary model. *Ecology* **75** (1994) 507–520.

Hurlbert, S. H., M. S. Mulla, H. R. Willson: Effects of an organophosphorus insecticide on the phytoplankton, zooplankton, and insect populations of freshwater ponds. *Ecol. Monogr.* **42** (1972) 269–299.

Hutchinson, G. E.: *A treatise on limnology. Vol. 1. Geography, physics, and chemistry.* Wiley, New York (1957).

Hutchinson, G. E.: Homage to Santa Rosalia, or Why are there so many kinds of animals. *Amer. Nat.* **93** (1958) 145–59.

Hutchinson, G. E.: The paradox of the plankton. *Am. Nat.* **95** (1961) 137–45.

Hutchinson, G. E.: *The ecological theater and the evolutionary play.* Yale Univ. Press., New Haven (1965).

Hutchinson, G. E.: *A treatise on limnology. Vol. 3. Limnological botany.* Wiley, New York (1975).

Hutchinson, G. E.: *An introduction to population ecology.* Yale Univ. Press, New Haven (1978).

Hynes, H. B. N.: *The ecology of running waters.* Univ. Toronto Press, Toronto (1970).

Hynes, H. B. N.: The stream and its valley. *Verh. Internat. Verein. Limnol.* **19** (1975) 1–15.

Illies, J.: Versuch einer allgemeinen biozönotischen Gliederung der Fließgewässer. *Int. Rev. ges. Hydrobiol.* **46** (1961) 205–13.

Ivlev, V. S.: Experimental ecology of the feeding of fishes (1955, Transl. by Yale Univ. Press 1961).

Jacobs, J.: Cyclomorphosis in *Daphnia galeata mendotae*, a case of environmentally controlled allometry. *Arch. Hydrobiol.* (1961) 7–71.

Jacobs, J.: Untersuchungen zur Funktion und Evolution der Zyklomorphose bei *Daphnia*, mit besonderer Berücksichtigung der Selektion durch Fische. *Arch. Hydrobiol.* **62** (1967) 467–541.

Jacobs, J.: Quantitative measurement of food selection. *Oecologia (Berlin)* **14** (1974) 413–17.

Jacobs, J.: Cyclomorphosis in *Daphnia*. In: *Daphnia*: R. H. Peters, R. DeBernardi (eds.). *Mem. Ist. Ital. Hydrobiol.* **45** (1987) 325–52.

Jacobs, J.: Microevolution in predominantly clonal populations of pelagic *Daphnia* (Crustacea, Phyllopoda)—selection, exchange, and sex. *J. Evol. Biol.* **3** (1990) 257–82.

Jacobsen, D., K. Sand-Jensen: Herbivory of invertebrates on submerged macrophytes from Danish freshwaters. *Freshw. Biol.* **28** (1992) 301–8.

Jakobsen, P. J., G. H. Johnsen: The influence of predation on horizontal distribution of zooplankton species. *Freshw. Biol.* **17** (1987) 501–7.

Janssen, J.: Alewives (*Alosa pseudoharengus*) and ciscoes (*Coregonus artedii*) as selective and non-selective planktivores. In: W. C. Kerfoot (ed.): *Evolution and ecology of zooplankton communities.* Univ. Press of New England, Hanover (1980) 580–86.

Jassby, A. D., T. Platt: Mathematical formulation of the relationship between photosynthesis and light for phytoplankton. *Limnol. Oceanogr.* **21** (1976) 540–47.

Jones, C. G., J. H. Lawton, M. Shachak: Organisms as ecosystem engineers. *Oikos* **69** (1994) 373–86.

Jörgensen, E. G.: The adaptation of plankton algae. IV. Light adaptation in different algal species. *Physiol. Plant.* **22** (1969) 1307–15.

Jürgens, K.: Impact of *Daphnia* on planktonic microbial food webs—A review. *Mar. Microb. Food Webs* **8** (1994) 295–324.

Jürgens, K., H. Güde: The potential importance of grazing- resistant bacteria in planktonic systems. *Mar. Ecol. Progr. Ser.* **112** (1994) 169–88.

Keddy, P. A.: Competition. Chapman and Hall, London (1989).

Keddy, P. A.: Competitive hierarchies and centrifugal organization in plant communities. In J. B. Grace & D. Tilman (eds.) Perspectives on plant competition. Academic Press, San Diego (1990) pp. 265–290.

Kerfoot, W. C.: Combat between predatory copepods and their prey: *Cyclops*, *Epischura*, and *Bosmina. Limnol. Oceanogr.* **23** (1978) 1089–102.

Kerfoot, W. C.: A question of taste: Crypsis and warning coloration in freshwater zooplankton communities. *Ecology* **63** (1982) 538–54.

Kerfoot, W. C.: Adaptive value of vertical migration: comments on the predation hypothesis and some alternatives. In: *Migration: Mechanisms and adaptive significance.* M. A. Rankin (ed.). Univ. of Texas, Port Aransas, *Contr. Mar. Sci.* **27** (1985) 91–113.

Kerfoot, W. C., W. R. DeMott, D. L. DeAngelis: Interactions among cladocerans: food limitation and exploitative competition. *Arch. Hydrobiol. Beih. Ergebn. Limnol.* **21** (1985) 431–51.

Klekowski, R. Z.: Neue Ergebnisse auf dem Gebiet der Bioenergetik und der physiologischen Ökologie der Tiere. *Verh. Internat. Limnol.* **18** (1973) 1594–609.

Klekowski, R. Z., A. Duncan: Physiological approach to ecological energetics. In: W. Grodzinski, R. Z. Klekowski, A. Duncan (eds.): *Methods for Ecological Bioenergetics.* IBP Handbook **24**. Blackwell, Oxford (1975) 15–64.

Koch, F., W. Wieser: Partitioning of energy in fish: Can reduction of swimming activity compensate for the cost of production? *J. exp. Biol.* **107** (1983) 141–46.

Koehl, M. A. R., J. R. Strickler: Copepod feeding currents: Food capture at low Reynolds numbers. *Limnol. Oceanogr.* **26** (1981) 1062–73.

Köpke, V., H. Schultz, R. Jarchow, V. Hornig, J. Peng: Analyse des Nahrungskonsums von Barschen (*Perca fluviatilis*) in der Talsperre Bautzen. *Limnologica (Berlin)* **19** (1988) 37–43.

Kozhov, M.: *Lake Baikal and its life.* Junk Publ., Den Haag (1963).

Krebs, C. J.: *Ecology.* 2nd ed. Harper & Row, New York (1985).

Kuhlmann, H.-W., K. Heckmann: Interspecific morphogens regulating predator-prey relationships in protozoa. *Science* **227** (1985) 1347–49.

Kusch, J., K. Heckmann: Isolation of the *Lembadion*-factor, a morphogenetically active signal, that induces *Euplotes* cells to change from their ovoid form into a larger lateral winged morph. *Develop. Genet.* **13** (1992) 241–46.

Lampert, W.: Untersuchungen zur Biologie und Populationsdynamik der Coregonen im Schluchsee. *Arch. Hydrobiol. Suppl.* **38** (1971) 257–314.

Lampert, W.: Climatic conditions and planktonic interactions as factors controlling the regular succession of spring algal bloom and extremely clear water in Lake Constance. *Verh. Int. Verein. Limnol.* **20** (1978) 969–74.

Lampert, W.: The measurement of respiration. In: J. A. Downing, F. H. Rigler (eds.): *A Manual on Methods for the Assessment of Secondary Production in Fresh Water.* IBP Handbook **17**, 2nd ed. Blackwell, Oxford (1984) 413–68.

Lampert, W.: Laboratory studies on zooplankton cyanobacteria interactions. *N. Z. J. Mar. Freshw. Res.* **21** (1987a) 483–90.

Lampert, W.: Predictability in lake ecosystems: the role of biotic interactions. In: Schulze, E. D., H. Zwölfer (eds.): *Potentials and Limitations of Ecosystem Analysis.* Ecological Studies **61**. Springer-Verlag, Heidelberg (1987b) 333–46.

Lampert, W.: The relationship between zooplankton biomass and grazing: A review. *Limnologica* **19** (1988a) 11–20.

Lampert, W.: The relative importance of food limitation and predation in the seasonal cycle of two *Daphnia* species. *Verh. Internat. Verein. Limnol.* **23** (1988b) 713–18.

Lampert, W.: The adaptive significance of diel vertical migration. *Functional Ecol.* **3** (1989) 21–27.

Lampert, W.: Zooplankton vertical migrations: Implications for phytoplankton-zooplankton interactions. *Arch. Hydrobiol. Beih. Ergebn. Limnol.* **35** (1992) 69–78.

Lampert, W.: Ultimate causes of diel vertical migration of zooplankton: New evidence for the predator-avoidance hypothesis. *Arch. Hydrobiol. Beih. Ergebn. Limnol.* **39** (1993) 79–88.

Lampert, W., C. Loose: Plankton Towers: Bridging the gap between laboratory and field experiments. *Arch. Hydrobiol.* **126** (1992) 53–66.

Lampert, W., P. Muck: Multiple aspects of food limitation in zooplankton communities: the *Daphnia-Eudiaptomus* example. *Arch. Hydrobiol. Beih.* **21** (1985) 311–22.

Lampert, W., H. G. Wolf: Cyclomorphosis in *Daphnia cucullata*: morphometric and population genetic analyses. *J. Plankton Res.* **8** (1986) 289–303.

Lancaster, J.: Predation and drift of lotic macroinvertebrates during colonization. *Oecologia* **85** (1990) 48–56.

Lancaster, J. A., A. G. Hildrew: Characterizig in-stream flow refugia. *Can. J. Fish. Aquat. Sci.* **50** (1993) 1663–75.

Lancaster, J., A. G. Hildrew, C. R. Townsend: Stream flow and predation effects on the spatial dynamics of benthic invertebrates. *Hydrobiologia* **203** (1990) 177–90.

Larsson, P.: The life cycle dynamics and production of zooplankton in Ovre Heimdalsvatn. *Holarctic Ecol.* **1** (1978) 162–218.

Larsson, P., S. I. Dodson: Chemical communication in planktonic animals. *Arch. Hydrobiol.* **129** (1993) 129–55.

Leah, R. T., B. Moss, D. E. Forrest: The role of predation in causing major changes in the limnology of a hyper-eutrophic lake. *Int. Rev. ges. Hydrobiol.* **65** (1980) 223–47.

Lean, D. R. S., C. Nalewajko: Phosphate exchange and organic phosphorus excretion by algae. *J. Fish. Res. Board Can.* **33** (1976) 1312–23.

Lehman, J. T.: Palearctic predator invades North American Great Lakes. *Oecologia* **74** (1987) 487–80.

Lewis, W. M.: Tropical limnology. *Ann. Rev. Ecol. Syst.* **18** (1987) 159–84.

Lieder, U.: Beiträge zur Kenntnis der Genus *Bosmina*. I. *Bosmina coregoni thersites* Poppe in den Seen des Spree-Dahme-Havelgebietes. *Arch. Hydrobiol.* **44** (1950) 77–122.

Likens, G. E., F. H. Bormann, R. S. Pierce, J. S. Eaton, N. M. Johnson: *Biogeochemistry of a forested ecosystem.* Springer, New York (1977).

Lim, E. L., L. A. Amaral, D. A. Caron, E. F. DeLong: Application of rRNA-based probes for observing marine nanoplanktonic protists. *Appl. Environ. Microbiol.* **59** (1993) 1647–55.

Loose, C. J.: *Daphnia* diel vertical migration behavior: Response to vertebrate predator abundance. *Arch. Hydrobiol. Beih. Ergebn. Limnol.* **39** (1993) 29–36.

Loose, C. J., P. Dawidowicz: Trade-offs in diel vertical migration by zooplankton: the costs of predator avoidance. *Ecology* **75** (1994) 2255–63.

Loose, C. J., E. von Elert, P. Dawidowicz: Chemically-induced diel vertical migration in *Daphnia*: a new bioassy for kairomones exuded by fish. *Arch. Hydrobiol.* **126** (1993) 329–37.

MacArthur, R. H., E. O. Wilson: *The theory of island biogeography.* Princeton Univ. Press, Princeton (1967).

MacIsaac, H. J., J. J. Gilbert: Discrimination between exploitative and interference competition between Cladocera and *Keratella cochlearis*. *Ecology* **72** (1991) 924–937.

Maltby, L.: Pollution as a probe of life-history adaptation in *Asellus aquaticus* (Isopoda). *Oikos* **61** (1991) 11–18.

Mann, H., U. Pieplow: Der Kalkhaushalt bei der Häutung der Krebse. *Sitzber. Ges. naturforsch. Freunde* (1938) 1–17.

Margalef, R.: *Perspectives in ecological theory.* Univ. Chicago Press, Chicago (1968).

Martens, K., B. Goodeeris, G. Coulter: Speciation in ancient lakes. *Arch. Hydrobiol. Beih. Ergebn. Limnol.* **44** (1994).

May, R. M.: Biological populations with nonoverlapping generations: Stable points, stable cycles and chaos. *Science* **186** (1974) 645–47.

May, R. M.: Simple mathematical models with very complicated dynamics. *Science* **261** (1976) 459.

McCauley, E., W. W. Murdoch: Cyclic and stable populations: plankton as paradigm. *Am. Nat.* **129** (1987) 97–121.

McCormick, P. V., D. Louie, C. Cairns, Jr.: Longitudinal effects of herbivory on lotic periphyton assemblages. *Freshw. Biol.* **31** (1994) 201–12.

McIntosh, A. R., C. R. Townsend: Interpopulation variaton in mayfly antipredator tactics: Differential effects of contrasting predatory fish. *Ecology* **75** (1994) 2078–90.

McQueen, D. J., M. R. S. Johannes, J. R. Post, T. J. Stewart, D. R. S. Lean: Bottom-up and top-down impacts on freshwater pelagic community structure. *Ecol. Monogr.* **59** (1989) 289–309.

Meulemans, J. T., F. Heinis: Biomass and production of periphyton attached to dead read stems in Lake Maarseveen. In: R. G. Wetzel (ed.): *Periphyton of freshwater ecosystems.* Junk, Den Haag (1983) 169–73.

Meyer, A., T. D. Kocher, P. Basasibwaki, A. C. Wilson: Monophyletic origin of Lake Victoria cichlid fishes suggested by mitochondrial DNA sequences. *Nature* **347** (1990) 550–53.

Milinski, M.: Do all members of a swarm suffer the same predation? *Z. Tierpsychol.* **45** (1977) 373–88.

Mills, E. L., D. M. Breen: Use of zooplankton size to assess the community structure of fish populations in freshwater lakes. *N. A. J. Fish. Man.* **7** (1987) 369–78.

Mills, E. L., R. O. Gorman, J. DeGisi, R. F. Heberger, R. A. House: Food of the alewife (*Alosa pseudoharengus*) in Lake Ontario before and after the establishment of *Bythotrephes cederstroemi*. *Can. J. Fish. Aquat. Sci.* **49** (1992) 2009–19.

Monod, J.: La technique de culture continue: theorie et applications. *Ann. Inst. pasteur Lille* **79** (1950) 390–410.

Morel, F: *Principles of aquatic chemistry.* Wiley, New York (1983).

Morre, D. R., P. A. Keddy, C. L. Gaudet, & I. C. Wisheu: Conservation of wetlands: Do infertile wetlands deserve a higher priority? *Biol. Conserv.* **47** (1989) 203–17.

Mort, M. A., H. G. Wolf: The genetic structure of large-lake *Daphnia* populations. *Evolution* **40** (1986) 756–66.

Muck, P., W. Lampert: Feeding of freshwater filter-feeders at very low food concentrations: Poor evidence for "threshold feeding" and "optimal foraging" in *Daphnia longispina* and *Eudiaptomus gracilis. J. Plankton Res.* **2** (1980) 367–79.

Müller-Navarra, D.: Evidence that a highly unsaturated fatty acid limits *Daphnia* growth in nature. *Arch. Hydrobiol.* **132** (1995) 297–307.

Mullin, M. M., E. Fuglister-Stewart, F. J. Fuglister: Ingestion by planktonic grazers as a function of concentration of food. *Limnol. Oceanogr.* **20** (1975) 259–62.

Mur, L., R. O. Bejsdorf: A model of the succession from green to blue-green algae based on light limitation. *Verh. Internat. Verein Limnol.* **20** (1978) 2314–21.

Neill, W. E.: Induced vertical migration in copepods as a defense against invertebrate predation. *Nature* **345** (1990) 524–26.

Neumann, D.: Ernährungsbiologie einer rhipidoglossen Kiemenschnecke. *Hydrobiologia* **7** (1961) 133–51.

Nilssen, J. P.: When and how to reproduce: A dilemma for limnetic cyclopoid copepods. In: W. C. Keerfoot (ed.): *Evolution and ecology of zooplankton communities.* University Press of New England, Hanover, N.H. (1980) 418–26.

O'Brien, W. J.: Planktivory by freshwater fish: Thrust and parry in the pelagia. In: W. C. Kerfoot, A. Sih (eds.): *Predation: direct and indirect impacts on aquatic communities.* University of New England Press, Hanover, N.H. (1987) 3–16.

O'Brien, W. J., B. Evans, C. Luecke: Apparent size choice of zooplankton by planktivorous sunfish—Exceptions to the rule. *Environ. Biol. Fish.* **13** (1985) 225–33.

Odum, E. P.: *Fundamentals of ecology,* 2nd edn. Saunders, Philadelphia (1959).

Odum, H. T: Trophic structure and productivity of Silver Springs, Florida. *Ecol. Monogr.* **27** (1957) 55–112.

Oksanen, L., S. D. Fretwell, J. Arruda, P. Niemala: Exploitation ecosystems in gradients of primary productivity. *Am. Nat.* **118** (1981) 240–61.

Ostwald, W.: Über eine neue theoretische Betrachtungsweise in d. Planktologie, insbes. Über die Bedeutung des Begriffs der inneren Reibung des Wassers für dieselbe. Forsch. Ber. Biol. Stat. Plön. (1903).

Overbeck, J.: Distribution pattern of uptake kinetic response in a stratified eutrophic lake. *Verh. Internat. Verein. Limnol.* **19** (1975) 2600–15.

Overrein, N. L., H. M. Seip, A. Tollan: Acid precipitation—effects on forest and fish. Final report of the SNSF-project 1972–1980. Norwegian Institute of Water Analysis, Oslo (1980).

Paerl, H. W.: Growth and reproductive strategies of freshwater blue- green algae (cyanobacteria). In: C. D. Sandgren (ed.): *Growth and survival strategies of freshwater phytoplankton.* Cambridge Univ. Press (1988) 261–315.

Paine, R. T.: A note on trophic complexity and community stability. *Am. Nat.* **103** (1969) 91–93.

Paloheimo, J. E.: Calculation of instantaneous birth rate. *Limnol. Oceanogr.* **19** (1974) 692–94.

Parent, S., R. D. Cheetham: Effects of acid precipitation on *Daphnia magna. Bull. Environm. Toxicol.* **25** (1980) 298–304.

Parker, M. S.: Size-selective predation on benthic macroinvertebrates by stream-dwelling salamander larvae. *Arch. Hydrobiol.* **128** (1993) 385–400.

Pastorok, R. A.: Prey vulnerability and size selection by *Chaoborus* larvae. *Ecology* **62** (1981) 1311–24.

Pechlaner, R.: Response of a eutrophicated lake (Piburger See) to reduced nutrient load and selective water renewal. *Arch. Hydrobiol. Beih. Ergebn. Limnol.* **13** (1979) 293–305.

Peckarsky, B. L.: A field test of resource depression by predatory stonefly larvae. *Oikos* **61** (1991) 3–10.

Peckarsky, B. L., C. A. Cowan, M. A. Penton, C. Anderson: Sublethal consequences of stream-dwelling predatory stonflies on mayfly growth and fecundity. *Ecology* **74** (1993) 1836–46.

Peckarsky, B. L., M. A. Penton: Mechanisms of prey selection by stream-dwelling stoneflies. *Ecology* **70** (1989) 1203–18.

Persson, L.: Competition induced switch in young of the year perch, *Perca fluviatilis*: an experimental test of resource limitation. *Env. Biol. Fish.* **19** (1987) 235–160.

Persson, L., G. Andersson, S. F. Hamrin, L. Johansson: Predator regulation and primary production along the productivity gradient of temperate lake ecosystems. In: S. R. Carpenter (ed.): *Complex interactions in lake communities.* Springer, New York (1988) 45–65.

Peters, R. H.: *The ecological implications of body size.* Cambridge Univ. Press (1983).

Peters, R. H.: Methods for the study of feeding, grazing and assimilation by zooplankton. In: J. A. Downing, F. H. Rigler: *A manual on methods for the assessment of secondary production in fresh waters.* IBP Handbook **17**, 2nd edn. Blackwell, Oxford (1984) 336–412.

Peters, R. H.: The role of prediction in limnology. *Limnol. Oceanogr.* **31** (1986) 1143–59.

Pfennig, N.: General physiology and ecology of photosynthetic bacteria. In: R. K. Clayton, W. R. Sistrom (eds.): *The photosynthetic bacteria.* Plenum, New York (1978) 3–18.

Porter, K. G.: The plant-animal interface in freshwater ecosystems. *Am. Sci.* **65** (1977) 159–70.

Power, M. E., W. J. Matthews, A. J. Stewart: Grazing minnows, piscivorous bass, and stream algae: dynamics of a strong interaction. *Ecology* **66** (1985) 1448–56.

Preijs, A., K. Lewandowski, A. Stancyzkowska-Piotrowska: Size- selective predation by roach (*Rutilus rutilus*) on zebra mussel (*Dreissena polymorpha*): field studies. *Oecologia* **83** (1990) 378–84.

Probst, L.: Sublitoral and profundal Oligochaeta fauna of Lake Constance (Bodensee-Obersee). *Hydrobiologia* **155** (1987) 277–82.

Prosser, C. L.: *Adaptational biology: Molecules to organisms.* Wiley, New York (1986).

Prus, T.: The assimilation efficiency of *Asellus aquaticus* L. (Crustacea, Isopoda). *Freshwat. Biol.* **1** (1971) 287–305.

Pütter, A.: Die Ernährung der Wassertiere durch gelöste organische Verbindungen. *Pflügers Arch. Physiol.* **137** (1911).

Pyke, G. H., H. R. Pulliam, E. L. Charnov: Optimal foraging: a selective review of theory and tests. *Quart. Rev. Biol.* **52** (1977) 137–54.

Reynolds, C. S.: *The ecology of freshwater phytoplankton.* Cambridge University Press, Cambridge (1984).

Reynolds, C. S.: Physical determinants of seasonal change in the species composition of phytoplankton. In: U. Sommer (ed.): *Succession in plankton communities.* Springer, Berlin (1989) 9–56.

Reynolds, C. S.: The ecological basis for the successful biomanipulation of aquatic communities. *Arch. Hydrobiol.* **130** (1994) 1–33.

Reynolds, C. S., P. A. Carling, K. J. Beven: Flow in river channels: new insights into hydraulic retention. *Arch. Hydrobiol.* **121,** (1991) 171–79.

Reynolds, C. S., A. E. Walsby, R. L. Oliver: The role of buoyancy in the distribution of *Anabaena* s. in Lake Rotongaio. *N. Z. J. Mar. Freshw. Res.* **21,** (1987) 525–26.

Reynolds, C. S., S. W. Wiseman, M. J. O. Clarke: Growth and loss rate responses of phytoplankton to intermittent artificial mixing and their potential application to the control of planktonic biomass. *J. Appl. Ecol.* **21** (1984) 11–39.

Reynoldson, T. B.: The population biology of Turbellaria with special reference to the triclads of the British Isles. *Adv. Ecol. Res.* **13** (1983) 235–326.

Richman, S.: The transformation of energy by *Daphnia pulex*. *Ecol. Monogr.* **28** (1958) 273–91.

Riemann, B., R. T. Bell: Advances in estimating bacterial biomass and growth in aquatic systems. *Arch. Hydrobiol.* **118** (1990) 385–402.

Riemann, B., K. Christoffersen: Microbial trophodynamics in temperate lakes. *Mar. Microb. Food Webs* **7** (1993) 69–100.

Riessen, H. P.: The other side of cyclomorphosis: Why *Daphnia* lose their helmets. *Limnol. Oceanogr.* **29** (1984) 1123–27.

Ringelberg, J.: Enhancement of the phototactic reaction in *Daphnia hyalina* by a chemical mediated by juvenile perch (*Perca fluviatilis*). *J. Plankton Res.* **13** (1991) 17–25.

Ringelberg, J., J. Van Kasteel, H. Servaas: The sensitivity of *Daphnia magna* Straus to changes in light intensity of various adaptation levels and its implication in diurnal vertical migration. *Z. vergl. Physiol.* **56** (1967) 397–407.

Rosemond, A. D., P. J. Mulholland, J. W. Elwood: Top-down and bottom-up control of stream periphyton: Effects of nutrients and herbivores. *Ecology* **74** (1994) 1264–80.

Rosenzweig, M. L., R. H. MacArthur, R. H.: Graphical representation and stability conditions of predator-prey interactions. *Am Nat.* **97** (1963) 209–223.

Rothhaupt, K. O.: Mechanistic resource competition theory applied to laboratory experiments with zooplankton. *Nature* **333** (1988) 660–62.

Rothhaupt, K. O.: Resource competition of herbivorous zooplankton: a review of approaches and perspectives. *Arch. Hydrobiol.* **118** (1990) 1–29.

Rothhaupt, K. O.: Stimulation of phosphorus limited phytoplankton by bacterivorous flagellates in laboratory experiments. *Limnol. Oceanogr.* **37** (1992) 750–59.

Rubenstein, D. I., M. A. R. Koehl: The mechanisms of filter feeding: some theoretical considerations. *Am. Nat.* **111** (1977) 981–94.

Sanders, R. W., K. G. Porter: Phagotrophic phytoflagellates. *Adv. Microb. Ecol.* **10** (1988) 167–92.

Sand-Jensen, K.: Environmental control of bicarbonate use among freshwater and marine macrophytes. In: R. M. Crawford (ed.): *Plant life in aquatic and amphibious habitats*. Blackwell, Oxford (1987) 99–112.

Santer, B., W. Lampert: Summer diapause in cyclopoid copepods: Adaptive response to fish predation or food bottleneck. *J. Animal Ecol.* xx (1995) xx.

Sas, H. (ed.): *Lake restoration by reduction of nutrient loading.* Academia Verlag Richarz, St. Augustin (1989).

Scheffer, M.: Should we expect strange attractors behind plankton dynamics - and if so, should we bother? *J. Plankton Res.* **13** (1991) 1291–1305

Scheffer, M., S. H. Hosper, M.-L. Meijer, B. Moss, E. Jeppesen: Alternative equilibria in shallow lakes. *TREE* **8** (1993) 275–79.

Schober, U.: Kausalanalytische Untersuchungen der Abundanzschwankungen des Crustaceen-Planktons im Bodensee. Ph.D. Thesis, University Freiburg/Brsg., Germany (1980).

Schwartz, S. S., G. N. Cameron: How do parasites cost their hosts? Preliminary answers from trematodes and *Daphnia obtusa*. *Limnol. Oceanogr.* **38** (1993) 602–12.

Seitz, A.: Are there allelopathic interactions in zooplankton? Laboratory experiments with *Daphnia*. *Oecologia* **62** (1984) 94–96.

Shannon, J. P., D. W. Blinn, L. E. Stevens: Trophic interactions and benthic animal community structure in the Colorado River, Arizona, U.S.A. *Freshw. Biol.* **31** (1994) 213–20.

Shapiro, J.: Blue-green dominance in lakes: The role and management significance of pH and CO_2. *Int. Revue ges. Hydrobiol.* **69** (1984) 765–80.

Shapiro, J.: Biomanipulation: the next phase—making it stable. *Hydrobiologia* **200/201** (1990) 13–27.

Shapiro, J., D. I. Wright: Lake restoration by biomanipulation: Round Lake, Minnesota, the first two years. *Freshw. Biol.* **14** (1984) 371–83.

Sheldon, R. W., A. Prakash, W. H. Sutcliffe: The size distribution of particles in the ocean. *Limnol. Oceanogr.* **17** (1972) 327–40.

Schindler, D. W.: The effect of fertilization with phosphorus and nitrogen versus phosphorus alone on eutrophication of experimental lakes. *Limnol. Oceanogr.* **25** (1980) 1149–52.

Siebeck, H. O.: Optical orientation of pelagic crustaceans and its consequences in the pelagic and littoral zone. In: W. C. Kerfoot (ed.): *Evolution and ecology of zooplankton communities.* Univ. Press of New England, Hanover (1980) 28–38.

Siebeck, O.: Der Königssee. Eine limnologische Projektstudie. Nationalpark Berchtesgaden, *Forschungsberichte* **5** (1982) 1–131.

Simon, M.: Specific uptake rates of amino acids by attached and free-living bacteria in a mesotrophic lake. *Appl. Environ. Microbiol.* **49** (1985) 1254–59.

Simpson, F. B., Neilands, J. B.: Siderochromes in cyanophyceae: isolation and characterization of schizokinen from *Anabaena* sp. *J. Phycol.* **12** (1976) 44–48.

Smith, V. H.: Nutrient dependence of primary productivity in lakes. *Limnol. Oceanogr.* **24** (1979) 1051–64.

Sommer, U.: Vertical niche separation between two closely related planktonic flagellate species (*Rhodomonas lens* and *Rhodomonas minuta* v.*nannoplanctica*). *J. Plankton Res.* **4** (1982) 137–42.

Sommer, U.: Sedimentation of principal phytoplankton species in Lake Constance. *J. Plankton Res.* **6** (1984) 1–14.

Sommer, U.: Comparison between steady state and non-steady state competition: experiments with natural phytoplankton. *Limnol. Oceanogr.* **30** (1985) 335.

Sommer, U.: Phytoplankton competition along a gradient of dilution rates. *Oecologia* **68** (1986) 503–6.

Sommer, U.: Factors controlling the seasonal variation in phytoplankton species composition—A case study for a deep, nutrient rich lake. *Progr. Phycol. Res.* **5** (1987) 123–78.

Sommer, U.: Does nutrient competition among phytoplankton occur in situ? *Verh. Int. Verein Limnol.* **23** (1988a) 707–12.

Sommer, U.: Phytoplankton succession in microcosm experiments under simultaneous grazing pressure and resource limitation. *Limnol. Oceanogr.* **33** (1988b) 1037–54.

Sommer, U.: Growth and reproductive strategies of planktonic diatoms. In: C. D. Sandgren (ed.): *Growth and survival strategies of freshwater phytoplankton.* Cambridge Univ. Press (1988c) 227–60.

Sommer, U.: The role of competition for resources in phytoplankton succession. In U. Sommer (ed.) *Plankton ecology: succession in plankton communities.* Springer, Berlin (1989) 57–106.

Sommer, U. (ed): *Plankton ecology: succession in plankton communities.* Springer, Berlin (1989).

Sommer, U.: Phytoplankton nutrient competition—from laboratory to lake. In: J. B. Grace, D. Tilman (eds.): *Perspectives on plant competition.* Academic Press, San Diego (1990) 193–213.

Sommer, U., U. Gaedke, A. Schweizer: The first decade of oligotrophication of Lake Constance. 2. The response of phytoplankton taxonomic composition. *Oecologia* **93** (1993) 276–84.

Sommer, U., Z. M. Gliwicz.: Long range vertical migration of *Volvox* in tropical lake Cahora Bassa (Mozambique). *Limnol. Oceanogr.* **31** (1986) 650–53.

Sommer, U., Z. M. Gliwicz, W. Lampert, A. Duncan: The PEG-model of seasonal succession of planktonic events in fresh waters. *Arch. Hydrobiol.* **106** (1986) 433–71.

Sorokin, Y. I., H. Kadota: *Microbial production and decomposition in fresh waters.* IBP Handbook **23.** Blackwell, Oxford (1972).

Spence, D. H. N., J. Chrystal: Photosynthesis and zonation of freshwater macrophytes. *New Phytol.* **69** (1970) 205–27.

Spindler, K. D.: Untersuchungen über den Einfluβ äuβerer Faktoren auf die Dauer der Embryonalentwicklung und den Häutungsrhythmus von *Cyclops vicinus. Oecologia* **7** (1971) 342–55.

Spitze, K.: Predator-mediated plasticity of prey life-history and morphology: *Chaoborus americanus* predation on *Daphnia pulex. Am. Nat.* **139** (1992) 229–47.

Sprules, W. G., H. P. Riessen, E. H. Jin: Dynamics of the *Bythotrephes* invasion of the St. Lawrence Great Lakes. *J. Great Lakes Res.* **16** (1990) 346–51.

Stadelmann, P.: Stickstoffkreislauf und Primärproduktion im mesotrophen Vierwaldstätter See und im eutrophen Rotsee, mit besonderer Berücksichtigung des Nitrats als limitierendem Faktor. *Schweiz. Z. Hydrol.* **33** (1971) 1–65.

Statzner, B., T. F. Holm: Morphological adaptation of shape to flow: Microcurrents around lotic macroinvertebrates with known Reynolds numbers at quasi-natural flow conditions. *Oecologia (Berl.)* **78** (1989) 145–57.

Stearns, S. C.: The evolutionary significance of phenotypic plasticity. *BioScience* **39** (1989) 436–45.

Stearns, S. C.: *The evolution of life histories.* Oxford Univ. Press, New York (1992).

Stemberger, R. S.: Reproductive costs and hydrodynamics benefits of chemically induced defenses in *Keratella testudo. Limnol. Oceanogr.* **33** (1988) 593–606.

Stemberger, R. S., J. J. Gilbert: Rotifer threshold food concentrations and the size-efficiency hypothesis. *Ecology* **68** (1987) 181–87.

Stenson, J. A. E.: Variation in capsule size of *Holopedium gibberum* (Zaddach): a response to invertebrate predation. *Ecology* **68** (1987) 928–34.

Sterner, R. W.: Herbivores' direct and indirect effects on algal populations. *Science* **231** (1986) 605–7.

Sterner, R. W.: The role of grazers in phytoplankton succession. In: U. Sommer (ed.): *Plankton Ecology. Succession in plankton communities.* Springer-Verlag, Heidelberg (1989) 107–70.

Sterner, R. W.: *Daphnia* growth on varying quality of *Scenedesmus*: Mineral limitation of zooplankton. *Ecology* **74** (1993) 2351–60.

Stibor, H.: Predator induced life-history shifts in a freshwater cladoceran. *Oecologia* **92** (1992) 162–65.

Stibor, H., Lüning, J.: Predator-induced phenotypic variation in the pattern of growth and reproduction in *Daphnia hyalina* (Crustacea, Cladocera). *Functional Ecol.* **8** (1994) 97–101.

Stich, H. B., W. Lampert: Predator evasion as an explanation of diurnal vertical migration by zooplankton. *Nature* **293** (1981) 396–98.

Streit, B.: Morphometric relationships and feeding habits of two species of *Chaetogaster, Ch. limnaei* and *Ch. diastrophus* (Oligochaeta). *Arch. Hydrobiol. Suppl.* **48** (1977) 424–37.

Strickland, J. D., T. R. Parsons: A practical handbook of seawater analysis. *Bull. Fish. Res. Bd. Can.* **169** (1968) 1–203.

Strickler, J. R.: Observation of swimming performances of planktonic copepods. *Limnol. Oceanogr.* **22** (1977) 165–70.

Swift, M. C., A. Y. Fedorenko: Some aspects of prey capture by *Chaoborus* larvae. *Limnol. Oceanogr.* **20** (1975) 418–25.

Talling, J. F.: The phytoplankton population as a compound photosynthetic system. *New Phytol.* **56** (1957) 133–49.

Taylor, B. E., Gabriel, W.: To grow or not to grow: optimal resource allocation for *Daphnia*. *Am. Nat.* **139** (1992) 248–66.

Thomasson, K.: Araucanian lakes. *Acta Phytogeographica Suecica* **47** (1963) 1–55.

Threlkeld, S. T., D. A. Chiavelli, R. L. Willey: The organization of zooplankton epibiont communities. *TREE* **8** (1993) 317–21.

Tilman, D.: Resource competition between planktonic algae: an experimental and theoretical approach. *Ecology* **59** (1977) 338–48.

Tilman, D.: *Resource competition and community structure.* Princeton University Press, Princeton (1982).

Tilman, D.: Plant strategies and the structure and dynamics of plant communities. Princeton (1988).

Tilman, D., S. S. Kilham, P. Kilham: Phytoplankton community ecology: the role of limiting nutrients. *Ann. Rev. Ecol. Syst.* **13** (1982) 349–72.

Tilzer, M. M.: The importance of fractional light absorption by photosynthetic pigments for phytoplankton productivity in Lake Constance. *Limnol. Oceanogr.* **28** (1983) 833–46.

Tilzer, M. M.: The quantum yield as a fundamental parameter controlling vertical photosynthetic profiles of phytoplankton in Lake Constance. *Arch. Hydrobiol. Suppl.* **69** (1984a) 169–98.

Tilzer, M. M.: Estimation of phytoplankton loss rates from daily photosynthetic rates and observed biomass changes in Lake Constance. *J. Plankton Res.* **6** (1984b) 309–24.

Tilzer, M. M., W. Geller, U. Sommer, H. H. Stabel: Kohlenstoffkreislauf und Nahrungsketten in der Freiwasserzone des Bodensees. *Konstanzer Blätter für Hochschulfragen* **73** (1982) 51–.

Tilzer, M. M., C. R. Goldman, E. de Amezaga: The efficiency of photosynthetic light energy utilization by lake phytoplankton. *Verh. Int. Verein. Limnol.* **19** (1975) 800–7.

Tollrian, R.: Predator-induced helmet formation in *Daphnia cucullata* (Sars). *Arch. Hydrobiol.* **119** (1990) 191–96.

Tollrian, R.: Fish-kairomone induced morphological changes in *Daphnia lumholtzi* (Sars). *Arch. Hydrobiol.* **130** (1994) 69–75.

Tollrian, R.: *Chaoborus crystallinus* predation on *Daphnia pulex*: can induced morphological changes balance effects of body size on vulnerability. *Oecologia* **101** (1995) 151–55.

Tollrian, R., E. von Elert: Enrichment and purification of *Chaoborus* kairomone from water: Further steps towards its chemical characterization. *Limnol. Oceanogr.* **39** (1994) 788–96.

Tschumi, P.-A.: Eutrophierung, Primärproduktion und Sauerstoffverhältnisse im Bielersee. *Gas-Wasser-Abwasser* **57** (1977) 245–52.

Uhlmann, D: *Hydrobiologie.* Fischer, Stuttgart (1975).

Underwood, G. J. C., Thomas, J. D.: Grazing interactions between pulmonate snails and epiphytic algae and bacteria. *Freshw. Biol.* **23** (1990) 505–22.

Van den Bund, W. J., D. Groenendijk: Seasonal dynamics and burrowing of littoral chironomid larvae in relation to competition and predation. *Arch. Hydrobiol.* **132** (1994) 213–25.

Van Donk, E.: The role of fungal parasites in phytoplankton succession. In: U. Sommer (ed.): *Plankton ecology, succession in plankton communities.* Springer, Berlin (1989) 171–94.

Van Donk, E., R. D. Gulati, M. P. Grimm: Food web manipulation in Lake Zwemlust: positive and negative effects during the first two years. *Hydrobiol. Bull.* **23** (1989) 19–34.

Van Donk, E., D. O. Hessen: Grazing resistance in nutrient-stressed phytoplankton. *Oecologia* **93** (1993) 508–11.

Vanni, M. J.: Effects of food availability and fish predation on a zooplankton community. *Ecol. Monogr.* **57** (1987) 61–88.

Vanni, M. J., C. Luecke, J. F. Kitchell, Y. Allen, J. Temte, J. J. Magnuson: Effects on lower trophic levels of massive fish mortality. *Nature* **344** (1990) 333–35.

Vannote, R. L., G. W. Minshall, K. W. Cumming, J. R. Sedell, C. E. Cushing: The river continuum concept. *Can. J. Fish. Aquat. Sci.* **37** (1980) 130–37.

Vareschi, E.: The ecology of Lake Nakuru (Kenya). III. Abiotic factors and primary production. *Oecologia* **55** (1982) 81–101.

Vareschi, E., J. Jacobs: The ecology of Lkae Nakuru. IV. Synopsis of production and energy flow. *Oecologia* **65** (1985) 412–24.

Vaughn, C. C., F. P. Gelwick, W. J. Matthews: Effects of algivorous minnows on production of grazing stream invertebrates. *Oikos* **66** (1993) 119–28.

Vogel, S.: *Life in moving fluids*; the physical biology of flow, 2nd ed., Princeton Univ. Press, Princeton 1994.

Vollenweider, R., J. Kerekes: *Eutrophication of waters, monitoring, assessment and control.* OECD, Paris (1982).

Wagner, G.: Simulationsmodelle der Seeneutrophierung, dargestellt am Beispiel des Bodensee-Obersees. *Arch. Hydrobiol.* **78** (1976) 1–41.

Walls, M., I. Kortelainen, J. Sarvala: Prey responses to fish predation in fresh-water communities. *Ann. Zool. Fenn.* **27** (1990) 183–99.

Walsby, A. E.: Mechanism of buoancy regulation by planktonic cyanobacteria with gas vesicles. In: Fay, P., C. van Baalen (eds.): *The cyanobacteria.* Elsevier, Amsterdam (1987) 377–92.

Washington, H. G.: Diversity, biotic and similarity indices. *Wat. Res.* **18** (1984) 653–94.

Weider, L. J.: Disturbance, competition and the maintenance of clonal diversity in *Daphnia pulex. J. Evol. Biol.* **5** (1992) 505–22.

Weider, L. J., P. D. Hebert: Ecological and physiological differentiation among low-arctic clones of *Daphnia pulex. Ecology* **68** (1987) 188–98.

Weider, J. L., W. Lampert: Differential response of *Daphnia* genotypes to oxygen stress: respiration rates, hemoglobin content and low oxygen tolerance. *Oecologia* **65** (1985) 487–91.

Werner, E. E., D. J. Hall: Optimal foraging and the size selection of prey by the bluegill sunfish (*Lepomis macrochirus*). *Ecology* **55** (1974) 1042–52.

Werner, E. E., D. J. Hall: Ontogenetic habitat shifts in Bluegill: the foraging rate-predation risk trade-off. *Ecology* **69** (1988) 1352–66.

Wetzel, R. G.: *Limnology,* 2nd edn. Saunders, Philadelphia (1983).

Wetzel, R. G., G. E. Likens: *Limnological analyses,* 2nd ed. Springer- Verlag, New York (1991).

Wieser, W.: Die Ökophysiologie der Cyprinidenfauna österreichischer Gewässer. *Österreichs Fischerei* **39** (1986) 88–93.

Williams, D. D., H. B. N. Hynes: Benthic community development in a new stream. *Can. J. Zool.* **55** (1977) 1071–1776.

Winberg, G.G. (ed.): *Methods for the estimation of production in aquatic animals.* Academic Press, London (1971).

Wright, R. F.: Input-output bdgets at Langtjern, a small acidified lake in southern Norway. *Hydrobiologia* **101** (1983) 1–12.

Wright, R. T., J. E. Hobbie: Use of gluctose and acetate by algae and bacteria in aquatic systems. *Ecology* **47** (1966) 447–64.

Zaika, V. E.: *Specific production of aquatic invertebrates.* Wiley, New York (1973).

Zaret, T. M., A. S. Rand: Competition in tropical stream fishes: support for the competitive exclusion principle. *Ecology* **52** (1971) 336–42.

Zelinka, M., P. Marvan: Zur Präzisierung der biologischen Klassifikation der Reinheit der fließenden Gewässer. *Arch Hydrobiol.* **57** (1961) 389–407.

Zevenboom, W.: *Growth and nutrient uptake kinetics of Oscillatoria agardhii*. Acad. Proefschrift, Univ. Amsterdam (1980).

Further Reading

Chapter 1
Futuyma, D. J.: *Evolutionary Biology.* Sinauer Assoc., Sunderland, MA (1986).
Mayo, O.: *Natural selection and its constraints.* Academic Press, London (1983).
Rigler, F. H., R. H. Peters: *Science and limnology.* Ecology Institute, Oldendorf/Luhe (1995).
Smith, J. M.: *Evolutionary genetics.* Oxford Univ. Press, Oxford (1986).
Williams, G. C.: *Adaptation and natural selection.* Princeton Univ. Press (1966).

Chapter 2
Hairston, N. G., Sr.: *Ecological experiments.* Cambridge Univ. Press, Cambridge (1989).
Pielou, E. C.: *The interpretation of ecological data.* Wiley, New York (1984).
Popper, K. R.: *Realism and the aim of science.* Rowman and Littlefield, Totowa (1983).
Wetzel, R. G., G. E. Likens: *Limnological analyses,* 2nd edn. Springer-Verlag, New York (1991).

Chapter 3
Stumm, W. (ed.): *Chemical processes in lakes.* Wiley, New York (1985).
Stumm, W., J. J. Morgan: *Aquatic chemistry,* Wiley, New York (1970).
Wetzel, R. G.: *Limnology,* 2. Aufl., Saunders, Philadelphia (1983).

Chapter 4
Hochachka, P. W., G. N. Somero: *Biochemial adaptation.* Princeton Univ. Press, Princeton (1984).
Sibly, R. M., P. Calow: *Physiological ecology of animals.* Blackwell, Oxford (1986).
Townsend, C. R., P. Calow: *Physiological ecology.* Blackwell, Oxford (1981).
Vogel, S.: *Life in moving fluids.* Princeton Univ. Press, Princeton (1981).

Chapter 5
Begon, M., M. Mortimer: *Population ecology.* Sinauer, Sunderland, MA (1981).
Crow, J. F.: *Basic concepts in population, quantitative and evolutionary genetics.* Freeman, New York (1986).
Ginzburg, L. R., E. M. Golenberg: *Lectures in theoretical population biology.* Prentice-Hall, Englewood Cliffs (1985).
Hazen, W. (ed.): *Readings in population and community ecology.* Saunders, Philadelphia (1975).
Hedrick, P. W.: *Genetics of population.* Bartlett, Boston (1985).

Hutchinson, G. E.: *An introduction to population ecology.* Yale Univ. Press, New Haven (1978).
Roughgarden, J.: *Theory of population genetics and evolutionary ecology: An introduction.* MacMillan, New York (1979).

Chapter 6
Allan, J. D.: *Stream ecology. Structure and function of running waters.* Chapman and Hall (1995).
Barnes, J. R., G. W. Minshall: *Stream ecology.* Plenum Press, New York (1983).
Kerfoot, W. C. (ed.): *Evolution and ecology of zooplankton communities.* Univ. Press of New England, Hanover (1980).
Kerfoot, W. C., A. Sih (eds.): *Predation. Direct and indirect impacts on aquatic communities.* Univ. Press of New England, Hanover (1987).
Meyers, D. G., J. R. Strickler (eds.): *Trophic interactions within aquatic ecosystems.* Westview Press, Boulder (1984).
Pontin, A. J.: *Competition and coexistence of species.* Pitman, Boston (1982).
Sommer, U. (ed.): *Plankton ecology: succession in plankton communities.* Springer, Berlin (1989).
Stearns, S. C.: *The evolution of life histories.* Oxford Univ. Press, New York (1992).
Tilman, D.: *Resource competition and community structure.* Princeton University Press, Princeton (1982).
Yodzis, P.: *Introduction into theoretical ecology.* Harper and Row, New York (1989).
Zaret, T. M.: *Predation and freshwater communities.* Yale Univ. Press, New Haven (1980).

Chapter 7
Carpenter, S. R. (ed.); *Complex interactions in lake communities.* Springer, Berlin (1989).
Cody, M. L., J. M. Diamond (eds.): *Ecology and evolution of communities.* Belknap Press, Harvard (1975).
Pimm, S. L.: *Food webs.* Chapman and Hall, London (1982).
Roughgarden, J., R. M. May, S. A. Levin (eds.): *Perspectives in ecological theory.* Princeton Univ. Press, Princeton (1989).

Chapter 8
Likens, G. E. (ed.): *An ecosystem approach to aquatic ecology.* Springer, New York (1985).
Likens, G. E.: *The ecosystem approach: its use and abuse.* Ecology Institute, Oldendorf/Luhe (1992).
Odum, H. T.: *Systems ecology.* Wiley, New York (1983).
Whittaker, R. H.: *Communities and ecosystems.x MacMillan, New York (1970).*

Index